Neuro-inspired Computing Using Resistive Synaptic Devices

Shimeng Yu
Editor

Neuro-inspired Computing Using Resistive Synaptic Devices

Springer

Editor
Shimeng Yu
School of Electrical, Computer and Energy Engineering
Arizona State University
Tempe, AZ, USA

ISBN 978-3-319-54312-3 ISBN 978-3-319-54313-0 (eBook)
DOI 10.1007/978-3-319-54313-0

Library of Congress Control Number: 2017937653

Printed on acid-free paper

This Springer imprint is published by Springer Nature
The registered company is Springer International Publishing AG
The registered company address is: Gewerbestrasse 11, 6330 Cham, Switzerland

Preface

Cognitive learning and intelligence have not yet been captured in today's computers which are based on the Boolean logic and the von Neumann architecture. As compared to biological systems, today's computers are less energy efficient by a factor of one million to one billion in dealing with the complex, real-world environmental problems such as image/speech recognition. The memory wall problem is one of the major limitations for further performance gain in the conventional computing paradigm. Therefore, new computing paradigm that emulates the brain, namely, neuro-inspired computing, is emerging which generally takes the advantage of massive parallelism in neural networks and aims at robustness, energy efficiency, and inherent error-tolerant capability. The key challenge for hardware implementation of neuro-inspired computing is the development of the solid-state synapse elements due to the requirements of high integration density, low energy consumption, and silicon CMOS fabrication compatibility. Recently, materials that show electrically triggered resistive switching phenomenon have been proposed as resistive synaptic device. Examples include the two-terminal emerging nonvolatile memories such as phase change memory and resistive memory. The programmable multilevel conductance states in such emerging memories hold the promises to efficiently emulate the tunable analog weights in the synapses in the neural networks. The research on developing resistive synaptic device-based large-scale neuromorphic system is beginning and will be blooming in the coming years.

Neuro-inspired computing has recently attracted a lot of interests in academic institutions and industrial research institutions, as reflected by the large-scale projects such as DARPA SyNAPSE, DARPA UPSIDE, NSF/SRC E2CDA, and DARPA/SRC JUMP in the USA, Human Brain Project (HBP) and NeuRAM3 in Europe, etc.

This book summarizes the latest research progresses in the field of neuro-inspired computing using resistive synaptic devices. The book is organized hierarchically: Following the overview of the field (Chap. 1), Part I is on the device-level synaptic characteristics such as the synaptic weight update behavior, in particular, device conductance as a function of programming pulses. Three representative resistive synaptic devices in the literature are introduced: phase change memory-based synaptic device (Chap. 2), PCMO-based synaptic device (Chap. 3), and TaO_x/TiO_2-based synaptic device (Chap. 4). Part II is on the array-level experimental demonstration of the key neuromorphic operations such as association (Chap. 5), inference and online (in situ) training (Chap. 6), convolution (Chap. 7) on the small-scale (e.g., 12 × 12) crossbar (or pseudo-crossbar with transistors) array, and spike-timing-dependent plasticity (STDP) on the medium-scale (e.g., 256 × 256) array. Part III is on the circuit-, architecture-, and algorithm-level design of resistive synaptic device-based system. Neuron peripheral circuits (Chap. 9) and architectural design automation tool of deep neural network (Chap. 10) are discussed. The impact of non-ideal synaptic device characteristics on system-level learning performance is discussed for different algorithms such as multilayer perceptron (Chap. 11) and sparse coding (Chap. 12). Applications including visual pattern recognition with convolutional neural network and real-time decoding (spike sorting) of complex brain signals are introduced (Chap. 13).

This book covers the topics ranging from the areas of materials engineering, nanotechnology, semiconductor devices, circuits and systems design, neuromorphic engineering, and machine/deep learning. Advances described in this book will motivate further interdisciplinary research in these areas. This book will serve as a good reference for faculty, students, researchers, and industry practitioners who wish to learn the basics, state of the art, challenges, prospects, and future trends in the field of neuro-inspired computing with emerging device technologies.

Tempe, AZ, USA Shimeng Yu

Acknowledgments

First of all, I wish to sincerely thank all the authors of the contributed chapters for their outstanding research works.

Second, I am especially grateful to my research group members (Pai-Yu Chen, Ligang Gao, Rui Liu, Xiaoyu Sun, etc.) for their hard work on this project and help in the graphics and proofreading of this book.

I also want to thank Charles Glaser (Springer) for offering the opportunity to publish this book and Dhanuj Nair (Springer) for the editorial support.

Finally, my research of neuro-inspired computing using resistive synaptic devices is sponsored by the National Science Foundation Faculty Early CAREER Award under the grant NSF-CCF-1552687.

Shimeng Yu

Contents

About the Editor

Shimeng Yu He received the B.S. degree in microelectronics from Peking University, Beijing, China, in 2009 and the M.S. degree and Ph.D. degree in electrical engineering from Stanford University, Stanford, CA, USA, in 2011 and in 2013, respectively. He joined Arizona State University, Tempe, AZ, USA, as an assistant professor of electrical engineering and computer engineering in 2013.

His research interests are emerging nano-devices and circuits with a focus on the resistive memories for different applications including neuro-inspired computing, monolithic 3D integration, hardware security, radiation-hard electronics, etc. He has published more than 50 journal papers and more than 90 conference papers with citations of more than 4000 and H-index of 28 according to Google Scholar.

Among his honors, he is a recipient of the Stanford Graduate Fellowship from 2009 to 2012, the IEEE Electron Devices Society Masters Student Fellowship in 2010, the IEEE Electron Devices Society Ph.D. Student Fellowship in 2012, the DOD-DTRA Young Investigator Award in 2015, and the NSF Faculty Early CAREER Award in 2016 on the topic of scaling up resistive synaptic arrays for neuro-inspired computing.

He did summer internship in IMEC, Belgium, in 2011 and the IBM TJ Watson Research Center in 2012. He held visiting faculty position in the Air Force Research Laboratory in 2016. He has been serving the Technical Committee of Nanoelectronics and Gigascale Systems, IEEE Circuits and Systems Society, since 2014.

Chapter 1
Introduction to Neuro-Inspired Computing Using Resistive Synaptic Devices

Shimeng Yu

1.1 The Demand for New Hardware Beyond von Neumann Architecture

In the era of big data, the data generation rate is far greater than our computing capability to analyze the data. For instance, every minute, 3 million GB of image data is generated, which includes 1.2 million GB of video surveillance, 100 h of YouTube video, and 200,000 photographs of individuals on Facebook [1]. It is estimated that more than 23% of the image data has great analytic value for scientific research, and yet only 3% has been tagged (identified as valuable for analysis), and only 0.5% has been actually analyzed [1]. To extract the useful information from the large-scale dataset, the neuro-inspired learning algorithms (e.g., deep learning with hierarchical neural network architecture [2]) have demonstrated its powerfulness in various tasks, including image classification, speech recognition, natural language processing, bioinformatics, etc. The most popular deep learning algorithms today are convolutional neural network (CNN) [3], stacked autoencoder [4], deep belief network [5], etc. These deep learning algorithms remarkably outperform the traditional artificial neural network or simple machine learning algorithms. However, deep learning typically requires huge computational recourses to train the weights in the network. For instance, Google's stacked autoencoder algorithm [4] was able to successfully identify faces of cats from 10 million random images taken from YouTube videos. Yet this task was accomplished on a cluster of 16,000 processor cores consuming ~100 kW power

S. Yu (✉)
School of Electrical, Computer and Energy Engineering, Arizona State University, 781 E Terrace Rd, ISTB4 room 591, Tempe, AZ 85287, USA
e-mail: shimengy@asu.edu

S. Yu (ed.), *Neuro-inspired Computing Using Resistive Synaptic Devices*,
DOI 10.1007/978-3-319-54313-0_1

and used 3 days to train the network. Therefore, hardware acceleration with orders of magnitude improvement in speed and energy efficiency is essential for both data center and mobile applications. This remains a grand challenge for the conventional hardware based on complementary metal-oxide-semiconductor (CMOS) technology and von Neumann architecture. On the one hand, the scaling of silicon CMOS transistors is approaching the physical limit [6]. On the other hand, the well-known "memory wall" problem that the data movement between the microprocessor and off-chip memory/storage has became the bottleneck of the entire system [7]. As the neuro-inspired learning algorithms extensively involve large-scale matrix operations, computing paradigms that take advantage of the parallelism at finer-grain level are attractive. One promising solution is the neuro-inspired architecture that leverages the distributed computing in the neurons and localized storage in the synapses [8]. Figure 1.1 shows such revolutional shift of the computing paradigm from the computation-centric (von Neumann architecture) to the data-centric (neuro-inspired architecture). The neurons are simple computing units, and the synapses are local memories that are massively connected via the communication channels. The ultimate goal of the hardware implementation of the neuro-inspired computing is to supplement (but not supplant) today's von Neumann architecture for application-specific tasks that require intelligence such as image/speech recognition, autonomous vehicles, etc.

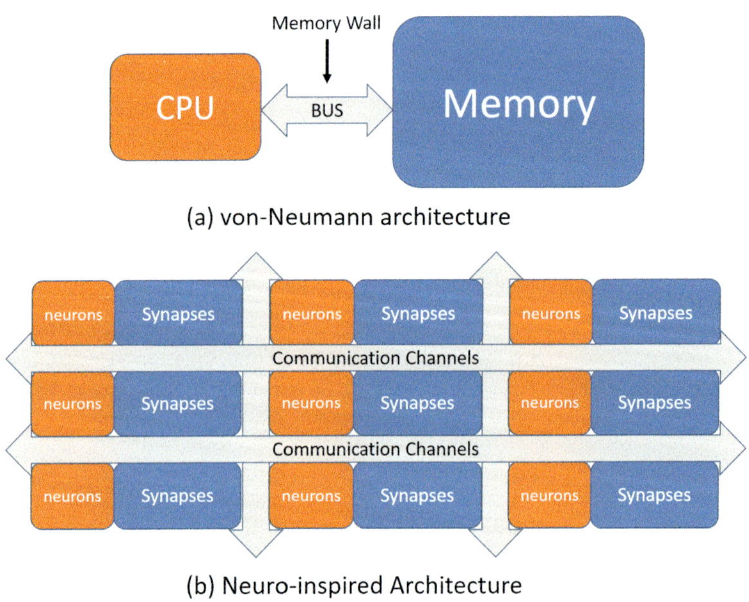

Fig. 1.1 A revolutional shift of the computing paradigm from the computation-centric (von Neumann architecture) to the data-centric (neuro-inspired architecture)

1.2 Neuromorphic Hardware Accelerators and Why Resistive Synaptic Devices?

Several hardware platforms with partial parallelism have been explored so far for implementing neuro-inspired learning algorithms. Generally, there are two approaches for neuromorphic hardware design depending on how to encode the information. The first approach stays on the digital (non-spiking) implementation while takes the inspirations from the neural network to maximize the parallel or distributed computation. As off-the-shelf technologies, graphics processing units (GPUs) [9] or field programmable gate arrays (FPGAs) [10] have been widely used for hardware acceleration for deep learning. To further improve the energy efficiency, CMOS-based application-specific integrated circuit (ASIC) accelerators such as those demonstrated in International Solid-State Circuit Conference (ISSCC) 2016 [11, 12] have been developed. Recently, Google also announced their custom-designed tensor processing unit (TPU) to accelerate the complex intelligent tasks such as AlphaGo [13]. The digital (non-spiking) approach aims to improve the computation efficiency in terms of the performance per watt per second (e.g., in the metric of GOPs/W/s). The second spiking approach aims to emulate the biological neural network more closely by adopting the spike representation of the information. Examples include Manchester's SpiNNaker with tens of thousands of ARM processors [14], as well as custom-designed CMOS-based neuromorphic chips (i.e., Heidelberg's HICANN [15], IBM's TrueNorth [16], etc.). Heidelberg's HICANN uses analog neurons similar as the leaky integrate-and-fire model and digital synapses made of 4-bit 6-transistor static random-access memory (SRAM) and 4-bit digital-to-analog converter (DAC) to interface with analog neurons, while IBM's TrueNorth uses digital neurons and digital synapses made of 1-bit transposable 8-transistor SRAM. In particular, the TrueNorth integrates 4096 neuro-synaptic cores with 1 million digital neurons and 256 million SRAM synapses that were fabricated in 28 nm node. The TrueNorth chip demonstrated 70 mW power consumption (@0.775 V) to perform real time (30 frames per second) object recognition on 480 $\times$ 240 input images. Although the TrueNorth can perform the inference at ultralow power, it can only perform offline training (i.e., synaptic weights are pretrained with software and then loaded into the SRAM arrays for inference only). Real-time operation actually means low speed with frequency (~ kHz), thus not suitable for acceleration of the online training (i.e., weights are trained on the fly). Therefore, such inference engine could not adapt to the constantly changing input data and/or learn new features during the runtime.

Table 1.1 summarizes the categories of different design options for hardware implementation of neuro-inspired computing. Here, the categories are roughly classified based on how the information is encoded and the availability of the hardware platforms. The digital representation uses binary bits or digital number of pulses (the spiking rate encoding) for encoding the neuron information, while the synapses can be either binary or multilevel (in an analog fashion). On the other hand, the spike representation encodes the information by the timing of the spikes.

Table 1.1 Categories of different design options for hardware implementation of neuro-inspired computing. Representative porotypes are shown

	Off-the-shelf technologies	CMOS ASIC	Emerging resistive synaptic devices
Digital representation	GPUs [9] FPGAs [10]	TPU [13] CNN accelerators [11, 12]	Analog synapses: UCSB's 12×12 crossbar array [18]
			Binary synapses: ASU/Tsinghua's 16 Mb RRAM macro [19]
Spike representation	SpiNNaker [14]	Analog neuron: HICANN [15]	IBM's 256×256 PCM array with STDP neuron circuits [20]
		Digital neuron: TrueNorth [16]	

One important biologically plausible learning rule to update the synaptic weight is the spike-timing-dependent plasticity (STDP) [17]. STDP learning rule states that if the presynaptic neuron fires earlier than the postsynaptic neuron, the conductance of the synapse (weight) will decrease and vice versa. The change of the weight is larger when the timing between the two neurons firing is closer. However, how to exploit such STDP learning rule (unsupervised and local to two adjacent neurons) to efficiently update the global synaptic network remains to be explored. Therefore, for solving today's practical supervised classification problems with high accuracy, the backpropagation learning rule is widely used where the training error is fed back from the output neuron layer to input neuron layer in a manner of layer by layer.

To overcome the aforementioned challenges with the CMOS platforms, the researchers are fascinated by exploiting the unique properties of emerging nonvolatile memory (NVM) technologies to better emulate the analog synapses in the neural networks. The goal is to replace the SRAM arrays with the resistive crossbar arrays to store and/or update the weights in a more parallel fashion. Compared to the binary SRAM cell with six or eight transistors, the emerging NVM cell occupies more than 30 times less area and can store multi-bit per cell, further increasing the integration density and thereby supporting a larger capacity on chip (for larger problem size or training dataset). Eliminating the off-chip memory access is critical to acceleration and reduction of the energy consumption from the entire system point of view. Thanks to the nonvolatility, these emerging NVM devices can be powered off and on instantly and consume no standby leakage. In addition, unlike SRAM array's sequential write and read, resistive crossbar array with emerging NVMs can do parallel programming and weighted sum to further speed up, potentially enabling the online training.

Generally speaking, emerging NVMs are mostly resistive memories that use *resistance* to represent and store data, including the spin-transfer torque magnetic random-access memory (STT-MRAM) [21], phase-change memory (PCM) [22], resistive random-access memory (RRAM) [23], and conductive bridge random-access memory (CBRAM) [24]. In some literature, resistive memories are also referred to as memristors [25]. In this book, we will focus on PCM, RRAM, and CBRAM technologies as they have demonstrated the multilevel states. These emerging NVMs are mostly pursued as the next-generation storage-class memory technologies by industrial research and development [26]. For example, Samsung has reported an 8 Gb PCM prototype chip in 20 nm node featuring 40 MB/s write bandwidth [27]. SanDisk/Toshiba has reported a 32 Gb RRAM prototype chip in 24 nm node [28]. Micron/Sony has reported a 16 Gb CBRAM prototype chip in 27 nm node featuring 200 MB/s write bandwidth and 1 GB/s read bandwidth. Panasonic has commercial products of microcontrollers with MB capacity embedded TaO_x RRAM [29]. These industrial works show that the emerging NVMs are viable technologies for large-scale integration.

In this book, resistive synaptic devices refer to a special subset of the resistive memories that can continuously tune the conductance into *multilevel* states, thereby emulating the analog synapses in the neural networks or the weights in the learning algorithms [30]. Recently, many resistive synaptic device candidates that are capable of tens to hundreds levels of conductance states have been demonstrated at single device level. Examples include $Ge_2Sb_2Te_5$-based PCM [31, 32]; Ag/a-Si [33]–, Ag/Ag_2S [34]–, and Ag/GeS_2 [35]-based CBRAM; $PrCaMnO_3$ [36], WO_x [37]-, TiO_x/HfO_x [38]–, and TiO_2/TaO_x [39]-based RRAM; etc. These resistive synaptic devices emulate the biological synapse in the sense that ion or defect migration in the solid-state dielectrics (e.g., in oxides) could modulate the conductance between the two electrodes, as the biological synapse modulates its conductance via the activation of voltage-gated calcium channels. Figure 1.2 shows this analogy between a biologic synapse and the resistive synaptic device.

1.3 Desirable Characteristics of Resistive Synaptic Devices

In this section, we will discuss the characteristics for resistive synaptic devices desirable for the applications where the synaptic weights need to be updated frequently. Table 1.2 summarizes the desirable performance metrics for resistive synaptic devices. It should be noted that many of the metrics are highly application dependent.

Device Dimensions The large-scale integration of neural network requires a compact synaptic device with a small device footprint. Resistive synaptic devices with scalability down to sub-10 nm regime are preferred. Today's RRAM, PCM, and CBRAM devices have proven such scalability; however, most of the demonstrations are for the digital memory application. Ultimately, the two-terminal synaptic device structure (possibly with a two-terminal selector) that is compatible with the crossbar array architecture and three-dimensional integration is desired.

Multilevel States Synaptic plasticity characteristics observed on biological synapses show an analog-like behavior with multilevel synaptic weight states. Most neuro-inspired algorithms also employ the analog synaptic weights to learn the patterns or extract features. In general, the more multilevel states (e.g., >hundreds of levels) could be translated into a better learning capability and an improved

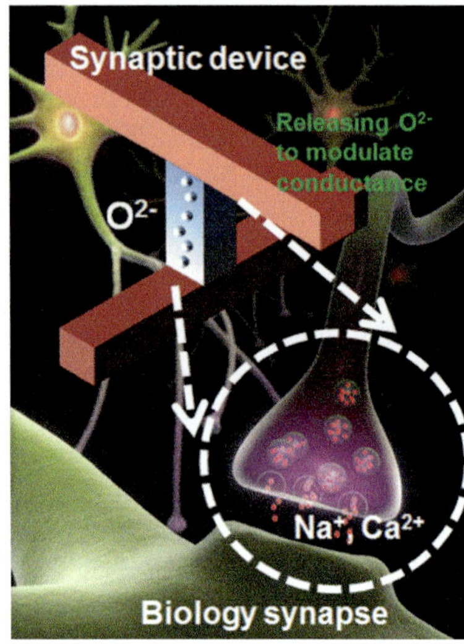

Fig. 1.2 An analogy between a biologic synapse and the resistive synaptic device

Table 1.2 Summary of the desirable performance metrics for synaptic devices

Performance metrics	Desired targets
Device dimension	<10 nm
Multilevel states' number	>100[a]
Energy consumption	<10 fJ/programming pulse
Dynamic range	>100[a]
Retention	>10 years[a]
Endurance	$>10^9$ updates[a]

Note: [a]These numbers are application dependent

network robustness. However, the weight precision requirement (i.e., the number of conductance states) remains strongly application dependent. If the multilevel states in the resistive synaptic devices are insufficient to meet the requirement, there are two alternative solutions: First, multiple devices could be grouped to represent higher precision at the expensive of area and energy [19]. Second, recent work shows that binary synaptic devices with stochastic weight update may equivalently provide the properties of analog synapses [35, 40].

Dynamic Range Dynamic range means the on/off ratio between the maximum conductance and minimum conductance. Most of the resistive synaptic device candidates exhibit a range of 3× to >100×. The larger the dynamic range is, the better mapping capability of the weights in the algorithms to the conductance in the devices, because the weights in the algorithms are typically normalized within a range (e.g., between 0 and 1). Considering the power consumption for parallel reading the weights in a large-scale integration of the neural network (e.g., with a matrix size 512 × 512 or above), a guideline of the desired range of a single device could be from 1 to 100 nS.

Linearity in Weight Update The linearity in weight update refers to the linearity of the curve between the device conductance and the number of identical programming pulses. Ideally, this should be a linear relationship for the direct mapping of the weights in the algorithms to the conductance in the devices. However, the resistive synaptic devices generally have the nonlinearity in weight update (see Fig. 1.3). The trajectory of the long-term-potentiation (LTP) process that increases the conductance differs from that of the long-term-depression (LTD) process that decreases the conductance. The weight tends to saturate at the end of LTP or LTD processes. This nonlinearity is undesired because the change of the weight (ΔW) depends on the current weight (W), or in other words, the weight update has a history dependence. Recent results have shown that this nonlinearity has caused the learning accuracy loss in the neural networks [41, 42].

Programming Energy Consumption The estimated energy consumption per synaptic event is around 1 ~ 10 fJ in biological synapses. Most RRAM/CBRAM devices show a programming energy around 100 fJ ~ 10 pJ, while most PCM devices may have even higher programming energy 10 ~ 100 pJ. The fundamental challenge is that it is much more difficult (thus paying more energy) to move the ions/defects in solid-state devices than moving calcium ions in the liquid environment in biological synapses. A back-of-envelope calculation is given as follows. In biological synapses, the spike voltage is ~10 mV, the ionic current ~1 nA, and the spike period ~1 ms; therefore, the energy is about 10 fJ. In resistive synaptic devices, the typical programming voltage is ~1 V, and the programming current is typically >μA; although the programming speed can be accelerated less than the real time to be <μs, still the energy is on the order of pJ. Further device engineering is thus needed to reduce the energy consumption.

Retention and Endurance During the online training, the weights are frequently updated, and the data retention requirement can be relaxed. When the training is complete, the resistive synaptic should behave as a long-term memory with a data retention in the order of 10 years at elevated temperature similarly as the

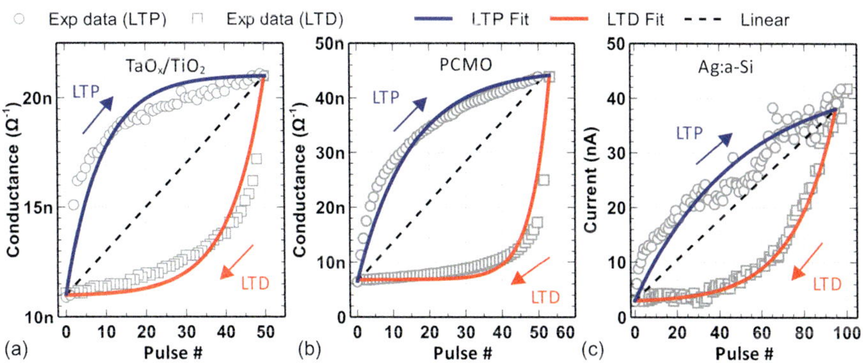

Fig. 1.3 The measured nonlinearity in the weight update reported from the literature: (**a**) TaO_x/TiO_2 device [39], (**b**) PCMO device [36], and (**c**) Ag:a-Si device [33]

requirement of NVM. The number of endurance is much application dependent, relying on how many weight updates are required in the training processes. For a relatively simple task (i.e., the MNIST handwritten digit recognition [43]), 60,000 training images with 50 training epochs (to repeated) give a maximum weight update possibility to be 3×10^6 updates. Actually not every synapse is updated in the training; thus, an endurance ~10^4 is sufficient for training MNIST dataset [19]. However, considering more challenging tasks (i.e., ImageNet challenge [44]), much more endurance may be required.

Uniformity and Variability Poor uniformity or significant variability in emerging NVMs is a major barrier for digital memory applications. In contrast, the neural networks promise robustness against device variations. The device variations could partially be tolerated by two mechanisms: the massive (thus maybe redundant) connections between neuron nodes by synaptic arrays and the iterative weight update process during the training. The degree of variations that can be tolerated at the system level strongly depends on the network architecture and the accuracy required by the target application. The device-algorithm co-simulations have shown the reasonable robustness against device variations in different neural networks [42, 45].

1.4 Crossbar Array Architecture for Accelerating Weighted Sum and Weight Update

The resistive crossbar array architecture has been proposed for implementing the weighted sum (or matrix-vector multiplication, dot product operation) [46], which is the most time-consuming step in the learning algorithms. As shown in Fig. 1.4, the crossbar array consists of perpendicular rows and columns with the resistive synaptic devices sandwiched at each cross-point. The weights in the neural network are then mapped to the conductance of the resistive synaptic devices.

The weighted sum operation is performed in a parallel fashion: read voltages are applied to all the rows, and then the read voltages are multiplied by the conductance of the synaptic devices at each cross-point, resulting in a weighted sum current in each column. Typically, the neuron circuits will be placed at the end of the column to convert this analog current to the digital output or spikes. The proposed crossbar array architecture only performs the analog computation in the array core, and the communication between arrays is still through digital fashion considering the signal integrity issues. It is also worth pointing out that the sneak path problem for the conventional crossbar memory does not exist here if all the rows and columns are activated during the weighted sum. This is because the conventional memory requires reading out data by bit or by row, while all the cells here participate in the computation essentially following the Kirchhoff law. The IR drop problem along long interconnect wires still exists here for a large-scale array, as the interconnect resistance may distort the weighted sum accuracy if it is a significant

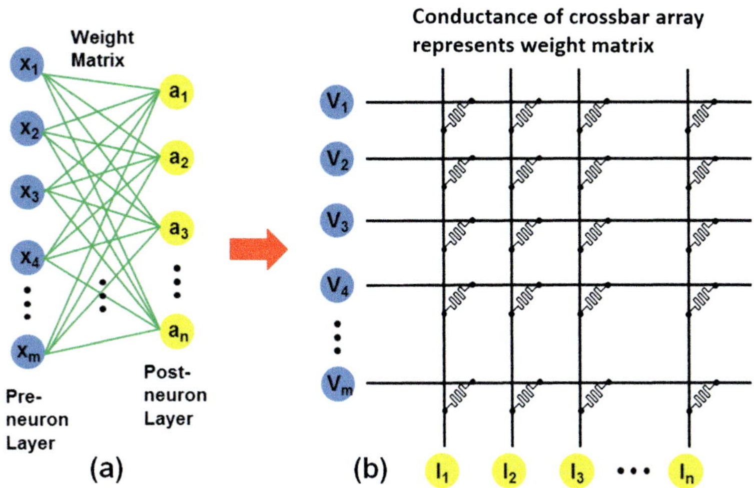

Fig. 1.4 (**a**) The weight matrix between neuron layers in the network. (**b**) The crossbar array consists of perpendicular rows and columns with the resistive synaptic devices sandwiched at each cross-point. The weights in the neural network are mapped to the conductance of the resistive synaptic devices

portion of the synaptic device resistance. The interconnect resistance effect can be mitigated by either relaxing the wire width [47] or remapping the weights of the weights in the algorithms to the conductance in the devices [48].

The weight update operation can be performed row by row (or column by column) in the crossbar array. In this case, selectors with nonlinear I-V relationship may be needed to minimize the leakage current in other unselected rows/columns. In principle, the weight update can be performed in a fully parallel fashion on the entire array as the programming voltages can be applied from both ends (row and column) to the synaptic device [49]. However, to program the entire array simultaneously usually demands a huge instant power from the peripheral circuits which seems not very feasible in practical designs.

Although various resistive synaptic device prototypes have been demonstrated in literature, most of these works still focus on the single device characterization and network simulation in software. So far, there have been a few experimental implementations of simple algorithms on small-scale arrays: 40 × 40 Ag:a-Si crossbar (loading weights without learning) [50], 12 × 12 TiO_x/Al_2O_3 crossbar (simple perceptron algorithm with software neurons) [18], 32 × 32 PCMO crossbar (off-chip neuron circuits on a board) [51], 10 × 10 PCM 1T1R array (simple recurrent network with software neurons) [52], IBM's 500 × 661 PCM 1T1R array (multilayer perceptron with software neurons) [41], and IBM's 256 × 256 PCM 1T1R array (with on-chip integrate-and-fire neurons) [20]. These demonstrations show the promises for future large-scale integration.

Nevertheless, design challenges arise from the device level up to the architecture level when the crossbar array size is scaled up to solve practical problems

[53]: Firstly, the resistive devices today are mostly engineered toward the NVM application, but the requirements of synaptic devices are quite different. For instance, synaptic devices need many more multilevel states (tens of hundreds of states) than NVM's 1 to 3 bit (eight levels) thereby requiring special materials and device engineering. Secondly, with the increase of the array size, issues associated with device yield, device variability, and array parasitics show up and may degrade the system performance, while the circuit or architectural mitigation techniques are yet to be developed. In addition to the array core, designs of the peripheral circuits are rarely explored. However, the CMOS neuron node (e.g., integrate-and-fire neuron circuits) at the periphery of the crossbar array is complex (with tens of transistors for each neuron), which is power-hungry and causes the column pitch matching problem (multiple columns have to share one neuron, thereby reducing the parallelism). To facilitate the design space exploration, development of a circuit-level benchmark simulator should be helpful to assess the trade-offs between the area efficiency, latency, and power. Furthermore, how to efficiently map various deep learning algorithms into the neuro-inspired architecture is an open question at the architecture level. Design automation tools are necessary for partitioning the array size and the number of layers given the constraints of area, latency, power, and learning accuracy. Lastly, the research in this field so far is mostly based on experimental work of single device and projection of large-scale array performance by simulations only. A large-scale chip demonstration with monolithic integration of resistive devices on top of CMOS peripheral circuits is critical to make a breakthrough in this field, as one can actually measure the speed and energy efficiency.

1.5 Challenges of Mapping Learning Algorithms to Neuromorphic Hardware

Most of the deep learning algorithms today generally rely on the availability of large datasets and the high-precision offline training to generate a huge set of model parameters, which are major limitations in mobile and dynamically varying applications. The high precision of data representation is needed by deep learning algorithms, which directly impacts the computation cost and energy efficiency. Traditionally, deep learning models are trained in the CPUs'/GPUs' environment using 32-bit (or even 64-bit) floating point, in order to satisfy the precision required by backpropagation or other gradient-based approaches; computations with such high-precision data consume significant amount of hardware resources and are impractical for emerging resistive synaptic devices. In the context of low-precision (i.e., fixed-point) training, recent research efforts have made the attempt, including binary weight storage [54] and stochastic rounding [55]. The adoption of low-precision weight is proposed for the classification (or inference) stage, due to the redundancy in the over-parameterized deep CNN models, such as the AlexNet [3] and VGG [56]. In a recent work [57], the minimum precision requirements of convolution and fully connected layer weights in CNN using the Caffe tool framework [58] were studied for inference stage. The pretrained models of the AlexNet (60 million parameters) and VGG (140 million parameters) were rounded and loaded on two Altera Stratix V FPGA platforms. The truncated models were tested using the ImageNet-2012 validation dataset of 50,000 images. The top-1 and top-5 image recognition accuracy did not drop until the weight precision reduced to 8 bits, as shown in Fig. 1.5.

Although recent efforts on network pruning and precision reduction show the promises for the classification (or inference) stage [59], high precision is still a must for the training stage due to the gradient descent in the error backpropagation or the decay of learning rate [54, 55] in the machine learning-driven algorithms. On the other hand, the biologically plausible algorithms may naturally tolerate the low precision and variations/noises in the emerging resistive synaptic devices even in the training stage. However, the biologically plausible algorithms have not yet demonstrated a competitively high learning accuracy for solving practical problems. Therefore, the research community should fundamentally rethink the algorithm and hardware co-optimization. Significant cross-layer research efforts are needed to develop new algorithms that can exploit the underlying unique device properties to realize a compact and energy-efficient mapping to the crossbar array

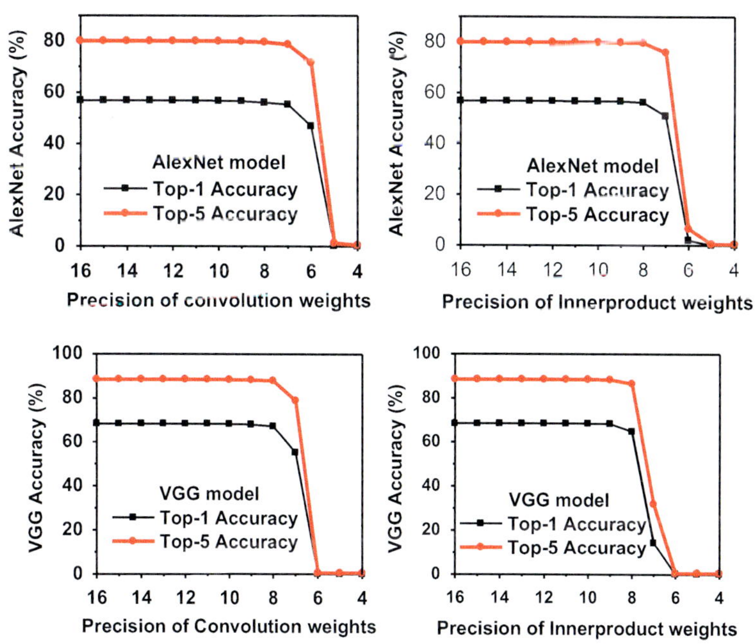

Fig. 1.5 The classification accuracy of ImageNet dataset as a function of with the weight precision in both convolution and inner-product layers in two CNN models, showing reduction to 8–10 bits is possible without accuracy loss [57]

architecture. This book aims to present state-of-the-art synaptic device properties, small-scale to medium-scale array integration, and preliminary exploration of device-architecture-algorithm co-design, with the hope of inspiring the research community for the future interdisciplinary collaborations on this exciting research topic.

References

1. J. Gantz, D. Reinsel, The Digital Universe in 2020: Big Data, Bigger Digital Shadows, and Biggest Growth in the Far East, IDC Bulletin: IDC IVIEW (2012)
2. Y. LeCun, Y. Bengio, G. Hinton, Deep learning. Nature **521**, 436–444 (2015)
3. A. Krizhevsky, I. Sutskever, G.E. Hinton, ImageNet classification with deep convolutional neural networks, in *Advances in Neural Information Processing Systems NIPS)* (2012)
4. Q. Le, M. Ranzato, R. Monga, M. Devin, K. Chen, G. Corrado, J. Dean, A. Ng, Building high-level features using large scale unsupervised learning, in *International Conference in Machine Learning (ICML)* (2012)
5. G.E. Hinton, S. Osindero, Y. Teh, A fast learning algorithm for deep belief nets. Neural Comput. **18**(7), 1527–1554 (2006)
6. K.J. Kuhn, Considerations for ultimate CMOS scaling. IEEE Trans. Electron Devices **59**(7), 1813–1828 (2012)
7. A. Sally, Reflections on the memory wall, in *Conference on Computing Frontiers* (2004)
8. C.-S. Poon, K. Zhou, Neuromorphic silicon neurons and large-scale neural networks: challenges and opportunities. Front. Neurosci. **5**(108), 1–3 (2011)
9. S. Chetlur, C. Woolley, P. Vandermersch, J. Cohen, J. Tran, cuDNN: Efficient Primitives for Deep Learning, arXiv:1410.0759 (2014)
10. G. Lacey, G.W. Taylor, S. Areibi, Deep Learning on FPGAs: Past, Present, and Future, http://arxiv.org/abs/1602.04283 (2016)
11. Y-H. Chen, T. Krishna, J. Emer, V. Sze, Eyeriss: an energy-efficient reconfigurable accelerator for deep convolutional neural networks, in *IEEE International Solid-State Circuits Conference (ISSCC)* (2016)
12. J. Sim, J-S. Park, M. Kim, D. Bae, Y. Choi, L-S. Kim, A 1.42TOPS/W deep convolutional neural network recognition processor for intelligent IoE systems, in *IEEE International Solid-State Circuits Conference (ISSCC)* (2016)
13. N. Jouppi, Google supercharges machine learning tasks with TPU custom chip (2016). [Online]. Available: https://cloudplatform.googleblog.com/2016/05/Google-supercharges-machine-learning-tasks-with-custom-chip.html
14. S.B. Furber, F. Galluppi, S. Temple, L.A. Plana, The SpiNNaker project. Proc. IEEE **102**(5), 652–665 (2014)
15. J. Schemmel, D. Bruderle, A. Grubl, M. Hock, K. Meier, S. Millner, A wafer-scale neuromorphic hardware system for large-scale neural modeling, in *IEEE International Symposium on Circuits and Systems (ISCAS)* (2010)
16. P.A. Merolla, J.V. Arthur, R. Alvarez-Icaza, A.S. Cassidy, J. Sawada, F. Akopyan, B.L. Jackson, N. Imam, C. Guo, Y. Nakamura, B. Brezzo, I. Vo, S.K. Esser, R. Appuswamy, B. Taba, A. Amir, M.D. Flickner, W.P. Risk, R. Manohar, D.S. Modha, A million spiking-neuron integrated circuit with a scalable communication network and interface. Science **345**(6197), 668–673 (2014)
17. G.Q. Bi, M.M. Poo, Synaptic modifications in cultured hippocampal neurons: dependence on spike timing, synaptic strength, and postsynaptic cell type. J. Neurosci. **18**(24), 10464–10472 (1998)

18. M. Prezioso, F. Merrikh-Bayat, B.D. Hoskins, G.C. Adam, K.K. Likharev, D.B. Strukov, Training and operation of an integrated neuromorphic network based on metal-oxide memristors. Nature **521**, 61–64 (2015)
19. S. Yu, Z. Li, P.-Y. Chen, H. Wu, B. Gao, D. Wang, W. Wu, H. Qian, Binary neural network with 16 Mb RRAM macro chip for classification and online training, in *IEEE International Electron Devices Meeting (IEDM)* (2016)
20. S. Kim, M. Ishii, S. Lewis, T. Perri, M. BrightSky, W. Kim, R. Jordan, G.W. Burr, N. Sosa, A. Ray, J.-P. Han, C. Miller, K. Hosokawa, C. Lam, NVM neuromorphic core with 64 k–cell (256-by-256) phase change memory synaptic array with on-chip neuron circuits for continuous in-situ learning, in *IEEE International Electron Devices Meeting (IEDM)* (2015)
21. B.J. Zhu, Magnetoresistive random access memory: the path to competitiveness and scalability. Proc. IEEE **96**(11), 1786–1798 (2008)
22. H.-S.P. Wong, S. Raoux, S. Kim, J. Liang, J.P. Reifenberg, B. Rajendran, M. Asheghi, K.E. Goodson, Phase change memory. Proc. IEEE **98**(12), 2201–2227 (2010)
23. H.-S.P. Wong, H.-Y. Lee, S. Yu, Y.-S. Chen, Y. Wu, P.-S. Chen, B. Lee, F.T. Chen, M.-J. Tsai, Metal–oxide RRAM. Proc. IEEE **100**(6), 1951–1970 (2012)
24. I. Valov, R. Waser, J.R. Jameson, M.N. Kozicki, Electrochemical metallization memories—fundamentals, applications, prospects. Nanotechnology **22**, 254003 (2011)
25. J.J. Yang, D.B. Strukov, D.R. Stewart, Memristive devices for computing. Nat. Nanotechnol. **8** (1), 13–24 (2013)
26. S. Yu, P.-Y. Chen, Emerging memory technologies: recent trends and prospects. IEEE Solid State Circuits Magazine **8**(2), 43–56 (2016)
27. Y. Choi, I. Song, M.-H. Park, H. Chung, S. Chang, B. Cho, J. Kim, Y. Oh, D. Kwon, J. Sunwoo, J. Shin, Y. Rho, C. Lee, M. Kang, J. Lee, Y. Kwon, S. Kim, J. Kim, Y.-J. Lee, Q. Wang, S. Cha, S. Ahn, H. Horii, J. Lee, K. Kim, H. Joo, K. Lee, Y.-T. Lee et al., A 20 nm 1.8 V 8Gb PRAM with 40 MB/s program bandwidth, in *IEEE International Solid-State Circuits Conference (ISSCC)* (2012)
28. T.-Y. Liu, T.H. Yan, R. Scheuerlein, Y. Chen, J.K. Lee, G. Balakrishnan, G. Yee, H. Zhang, A. Yap, J. Ouyang, T. Sasaki, S. Addepalli, A. Al-Shamma, C.-Y. Chen, M. Gupta, G. Hilton, S. Joshi, A. Kathuria, V. Lai, D. Masiwal, M. Matsumoto et al., A 130.7 mm2 2-layer 32Gb ReRAM memory device in 24 nm technology, in *IEEE International Solid-State Circuits Conference (ISSCC)* (2013)
29. A. Kawahara, R. Azuma, Y. Ikeda, K. Kawai, Y. Katoh, K. Tanabe, T. Nakamura, Y. Sumimoto, N. Yamada, N. Nakai, S. Sakamoto, Y. Hayakawa, K. Tsuji, S. Yoneda, A. Himeno, K. Origasa, K. Shimakawa, T. Takagi, T. Mikawa, K. Aono, An 8 Mb multi-layered cross-point ReRAM macro with 443 MB/s write throughput, in *IEEE International Solid-State Circuits Conference (ISSCC)* (2012)
30. D. Kuzum, S. Yu, H.-S.P. Wong, Synaptic electronics: materials, devices and applications. Nanotechnology **24**, 382001 (2013)
31. D. Kuzum, R.G.D. Jeyasingh, B. Lee, H.-S.P. Wong, Nanoelectronic programmable synapses based on phase change materials for brain-inspired computing. Nano Lett. **12**(5), 2179–2186 (2012)
32. B.L. Jackson, B. Rajendran, G.S. Corrado, M. Breitwisch, G.W. Burr, R. Cheek, K. Gopalakrishnan, S. Raoux, C.T. Rettner, A. Padilla, A.G. Schrott, R.S. Shenoy, B.N. Kurdi, C.H. Lam, D.S. Modha, Nanoscale electronic synapses using phase change devices. ACM J. Emerg. Technol. Comput. Syst. **9**(2), 12 (2013)
33. S.H. Jo, T. Chang, I. Ebong, B.B. Bhadviya, P. Mazumder, W. Lu, Nanoscale memristor device as synapse in neuromorphic systems. Nano Lett. **10**(4), 1297–1301 (2010)
34. T. Ohno, T. Hasegawa, T. Tsuruoka, K. Terabe, J.K. Gimzewski, M. Aono, Short-term plasticity and long-term potentiation mimicked in single inorganic synapses. Nat. Mater. **10** (8), 591–595 (2011)
35. S. Manan, O. Bichler, D. Querlioz, G. Palma, E. Vianello, D. Vuillaume, C. Gamrat, B. DeSalvo, CBRAM devices as binary synapses for low-power stochastic neuromorphic

systems: auditory (cochlea) and visual (retina) cognitive processing applications, in *IEEE International Electron Devices Meeting (IEDM)* (2012)
36. S. Park, A. Sheri, J. Kim, J. Noh, J. Jang, M. Jeon, B. Lee, B.R. Lee, B.H. Lee, H. Hwang, Neuromorphic speech systems using advanced ReRAM-based synapse, in *IEEE International Electron Devices Meeting (IEDM)* (2013)
37. T. Chang, S.-H. Jo, W. Lu, Short-term memory to long-term memory transition in a nanoscale memristor. ACS Nano **5**(9), 7669–7676 (2011)
38. S. Yu, B. Gao, Z. Fang, H.Y. Yu, J.F. Kang, H.-S.P. Wong, A low energy oxide-based electronic synaptic device for neuromorphic visual system with tolerance to device variation. Adv. Mater. **25**(12), 1774–1779 (2013)
39. I.-T. Wang, Y.-C. Lin, Y.-F. Wang, C.-W. Hsu, T.-H. Hou, 3D synaptic architecture with ultralow sub-10 f. energy per spike for neuromorphic computation, in *IEEE International Electron Devices Meeting (IEDM)* (2014)
40. S. Yu, B. Gao, Z. Fang, H.Y. Yu, J.F. Kang, H.-S.P. Wong, Stochastic learning in oxide binary synaptic device for neuromorphic computing. Front. Neurosci. **7**, 186 (2013)
41. G.W. Burr, R.M. Shelby, C.D. Nolfo, J.W. Jang, R.S. Shenoy, P. Narayanan, Experimental demonstration and tolerancing of a large-scale neural network (165,000 synapses), using phase-change memory as the synaptic weight element, in *IEEE International Electron Devices Meeting (IEDM)* (2014)
42. P.-Y. Chen, B. Lin, I.-T. Wang, T.-H. Hou, J. Ye, S. Vrudhula, J.-S. Seo, Y. Cao, S. Yu, Mitigating effects of non-ideal synaptic device characteristics for on-chip learning, in *IEEE/ACM International Conference on Computer-Aided Design (ICCAD)* (2015)
43. MNIST Handwritten Digits Dataset, [Online]. Available: http://yann.lecun.com/exdb/mnist/
44. "ImageNet," [Online]. Available: http://www.image-net.org/
45. D. Garbin, O. Bichler, E. Vianello, Q. Rafhay, C. Gamrat, L.Perniola, G. Ghibaudo, B. DeSalvo, Variability-tolerant convolutional neural network for pattern recognition applications based on OxRAM synapses, in *IEEE International Electron Devices Meeting (IEDM)* (2014)
46. M. Hu, H. Li, Q. Wu, G.S. Rose, Hardware realization of BSB recall function using memristor crossbar arrays, in *IEEE/DAC Design Automation Conference (DAC)* (2012)
47. P.-Y. Chen, D. Kadetotad, Z. Xu, A. Mohanty, B. Lin, J. Ye, S. Vrudhula, J.-S. Seo, Y. Cao, S. Yu, Technology-design co-optimization of resistive cross-point array for accelerating learning algorithms on chip, in *IEEE/ACM Design, Automation & Test in Europe (DATE)* (2015)
48. B. Liu, H. Li, Y. Chen, X. Li, T. Huang, Q. Wu, M. Barnell, Reduction and IR-drop compensations techniques for reliable neuromorphic computing systems, in *IEEE/ACM International Conference on Computer-Aided Design (ICCAD)* (2014)
49. L. Gao, I.-T. Wang, P.-Y. Chen, S. Vrudhula, J.-S. Seo, Y. Cao, T.-H. Hou, S. Yu, Fully parallel write/read in resistive synaptic array for accelerating on-chip learning. Nanotechnology. **26**, 455204 (2015)
50. K.-H. Kim, S. Gaba, D. Wheeler, J.M. Cruz-Albrecht, T. Hussain, N. Srinivasa, W. Lu, A functional hybrid memristor crossbar-array/CMOS system for data storage and neuromorphic applications. Nano Lett. **12**(1), 389–395 (2011)
51. S. Park, M. Chu, J. Kim, J. Noh, M. Jeon, B.H. Lee, H. Hwang, B. Lee, B.-G. Lee, Electronic system with memristive synapses for pattern recognition. Sci. Rep. **5**, 10123 (2015)
52. S.B. Eryilmaz, D. Kuzum, R.G.D. Jeyasingh, S. Kim, M. BrightSky, C. Lam, H.-S.P. Wong, Experimental demonstration of array-level learning with phase change synaptic devices, in *IEEE International Electron Devices Meeting (IEDM)* (2013)
53. S. Yu, P.-Y. Chen, Y. Cao, L. Xia, Y. Wang, H. Wu, Scaling-up resistive synaptic arrays for neuro-inspired architecture: challenges and prospect, in *IEEE International Electron Devices Meeting (IEDM)* (2015)

54. M. Courbariaux, Y. Bengio, J.P. David, BinaryConnect: Training Deep Neural Networks with binary weights during propagations, in *Advances in Neural Information Processing Systems (NIPS)* (2015)
55. S. Gupta, A. Agrawal, K. Gopalakrishnan, P. Narayanan, Deep learning with limited numerical precision, in *International Conference on Machine Learning (ICML)* (2015)
56. K. Simonyan, K. Zisserman, Very deep convolutional networks for large-scale image recognition, in *International Conference on Learning Representations (ICLR)* (2015)
57. N. Suda, G. Dasika, V. Chandra, A. Mohanty, Y. Ma, S. Vrudhula, J. Seo, Y. Cao, Throughput-optimized openCL-based FPGA accelerator for large-scale convolutional neural networks, in *ACM/SIGDA International Symposium on Field-Programmable Gate Arrays (FPGA)* (2016)
58. Caffe, [Online]. Available: http://caffe.berkeleyvision.org/
59. S. Han, H. Mao, W.J. Dally, A deep neural network compression pipeline: pruning, quantization, huffman encoding, in *International Conference on Learning Representations (ICLR)* (2016)

Part I
Device-Level Demonstrations of Resistive Synaptic Devices

Chapter 2
Synaptic Devices Based on Phase-Change Memory

Yuhan Shi, Scott Fong, H.-S. Philip Wong, and Duygu Kuzum

2.1 PCM Basics for Synaptic Devices

The phenomenon of resistive switching upon a change in phase of chalcogenide materials was first discovered by Stanford Ovshinsky (1922–2012) in the 1960s [2]. Phase-change materials own superior features such as scalability, high reliability, long endurance, multiple programming resistance states, and low device-to-device variation [3–7]. In the last two decades, phase-change materials have been actively explored for nonvolatile memory applications. Since 2010, device concepts based on phase-change materials have been investigated for neuromorphic computing applications [1, 8]. Nonvolatile memory cell based on phase-change material is called phase-change memory (PCM). PCM stores the information using large resistivity contrast between amorphous (high resistivity) and polycrystalline (low resistivity) states of phase-change material by applying electric pulses. The resistivity of the materials can also be modulated by applying voltage pulses with specific amplitude and duration, leading to multiple states between the high- and low-resistivity states. This unique switching behavior enables building nanoelectronic synapses using PCM. PCM based synaptic devices can emulate synaptic plasticity observed in biological synapses. The synaptic plasticity has been suggested to be the basis of learning and information storage in the biological brain [9].

Y. Shi • D. Kuzum (✉)
Department of Electrical and Computer Engineering, University of California, San Diego, 9500 Gilman Drive, La Jolla, CA 92093, USA
e-mail: dkuzum@ucsd.edu

S. Fong • H.-S.P. Wong (✉)
Department of Electrical Engineering, Stanford University, 420 Via Palou, Stanford, CA 94305, USA
e-mail: hspwong@stanford.edu

S. Yu (ed.), *Neuro-inspired Computing Using Resistive Synaptic Devices*,
DOI 10.1007/978-3-319-54313-0_2

2.1.1 Device Operation

PCM exhibits a reversible phase-change phenomenon between the amorphous and the crystalline state. In the amorphous state, the material has high resistivity. In contrast, in the crystalline state, the material has low resistivity. To understand this phase transformation, a graphical representation of the PCM device structure is shown in Fig. 2.1a [10].

To fabricate a phase-change memory cell, a layer of phase-change material is sandwiched between the top electrode contact and bottom electrode contact. The programming region and the heater element form a mushroom cell, so named because of the dome shape of the programmable region. The cell is programmed and read using voltage pulses, which generate Joule heating to make phase transformation occur. Set and reset state of PCM refers to low- and high-resistance state, respectively. To reset the PCM cell into the amorphous phase, the programming region is first melted and then quenched rapidly by applying a large electrical current pulse to bottom electrode contact for a short time period (typically <50 ns). To set the PCM cell into the crystalline phase, a medium electrical current pulse is applied to anneal the programming region at a temperature between the crystallization temperature and the melting temperature for a time period in the range of 100 ns–1 ms to crystallize. To read the state of the cell, a small electrical current is applied to measure the resistance but does not disturb the current state of the cell. The resulting temperatures for the given programming pulses used for set, reset, and read of the PCM are summarized in Fig. 2.1b [1, 10].

The ratio between the highest-resistance state and the lowest-resistance state is called dynamic range for the phase-change device. The biological synapses exhibit a cumulative weight change with a maximum of 100% and a minimum 50% during

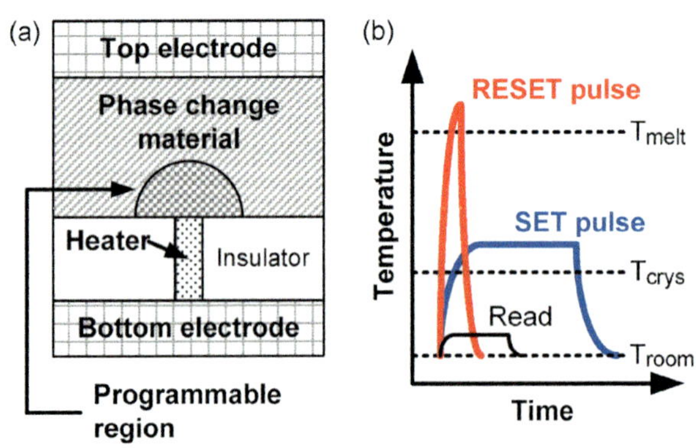

Fig. 2.1 Phase-change memory operation principle. (**a**) The cross-sectional schematic of the conventional mushroom-type PCM cell. The electrical current passes through the phase-change material between the top electrode and heater. Current crowding at the "heater" to phase-change material contact results in a programmed region illustrated by the mushroom boundary. (**b**) PCM cells are programmed and read by applying electrical pulses, which change temperature accordingly (Reprinted with permission from Wong et al. [10])

the implementation of biological synaptic plasticity. Thus, having a dynamic range of more than 4 is assumed to be sufficient to implement biological plasticity in phase-change synaptic device. However, a larger dynamic range is more desirable for implementation of robust learning using neural networks consisting of PCM synapses. Phase-change materials exhibit very high resistivity (10^3 Ohm-cm) in their amorphous state and a very low resistivity (10^{-3} Ohm-cm) in their crystalline state for the bulk form [11]. This wider programmable resistance range provides flexibility in choosing the resistance range for operation as well as allows having the multilevel states for information storage as synaptic connectivity strength.

2.1.2 Phase-Change Materials

Phase-change materials have the following three properties crucial for synaptic device applications: bistability of the amorphous and crystalline states, ability to rapidly switch between the two states, and high electrical resistance contrast between the two states. Figure 2.2 [9] shows the resistivity change with temperature for different classes of phase-change materials. The electrical resistivity contrast between the crystalline and amorphous (as deposited) state in phase-change materials can be up to six orders of magnitude [12]. A class of materials known as chalcogenides is widely used in current PCM technology because of its strong resistance contrast, fast crystallization, and high-crystallization temperature. More

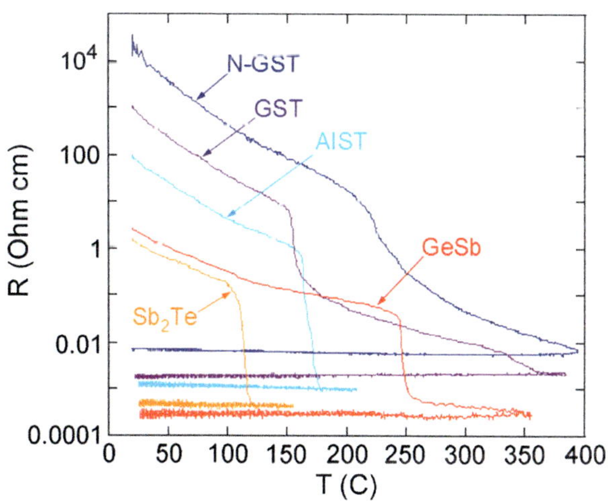

Fig. 2.2 Resistivity as a function of temperature during a heating cycle at 1 K/s for initially amorphous, as-deposited 50 nm thick films of various phase-change materials. Initially, the thin films have a high resistance that drops sharply when the crystallization temperature is reached, and it stays low upon cooling. GST-$Ge_2Sb_2Te_5$, N-GST-7 at. % N-doped GST, GeSb–Ge/Sb ratio 15:85, AIST-Sb_2Te doped with 7 at. % Ag and 11 at. % (Reprinted with permission from Raoux et al. [12])

specifically, GST ($Ge_2Sb_2Te_5$), which is located in the pseudobinary line between GeTe and Sb_2Te_3 in phase diagram, is one of the commonly used materials for memory and synaptic device applications [13].

Recently, a wide variety of materials have been developed to improve phase-change switching speeds, write energy, retention, and endurance. The most promising ones are doped GST; new materials using some of the Ge, Sb, and Te alloys; PVD-based materials; and most recently superlattice PCM. The detailed properties of these materials will be covered in Sect. 2.3.2.

2.1.3 Threshold Switching Mechanism

Threshold switching is a key mechanism for PCM operation. The current-voltage (I-V) curve for threshold switching observed in a PCM device is shown in Fig. 2.3 [10]. In the voltage below threshold switching voltage (V_{th}) region, the set and reset states have large resistance contrast. The cell in the reset state has a high resistance below V_{th} and shows electronic threshold switching behavior at V_{th}, i.e., a negative differential resistance. If the voltage pulse is removed quickly after reaching the threshold switching voltage, the PCM will return to its original amorphous state. But if the voltage above V_{th} is applied to the cell for a sufficient amount of time, it will lead the PCM cell switch into the low resistivity, namely, crystalline state. The electronic threshold switching mechanism is the most critical process that enables programming of the PCM cell [10, 14].

There are several different models proposed to explain threshold switching. In terms of thermal mechanisms, the Joule heating model is based on a simple observation that the current through the phase-change material increases exponentially due to the temperature-dependent conductivity of the phase-change material as temperature increases [15]. In terms of electronic mechanisms, the impact

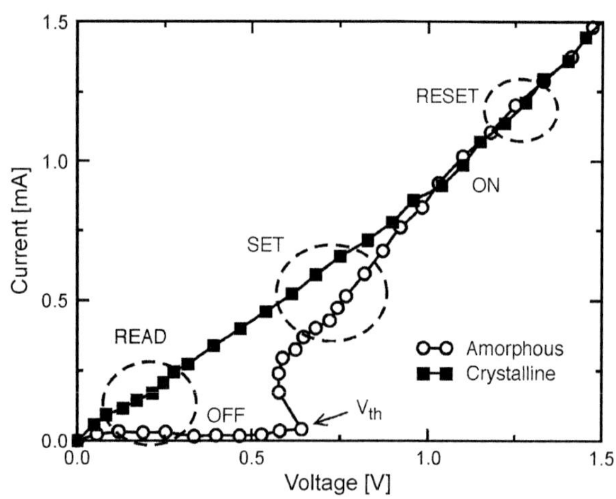

Fig. 2.3 I–V characteristics of set and reset state of phase-change memory. The reset state shows switching behavior at the threshold switching voltage V_{th}. The reset state stays in the high-resistance state below V_{th} (subthreshold region) and switches to the low-resistance state at V_{th} (Reprinted with permission from Pirovano et al. [20])

ionization model explains the threshold switching is due to strong carrier generation via impact ionization caused by high electric field and large carrier density [16–18]. However, another electronic model explains threshold switching based on the balance between electron energy gain and relaxation in the hopping transport [14]. The field-induced nucleation model attributes threshold switching to crystalline nuclei formation which is facilitated by the electric field [19]. There is not a comprehensive model that completely explains the threshold switching mechanism of the PCM. Based on different operating conditions and variations of the devices, there might be more than one model that can be used to describe the experimental observations.

2.1.4 *Resistance Drift*

The resistance drift of PCM devices has long been identified as one of the biggest challenges toward storing multiple bits per cell by programming it to different resistance values. Resistance drift occurs immediately after the PCM is programmed to the high-resistance amorphous phase. As the cell releases stress created from the volume expansion induced by the phase transformation, it undergoes a gradual resistance increase with time [21–23]. Figure 2.4 shows the conductivity of a PCM cell in the amorphous and crystalline state over time [10, 24]. Although the conductivity in the crystalline state remains constant over the time, the amorphous conductivity drifts logarithmically with time to a higher

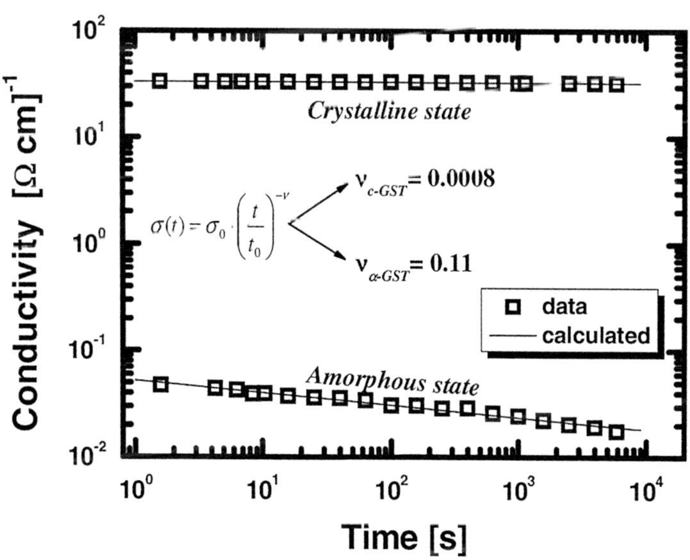

Fig. 2.4 Conductivity in the amorphous and crystalline phase of $Ge_2Sb_2Te_5$-based PCM as a function of time (Reprinted with permission from Boniardi et al. [24])

value immediately after the reset pulse is removed following the power law $R = R_0 \times (t/t_0)^{\nu}$ where R and R_0 are the present and initial resistances, t and t_0 are the present and initial times, and ν is the drift exponent [20]. The drift strongly depends on temperature [25] and the volume of the amorphous region [26]. Typically, higher temperatures and larger amorphous volumes have larger drift coefficient [27]. The cause of the drift includes relaxation of structural defects (distorted bonds, wrong bonds, vacancies, etc.), stress release [20, 24, 28], shift of the Fermi level, and increase of the band gap [20].

Fine control of resistance is required to emulate the analog transition of biological synapse states. Therefore, a fundamental requirement for any nonvolatile memory to be used as a synaptic device is the ability to be programed into a large number of resistance states with predictable read resistances. Resistance drift observed in PCM will only be a concern for implementation of multilevel synaptic states if the synaptic device is operated in the high-resistance regime, where the drift is most dominant.

2.1.5 Scaling

In the human cortex, the density of synapses is $> \sim 10^9$ mm^{-3} [29]. Thus, the size of the synaptic device is one of the important criteria to be considered in synaptic device design. PCM is highly scalable to small sizes. Sub-10 nm PCM devices have already been demonstrated for memory applications [30]. It has been also shown that phase-change memory devices with carbon nanotube electrodes (bottom electrode, ~2 nm) can exhibit low power switching [31]. In the array level, functional arrays of phase-change memory cells have already been demonstrated in 20 nm technology node [32].

2.1.6 Array Architecture and Memory Cell Selector

In this section, we focus on discussing possible choices of array architectures, which are suitable for stacking many layers of electronic synapses to mimic the parallelism and connectivity of the brain. The cross-point structure is one of the most attractive and simplest array architectures as it has the highest possible nonvolatile memory density. However, cross-point arrays consisting of only memory devices are prone to parallel leakage paths formed by unselected devices, so-called sneak paths [33]. Various forms of selector devices connected in series with the memory device have been proposed to prevent sneak path problem [34–36]. Array architectures based on the selector device choice can be categorized into three groups: 1T1R (one-transistor-one-resistor), 1D1R (one-diode-one-resistor), and 1S1R (one-selector-one-resistor).

The 1T1R array architecture consists of one transistor and one resistor. The memory element (1R) can only be accessed by controlling the on/off of the transistor (1T). When the voltage equals to V_{dd} applied to the gate of the transistor in word line, the memory cell can be programmed. The states of memory cell, either 1 (low-resistance state) or 0 (high-resistance state), can be read by sense amplifiers. For PCM cell, 1T1R cell can effectively avoid the leakage currents through the system because every single memory cell has its own dedicated transistor. However, phase-change memory requires large current levels for programming resulting in a need for large transistors to supply enough current for the memory cell. Thus, the scaling of transistors is still a challenge for a massively parallel system with ultra-small synaptic devices.

The 1D1R has been proposed to solve the problems caused by sneak leakage by allowing the electric current to pass in one direction. Different types of diodes have been proposed such as a poly-Si diode [37], binary oxide diodes [38, 39], a hybrid organic/inorganic diode [40], and Schottky diodes [41], which can bring the memory cell size down to $4F^2$. The cross-point PCM fabricated with the selection diode made of poly-Si has been developed to make the set/reset operation possible in $4F^2$ cell [37].

The 1S1R contains a resistor and a nonlinear device as the memory selector. The memory cell selectors are not only limited to transistor or diode but also include the use of metal-insulator transition (MIT) [42] and the ovonic threshold switch (OTS) [43]. Recently, significant improvements have been made using OTS as a selector device, and it will be discussed in Sect. 2.3.3. In addition, a detailed review of state-of-the art selector devices can be found in reference [44].

2.1.7 *Power Consumption*

The biological brain is a highly energy efficient system. There are approximately 10^{11} neurons and 10^{15} synapses in the brain. Neurons produce spikes with a frequency of 10 Hz, and only 1% of the neurons are active at the same time [45]. Total power consumption of human brain is ~20 W. If we assume that the total power consumed by synapses is of the order of 10 W (with the rest of the power consumption taken up by the interconnection network and the neurons), then the power consumption per synaptic event can be estimated to be 10^{-12} W. Each synaptic event has a duration of ~100 ms, leading to an energy consumption requirement per synaptic event of 100 fJ. That sets the approximate biological limit for energy consumption per synaptic event. However, the metrics for synaptic energy consumption for hardware implementation depends on the scale of the system and total number of synapses. If one targets to build a system consisting of 10^{14} synapses with a power budget of 10 W, the synapse energy consumption requirement can be stretched to 1 pJ per synaptic event. In summary minimizing the energy consumption of the synaptic devices in a neuromorphic system is a high priority metric for building energy-efficient, massively parallel systems with a large number of synaptic devices.

High power consumption of PCM has been a challenge for memory applications. Since PCM works by melting and quenching the phase-change layer, high energies and high temperatures, particularly for reset operations, are required to heat up the phase-change layer. For synaptic device applications, energy consumption of PCM could be minimized by carefully choosing the operation regime. As an illustrative example, energy consumption of PCM synaptic devices has been investigated in high-resistance (high-R) and low-resistance (low-R) regimes. Figure 2.5 shows the applied pulses and the current flowing across the synapse during gradual reset and set programming, respectively, in low-R (blue) and high-R (red) regimes. The reset current is observed to be rather constant (~2 mA) or weakly dependent on reset pulse amplitude, while the set current strongly depends on the set pulse amplitude, varying in the range of 40–500 μA. Minimum reset and set energy are 50 pJ (Iow-R regime) and 500 fJ. (high-R regime), respectively (Fig. 2.6). Energy consumption can be optimized by operating the synapse in high-R or low-R regimes depending on the desired plasticity characteristics [46]. For PCM, scaling devices, improving the heating efficiency, and using new materials or superlattice structures are the major approaches explored for reducing power. PCM device with carbon nanotube contacts have shown energy consumption as low as 210 fJ/operation [30, 47]. Here, we will focus on the first two approaches and further discuss the new materials in Sect. 2.3.2.

Since PCM operates by storing the information in the phase of the material, as the device dimensions scale, so does the effective volume which is melted and thus

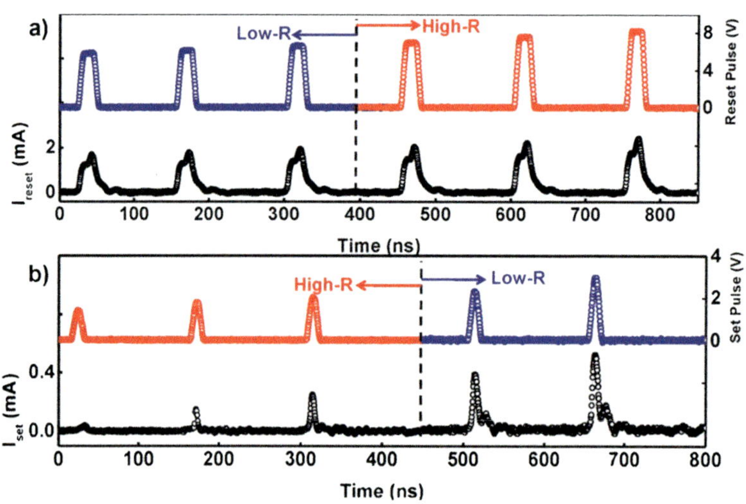

Fig. 2.5 Current consumption in low-R and high-R. Applied pulses and current flowing across the synapse measured by oscilloscope during (**a**) gradual reset and (**b**) gradual set programming are shown, respectively, in low-R (*blue*) and high-R (*red*) regimes. (**a**) Reset pulse rise time, pulse width, and fall time are 5 ns, 20 ns, and 5 ns, respectively, and (**b**) set pulse rise time, pulse width, and fall time are 5 ns, 10 ns, and 5 ns, respectively (Reprinted with permission from Kuzum et al. [46])

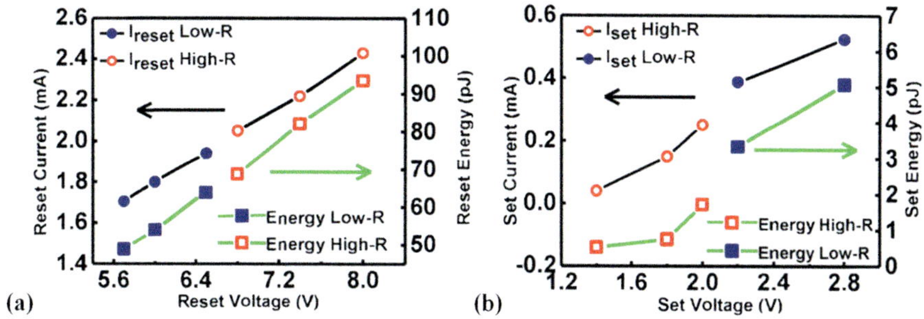

Fig. 2.6 Reset and set power consumption profile. (**a**) Reset energy consumption is in 70–90 pJ range for high-R (*red*) regime and in 50–70 pJ range for low-R (*blue*) regime. (**b**) Set energy consumption is 500 fJ for high-R (*red*) regime; it goes up a maximum of 5 pJ for low-R (*blue*) regime (Reprinted with permission from Kuzum et al. [46])

Fig. 2.7 Scaling trends of PCM. Data showing the reset current as a function of contact area and equivalent diameter. Red signifies data collected in [10], and black signifies data collected with CNT-based contacts in [50] (Reprinted with permission from Xiong et al. [50])

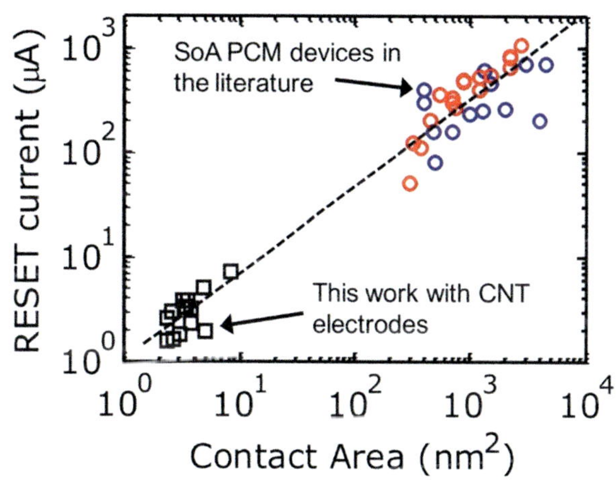

the required power to program. As shown in Fig. 2.7, the current required for the reset operation consistently decreases as the contact area of the PCM decreases. The scaling trend shows that the programming current and thus the power required will scale at a rate of about 1/k, where k is the feature size for a wide variety of structures [48]. This trend has been confirmed even for very small feature sizes with patterned devices of 7.5 × 20 nm [49] and CNT-contacted devices scaling down to 2 nm widths [30, 31, 50]. These works together show that as the device feature size is scaled to sub-10 nm^2 contact areas, there is an expected power reduction toward the single ~pJ range [44].

Further reduction in power can be attained by minimizing the heat loss to the surrounding structure. In many thermally unoptimized devices, >99% of the heat is lost to the surrounding structure [51, 52]. This means that most of the heat generated into the cell ends up heating up the nearby oxide and metal lines. Figure 2.8 shows the heat lost through various paths in mushroom and confined cells. This heat loss

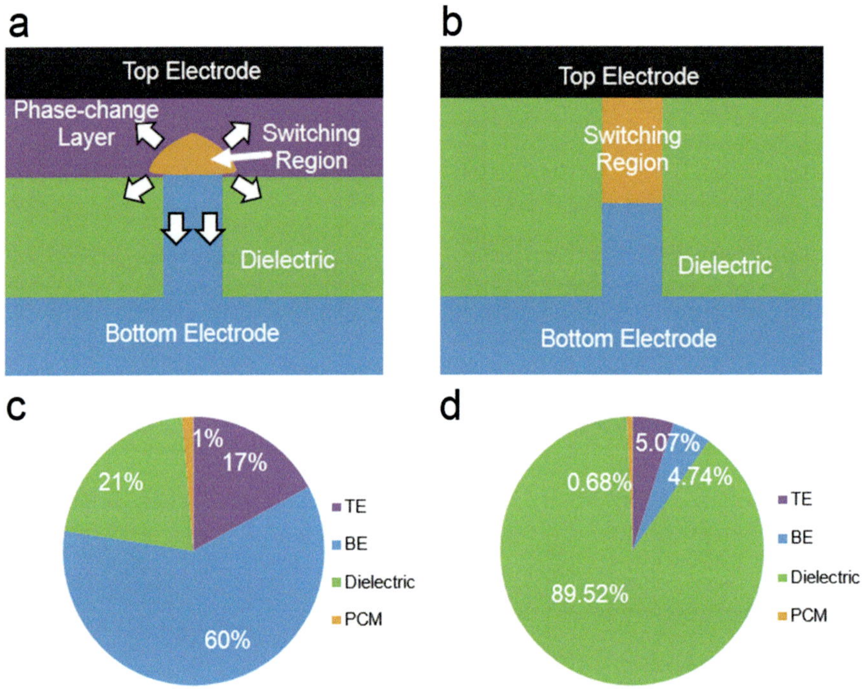

Fig. 2.8 Thermal model of PCM. Schematic of a typical (**a**) mushroom and (**b**) confined cell. Both have various heat loss paths as labeled. (**c**, **d**) Relative heat loss through each of the paths as described in the respective cells (Reprinted with permission from Sadeghipour et al. [51] and Fong et al. [52])

can be reduced by using thermally resistive electrodes and oxides to retain heat within the cell and minimize heat wasted. To this end, various electrodes such as TiN/TaN multilayers [53], fullerenes [54], and graphene [55] have all been used. Additionally, dielectric multilayers are currently being investigated for replacing the dielectric [52]. Each of these works has shown reduced power consumption by improving the thermal resistances of the structure. As devices are scaled more aggressively, the thermal boundary resistances between layers contribute a larger component of the total thermal resistance [56, 57].

2.2 PCM Synaptic Device Implementations

2.2.1 PCM as an Electronic Synapse

PCM, as one of the most mature emerging nonvolatile memory, has gained considerable attention over the years for use as electronic synapses in biologically inspired neuromorphic systems. A single-element phase-change electronic synapse with the capability of both the modulation of the time constant and the realization of the different STDP types has been first demonstrated by *Kuzum* et al. in 2011 [8]. In the same year, the group of *Wright* et al. used PCM to execute arithmetic processes such as addition, multiplication, division, and subtraction by programming it optically [58]. Later, *Suri* et al. developed 2-PCM synapses to implement a simplified rule where long-term potentiation and long-term depression can both be produced with a single invariant crystallizing pulse. They also showed a real-world application of extracting complex patterns from recorded video data by simulating two-layer feedforward spiking neural networks (SNN) [59, 60]. *Jackson* et al. investigated different programming schemes to reproduce biological synaptic plasticity in 10 nm and 40 nm PCM-based synaptic devices [61]. Additionally, various novel pulse programming schemes have been proposed by different groups to reduce the complexity and power consumption of neuromorphic circuits using PCM devices [62–64].

In addition to building single synapses using PCM, building array level systems have been widely investigated. *Eryilmaz* et al. experimentally employed a Hopfield network consisting of 100 synaptic devices and 10 recurrently connected neurons for implementation of brain-like associative learning [65]. *Burr* et al. used two PCM devices per synapse, a three-layer perceptron with 164,885 synapses to build large-scale neural network, which was trained with backpropagation [66]. A 64k-cell (256-by-256) PCM synaptic array with on-chip neuron circuits for continuous in situ learning was developed by *Kim* et al. The novel 2T1R (two transistors one resistor) structure enables both LIF and STDP learning to operate asynchronously within the array [67]. STDP-based unsupervised learning was first shown using 1T1R PCM arrays by *Ambrogio* et al. [68]. Additionally, while most of the PCM synaptic device implementations have used $Ge_2Sb_2Te_5$ as the phase-change material, a recent report proposed $Ge_1Cu_2Te_3$ (GCT) as a promising candidate for higher-density PCM arrays [69]. In order to compare different synaptic devices mentioned above, Table 2.1 summarizes past and recent advances reported in the literature. The detailed explanations in terms of working principle and pulse schemes for plasticity will be covered in Sect. 2.2.3.

2.2.2 Synaptic Plasticity and Learning

The two basic computational units in the brain are neurons and synapses. Each neuron is connected through 1000–10,000 synapses to other neurons. A neuron integrates the inputs coming from other neurons and generates action potentials (spikes) to perform computation. The synapses contribute to the computation by changing their connection strength as a result of neuronal activity, which is known as synaptic plasticity [1]. Synaptic plasticity has been found in both excitatory and inhibitory synapses, and it is the foundation of learning and memory in the human brain [73]. An excitatory synapse in a presynaptic neuron can increase the

Table 2.1 A representative list of PCM synaptic device characteristics

PCM synaptic device	Dimensions D or W × L	Energy consumption	Programming time	Multilevel states	Max dynamic range	Achievable retention/endurance
$Ge_2Sb_2Te_5$ (Kuzum et al. [8])	$D = 75$ nm	2–50 pJ	~50 ns	100	~1000	Years/10^{12}
$Ge_2Sb_2Te_5$ (Suri et al. [60])	$D = 300$ nm	121–1552 pJ	~50 ns	30	~100	N/A
$Ge_2Sb_2Te_5$ (Jackson et al. [61])	$D = 10$ nm $D = 40$ nm	5 pJ	~50 ns	>20	~100	N/A
$Ge_2Sb_2Te_5$ (Eryilmaz et al. [65, 70])	$D = 90$ nm [71, 72]	52.8 nJ	~50 ns reset ~300 ns set	9	~300	N/A
Doped GeSbTe (Kim et al. [67])	$D = 90$ nm	~0.9 pJ	~20–164 ns	N/A	~1000	N/A
$Ge_2Sb_2Te_5$ (Kang et al. [63])	$D = 39$ nm	~200 pJ–300 pJ	~80 ns–250 ns	N/A	~300	N/A
$Ge_1Cu_2Te_3$ (An et al. [69])	$D = 50$ nm	N/A	~50 ns	N/A	~100	3×10^3 cycles
$Ge_2Sb_2Te_5$ (Ambrogio et al. [68])	$D = 45$ nm	1 pJ	~40–250 ns	3	~10^3	N/A
$Ge_2Sb_2Te_5$ (Zhong et al. [64])	$D = 125$ nm	1.9 pJ–13.9 pJ	~40 ns reset ~500 ns set	<10	<10	N/A

probability of an action potential occurring in a postsynaptic cell, and an inhibitory synapse can decrease this probability. In addition, the concept of synaptic plasticity has been heavily influenced by Hebb's postulate stating that the connection strength between neurons is modified based on neural activities in presynaptic and postsynaptic cells [74, 75] (Fig. 2.9a). The synaptic plasticity can be categorized as short-term plasticity and long-term plasticity depending on the duration of the potentiation or depression event. Here, we focus on use of PCM technology for implementing long-term plasticity. A form of long-term plasticity, so-called spike-timing-dependent plasticity (STDP), depends on the relative timing of pre- and postsynaptic spikes. The synapse potentiates (increases in synaptic weight or conductance) if a presynaptic spike precedes a postsynaptic spike repeatedly, and the synapse depresses (decreases in synaptic weight or conductance) if a postsynaptic spike precedes a presynaptic spike repeatedly (Fig. 2.9b). Different forms of STDP showing different temporal dependency on spike timing and order have been observed in biological synapses (Fig. 2.9c) [1, 76–80]. As with other forms of synaptic plasticity, STDP attracts significant attention because it underlies learning

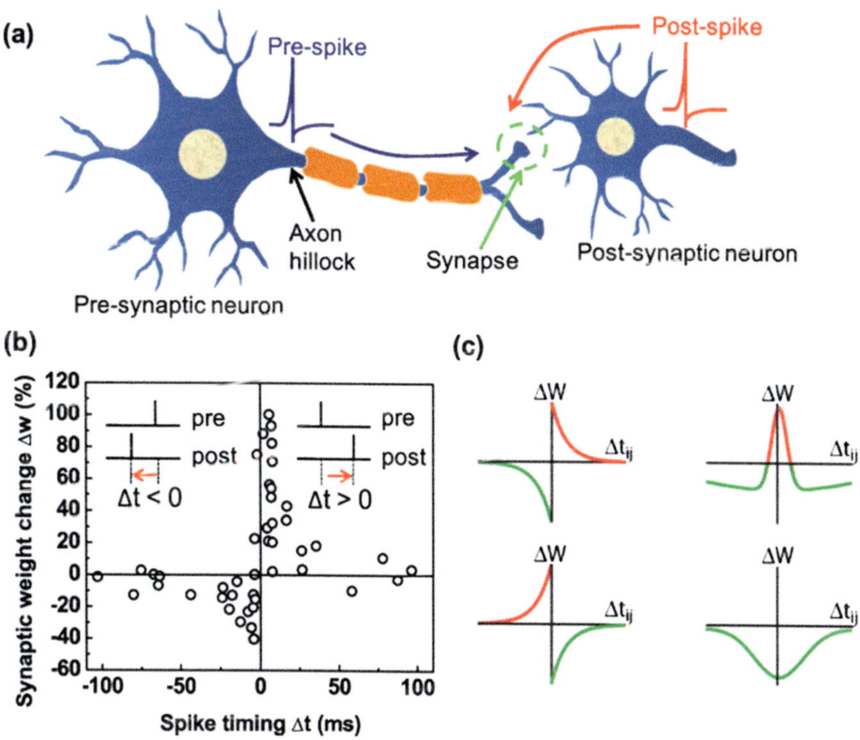

Fig. 2.9 Synaptic plasticity. (**a**) Relative timings of neuronal spikes from the presynaptic neuron and the postsynaptic neuron determine the weight change in synapse. (**b**) Synaptic weight change is plotted as a function of relative timing of pre- and post-spikes (Reprinted with permission from Bi and Poo [82]) (**c**) Diverse forms of STDP (Reprinted with permission from Shouval et al. [76])

and information storage in the brain, as well as the development and refinement of neuronal circuits during brain development [80, 81].

In general, STDP can be summarized as a mechanism, which regulates the connection strength of neurons depending on their coactivity and spike timing. Hence, the most important requirement to emulate strengthening or weakening a synapse is the gradual modulation of the conductance (or resistance) of the PCM device. The increase in synaptic strength (or potentiation) can be implemented by decreasing the resistance of the PCM synapse, in other words, by creating crystalline regions inside the chalcogenide material. Similarly, the decrease in synaptic strength (or depression) can be implemented by increasing the PCM resistance or creating amorphous regions in the chalcogenide [83]. To realize fine control of resistance, which is close to 1% change per synaptic activity, gradual programming has been demonstrated by using various schemes including increasing voltage pulse amplitude, number, or duration as discussed in Sect. 2.2.3.

2.2.3 *Pulse Schemes for Plasticity*

Gradual programming and a continuous transition between intermediate resistance states is the first requirement to achieve the analog nature of synaptic weight change in biological synapses. Various ways of programming PCM devices have been proposed to achieve gradual switching. *Kuzum* et al. have used gradually increasing voltage pulses with a custom engineered timing to probe the intermediate resistance levels and to maintain continuous transitions between adjacent levels (Fig. 2.10a) [8]. Gradual reset of the cell resistance was implemented by using pulses with increasing amplitude in the 2–4 V voltage range with 20 mV voltage steps. The pulse width, rise, and fall times of the reset pulses were 75 ns, 25 ns, and 25 ns, respectively. Gradual set of the cell resistance was achieved by using staircase pulses with an increasing step of 0.1 V. Each voltage value in the step was repeated continuously for 20 pulses, i.e., 20 pulses for each voltage 0.5, 0.6, 0.7, 0.8, and 0.9 V. The pulse width, rise, and fall times of the set pulses were 5 μs, 500 ns, and 500 ns, respectively. An order of magnitude change in the phase-change cell resistance was achieved through 100 steps for both the set and reset transitions. This way of programming using voltage pulses with increasing amplitude has resulted in a smooth transition with minimal variations. *Zhong* et al. have also demonstrated gradual programming using voltage pulses with increasing amplitude (Fig. 2.10b) [64]. In this case, choice of larger voltage steps (50 mV) has resulted in less number of resistance levels and higher variation in gradual switching characteristics. *Suri* et al. first examined the response of PCM synaptic devices by applying identical potentiating and depressing pulses (Fig. 2.10c) [83]. For the synaptic potentiation experiment, the device was initialized to a high-resistance state (110 kΩ) by applying a strong reset pulse (8 V, 500 ns). After that, 30 identical potentiating pulses (all equal to 2 V, with a fixed pulse width) were applied. Similarly, for the synaptic depression experiment, the device was first initialized

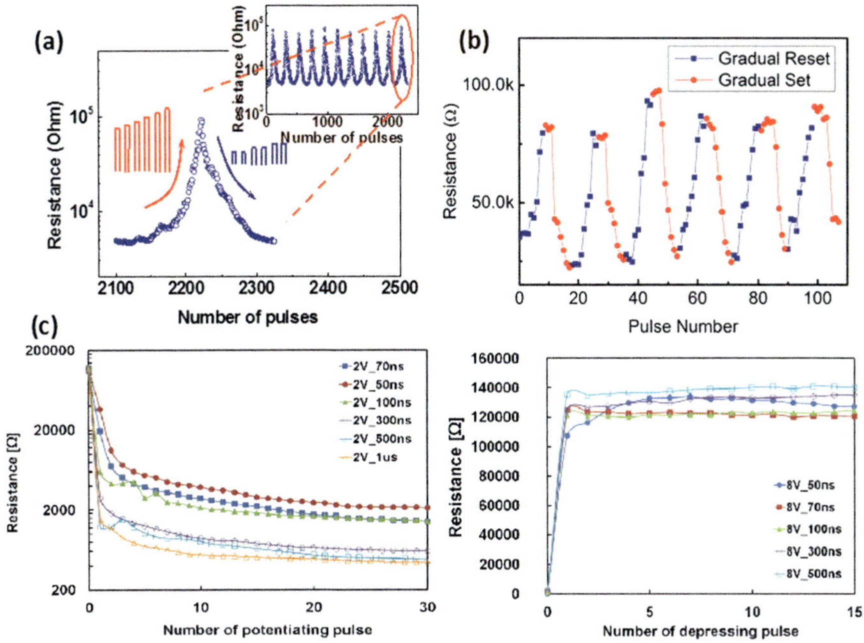

Fig. 2.10 (**a**) Gradual reset of the cell resistance is implemented by using pulses with increasing amplitude in 2–4 V range with 20 mV voltage steps. Gradual set of the cell resistance is achieved by using staircase pulses with an increasing step of 0.1 V. There are 20 pulses for each voltage step, 0.5, 0.6, 0.7, 0.8, and 0.9 V (Reprinted with permission from Kuzum et al. [8]). (**b**) Gradual set of the resistance realized by using pulses with amplitude in the 0.75–1.25 V range with 50 mV increasing steps; the pulse width, rise, and fall times of the set pulses are 500 ns, 10 ns, and 10 ns, respectively. Gradual reset of the resistance realized by using pulses with amplitude to the 1.5–2 V range with 50 mV increasing steps; the pulse width, rise, and fall times of the reset pulses are 40 ns, 10 ns, and 10 ns, respectively (Reprinted with permission from Zhong et al. [64]). (**c**) Identical potentiating and depressing pulses. *Left*: Experiment showing emulation of synaptic potentiation behavior using lance-type GST PCM devices. Conductance can be modulated as desired by choosing the right potentiating pulse width. The device was initialized with a strong reset pulse before starting the test. *Right*: Experiment showing emulation of synaptic depression behavior using lance-type GST PCM devices. The depression behavior is more abrupt due to physics governing the amorphization. The device was initialized with a long set pulse before starting the test (Reprinted with permission from Suri et al. [84])

to a very low resistance (around 400 Ω) by applying a potentiation pulse (2 V, 1 μs). After the application of the potentiation pulse, 15 identical depression pulses (each of 8 V, with a fixed pulse width) were applied. To read the resistance, a small read pulse of 0.1 V was applied between two consecutive potentiation pulses. The process was repeated with different pulse widths. The result of resistance measurements for long-term potentiation (LTP) and long-term depression (LTD) is shown in Fig. 2.10c. The results indicate that the resistance gradually decreases with each potentiating pulse, until it reaches a saturating minimum value. However, synaptic depression in PCM does not exhibit a gradual change, and it rather shows an abrupt increase in resistance. The explanation for that is that the impact of applying several short crystallizing pulses is almost the same as applying a single long crystallizing pulse of the same amplitude because PCM is an energy cumulative process. For amorphization, the volume of the molten region created inside the chalcogenide layer increases with the amplitude of the applied reset pulse. Changing the pulse duration does not significantly change the programmed resistance since the entire volume is melted and quenched during all of the pulse durations. This problem can be addressed by applying voltage pulses with increasing amplitude as shown by *Kuzum* et al. [8].

Once gradual programming is achieved, the next step requires development of pulse schemes for implementation of spike-timing-dependent plasticity. To emulate the STDP, pre-spikes and post-spikes are applied to the top electrode and bottom electrode of the cell, respectively. The spike scheme translates the arrival timing relationship of the pre- and the post-spikes into programming pulses (amplitude, duration, and number of pulses) for the phase-change synaptic device. The choice of pulse scheme is important to reduce the complexity of the neuromorphic system and decrease the total power consumption.

The initial scheme proposed to emulate STDP was a multi-pulse scheme, in which the pre-spike consists of set and reset pulses with changing amplitudes [8]. In this scheme, the pre-spike was a pulse train, consisting of depression pulses with increasing amplitudes and potentiation pulses with decreasing amplitude. Depression and potentiation correspond to reset and set states, respectively. Total duration of pre-spike was 120 ms. The pulse width, rise, and fall times of depression pulses were 50 ns, 10 ns, and 10 ns, respectively, while they are 1 μs, 100 ns, and 100 ns for potentiation pulses. In this scheme, the post-spike serves as a gating function for the pre-spike. It is a low amplitude, continuous pulse with 120 ms duration with a short negative amplitude pulse of 8 ms at the center. Depending on the overlap of pre-spike and post-spike in time, the difference between the amplitudes (Vpre – Vpost) defines the net programming voltage applied across the electronic synapse at each point in time. Depending on the different spike time of the pre-neuron and post-neuron, the overlapping parts of the pre-spike and post-spike results in different changes in synaptic weight. For example, when the post-neuron spikes before the pre-neuron, it generates a negative synaptic weight change. In the opposite scenario when the post-neuron spikes after the pre-neuron, it generates a positive synaptic weight change. This multi-pulse scheme and the STDP measurement results are shown in Fig. 2.11. STDP results were consistent with the biological data measured in the hippocampal glutamatergic synapses proposed by Bi and Poo [82]. More interestingly, using the same pulse scheme technique, the symmetric STDP has been demonstrated by simply changing the amplitude and the order of pulses in pre-spike as shown in Fig. 2.12 [85].

The disadvantage of the multi-pulse scheme is undesired disturbance generated by additional pulses during the programming and excess energy consumption associated with the charging of interconnected wires of a large-scale synapse array. To address these issues, single-pulse scheme has been proposed [85]. In this scheme, the timing of the spike can be tracked by the neuron circuitry. Every

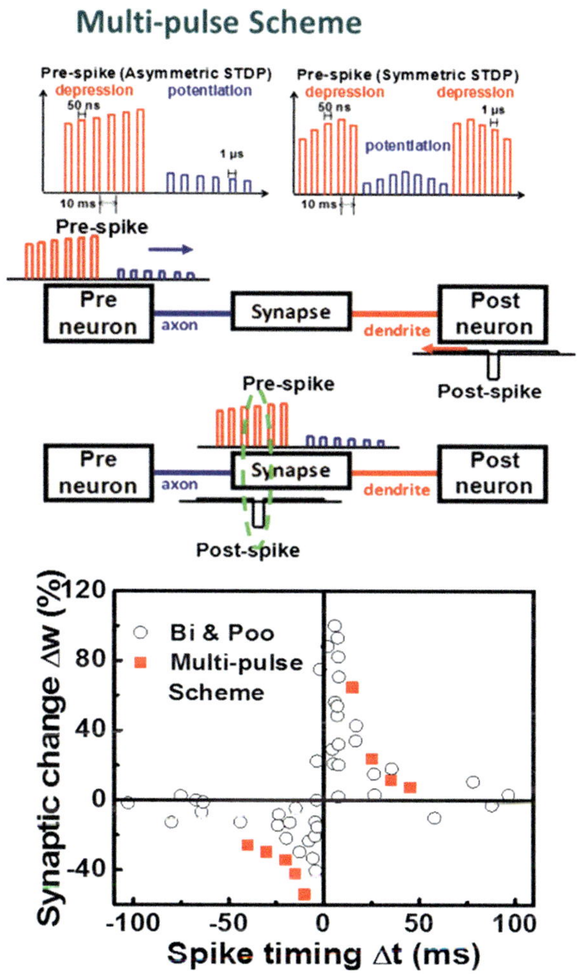

Fig. 2.11 Multi-pulse scheme: the pre-spike consists of set and reset pulses with changing amplitudes, and the post-spike is a single pulse acting as gating function. At the top, pre-spikes for asymmetric and symmetric STDP are shown. This scheme translates timing difference into the pulse amplitude and order difference. Depending on the timing difference, one of the set or reset pulses in the pre-spike overlaps with the post-spike, resulting in a change in synaptic weight (Reprinted with permission from Kuzum et al. [1])

time neurons spike, they generate a communication signal, a low amplitude square pulse. The amplitude of the communication signal is significantly below the set threshold. Hence, it does not cause any disturbance in cell resistance. The communication signal is transmitted through the axons and dendrites of the neuron (Fig. 2.13) to all the neurons, which have synapses with spiking neuron. Once communication signal is received, the receiving neuron samples its internal neuron voltage, generates a response spike, and initializes internal neuron voltage. Internal neuron voltage is a function of spike time, and the timing between last time it spiked and the time it received communication signal determines the magnitude of response spike. The neuron, which has sent out communication signal, also generates a gating spike after a specific timing delay. The response spike from the neuron (postsynaptic neuron in Fig. 2.13), which has received communication signal, overlaps with gating spike from the neuron (presynaptic neuron in Fig. 2.13),

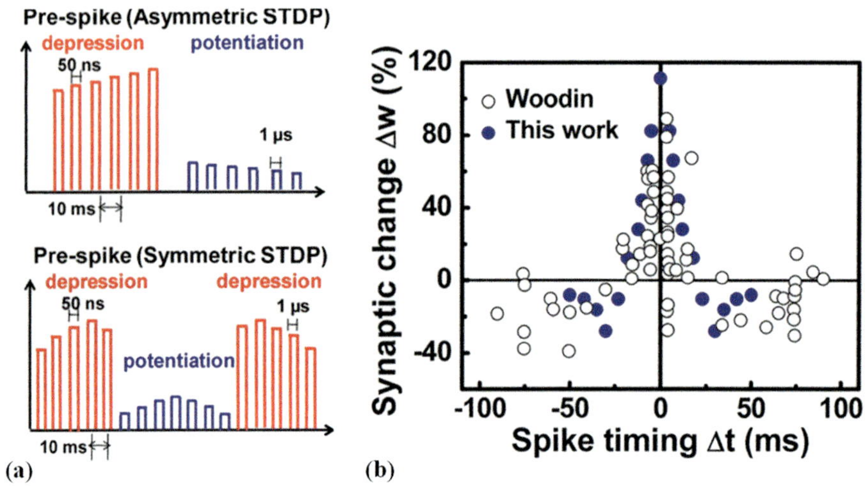

Fig. 2.12 (**a**) Spikes for asymmetric and symmetric STDP are shown (Reprinted with permission from Kuzum et al. [85]). (**b**) Asymmetric STDP behavior is obtained (Reprinted with permission from Woodin et al. [86])

which has transmitted communication signal initially. That overlap results in relative synaptic weight change. Use of gating spike enables that only the synapses between the spiking neuron and the associated receiving neuron are programmed. STDP characteristics implemented using single-pulse scheme is shown in Fig. 2.13.

Using these pulse schemes, four different forms of STDP observed in biological synapses can be demonstrated with PCM synapses (Fig. 2.14a). Another scheme has utilized different time overlap of single voltage pulses to achieve different forms of STDP [63] (Fig. 2.14b) similar to the single-pulse scheme used in reference [85].

In order to address abrupt transitions for the depression programming, a two PCM synapse device has been proposed. *Suri* et al. showed a system containing two PCM devices: one is used to implement synaptic potentiation (LTP-device), while the other one is used to implement synaptic depression (LTD-device) [59]. In both cases, the device undergoes crystallization process. Although with this scheme, the conductance of both PCM devices keeps increasing upon undergoing LTP and LTD, the contribution of the currents through the LTP device is positive, while the contribution through the LTD device is negative. The negative current through the LTD device acts like synaptic depression, because the current flowing through it is subtracted in the post neuron. The operation principle of the two PCM synapse device is shown in Fig. 2.15 [60].

While initial work of PCM synapses focused on single device demonstrations and fine-tuning of device characteristics, the interest in developing synaptic arrays for neuromorphic computation has increased the efforts toward PCM architectures with selector devices. The use of a selector device or multiple devices in series with the PCM synapse has brought more flexibility in designing pulse schemes for

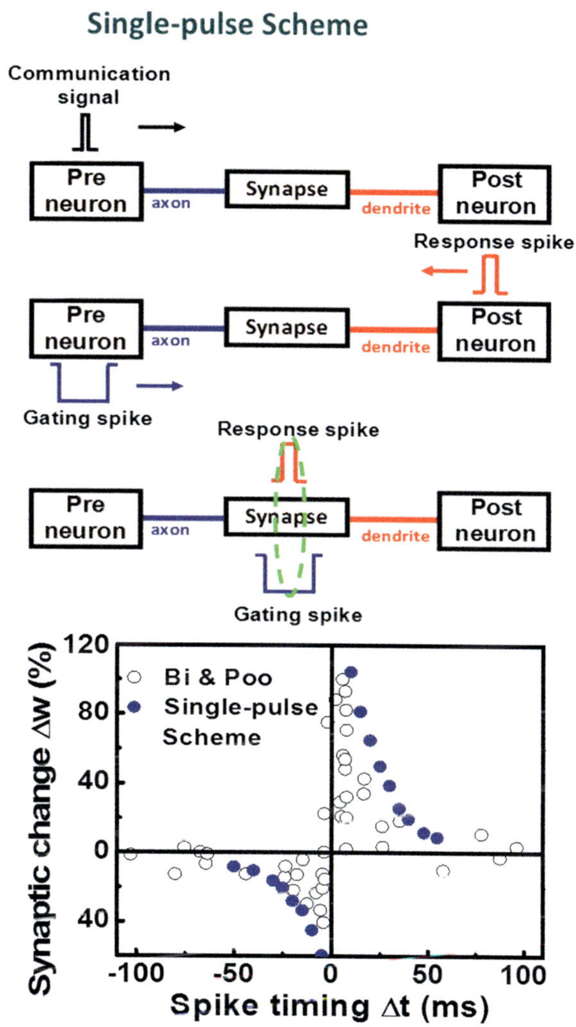

Fig. 2.13 Single-pulse scheme: the single-pulse scheme makes use of a communication signal between the presynaptic neuron and postsynaptic neuron, and the timing of the spike is tracked by the neuron circuitry. Whenever a neuron spikes, it sends a communication signal through its axon. A neuron, which receives a communication signal, samples its internal potential and generates a response spike. The overlap of gating spike and response spike across the synapse results in synaptic weight change (Reprinted with permission from Kuzum et al. [1])

synaptic plasticity. An example of synaptic plasticity implementation with 1T1R PCM devices was demonstrated by [68]. The 1T1R synapse contains three terminals: the gate electrode of the transistor, the top electrode of the PCM, and the bottom electrode, consisting of the transistor channel contact not connected to the PCM. The gate of the transistor is driven by the presynaptic neuron (PRE), which applies a sequence of rectangular spikes. The positive gate voltage activates a current spike in the synapse, which is fed into the postsynaptic neuron (POST). After the integrate-and-fire stage, the POST delivers a pulse back to the top electrode to update the weight of the synapse. The top electrode spike contains two rectangular pulses, one for set and the other for reset. Depending on the timing difference between PRE and POST, potentiation or depression is induced as

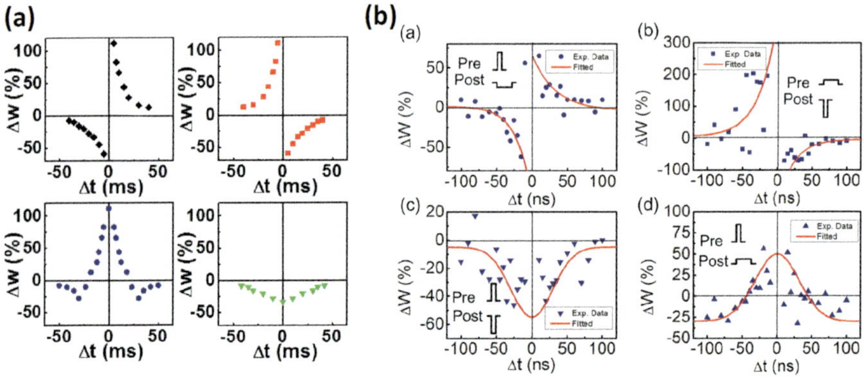

Fig. 2.14 (**a**) Different forms STDP learning rules. Upper left is asymmetric STDP with potentiation at positive Δt. Upper right is asymmetric STDP with potentiation at negative Δt. Lower left is symmetric STDP with potentiation at positive $\Delta t = \sim 0$. Lower right is symmetric STDP with depression at all Δt (Reprinted with permission from Kuzum et al. [8]). (**b**) Implementation of STDP with square pulse as the pre- and postsynaptic spike with (**a**) the antisymmetric Hebbian learning rule with potentiation at positive t, (**b**) the antisymmetric anti-Hebbian learning rule with potentiation at negative Δt, (**c**) the symmetric antisymmetric Hebbian learning rule with depression at all Δt, and (**d**) the symmetric Hebbian learning rule with potentiation at $\Delta t \sim 0$ (Reprinted with permission from Zhong et al. [64])

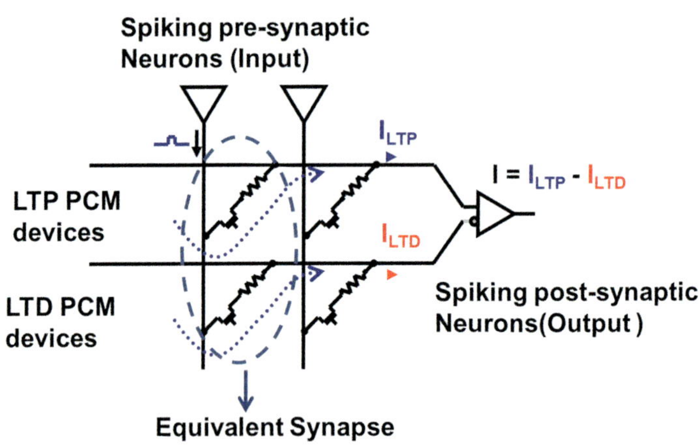

Fig. 2.15 Schematic showing the concept of 2-PCM synapse. The contribution of the current from the LTD device is subtracted at the postsynaptic neuron (Reprinted with permission from Suri et al. [60])

explained in Fig. 2.16b. The working principles of the 1T1R structure and pulse scheme are shown in Fig. 2.16a, b, respectively.

Another different pulse scheme called time-modulated voltage (TMV) scheme has been implemented in two PCM devices by *Kang* et al. [63]. The concept was to use two resistive memory cells connected to presynaptic and postsynaptic neurons

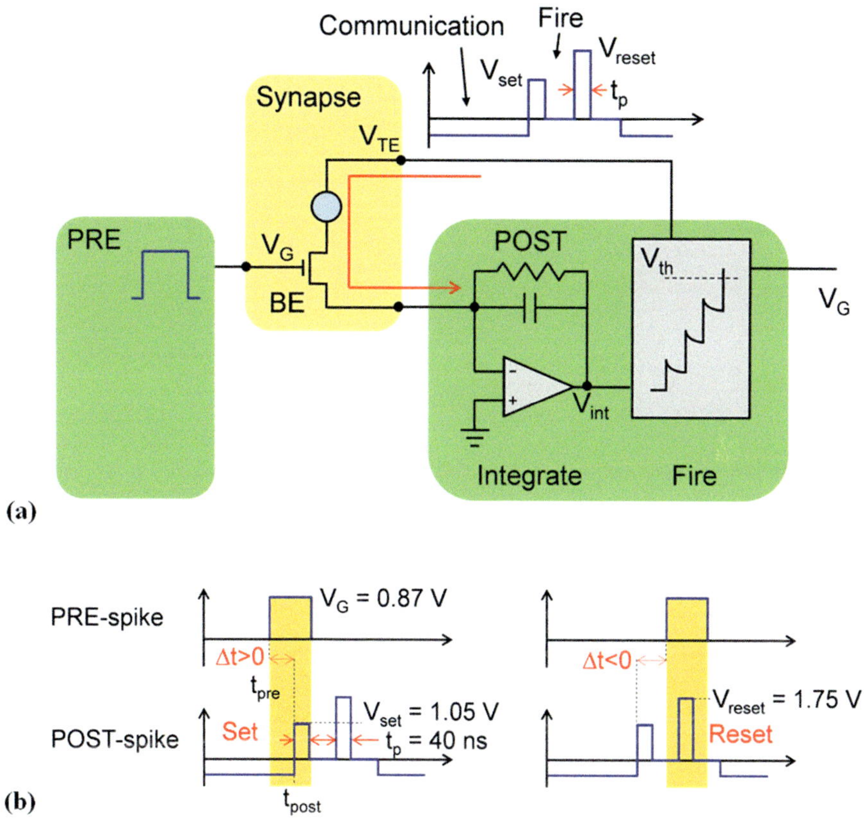

Fig. 2.16 1T1R synapse. (**a**) Schematic illustration of the neuromorphic network with a 1T1R synapse. The PRE drives the MOS transistor gate voltage VG, thus, activating a current spike due to the low negative TE voltage (VTE = −30 mV) set by the POST. The current spikes are fed into the POST, which eventually delivers a VTE spike back to the synapse as the internal voltage Vint exceeds threshold Vth. The VTE spike includes a set and reset pulse to induce potentiation/depression according to the STDP protocol. (**b**) Scheme of the applied pulses from the PRE and POST neurons to the 1T1R synapse. In the case of small positive delay Δt, when the PRE spike is applied just before the POST spike, the PCM receives a potentiating pulse with voltage Vset inducing set transition. On the other hand, for small negative delay Δt, when the PRE spike is applied just after the POST spike, the PCM receives a depressing pulse with voltage Vreset inducing reset transition. For positive/negative delays larger than 10 ms, there is no overlap between PRE and POST spikes; thus, no potentiation/depression can take place (Reprinted with permission from Ambrogio et al. [68])

through their respective bit lines, which is shown in Fig. 2.17a. One cell is a reference cell with fixed resistance, and the other acts as a control cell with variable resistance. The V_{ref} applies to the reference PCM cell to make it always have an intermediate resistance, whereas the set voltage to control cell (V_{cont}) is modulated to have various resistance levels so that the two cells have different resistance values with respect to the time interval between the spikes of the presynaptic and

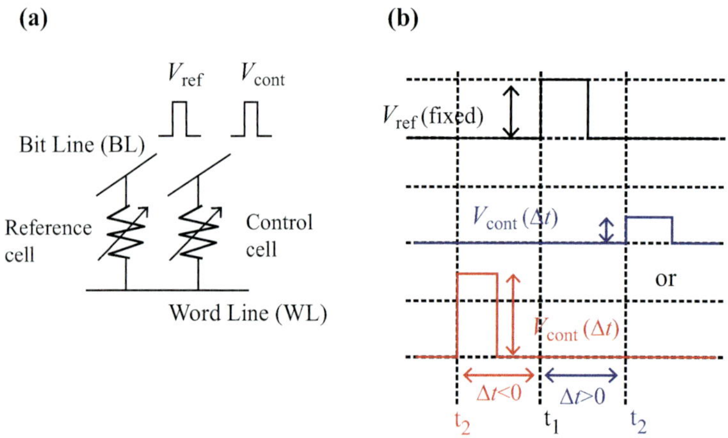

Fig. 2.17 Time-modulated voltage (TMV) scheme. (**a**) A circuit diagram of 2-PCM cells with two-channel pulse generators. (**b**) The voltage timing diagram, where *blue* and *red* texts, respectively, correspond to LTP and LTD cases. If V_{ref} is applied to the reference cell at a specific time (t_0), then V_{cont} is applied to the control cell at $t_0 + \Delta t$ (Reprinted with permission from Kang et al. [63])

postsynaptic neurons (Δt). To obtain the STDP characteristics, the pulse scheme and its programming algorithm are shown in Fig. 2.17b. The reciprocal of the resistance difference of the two cells is considered as the conductance change (ΔC), and it is plotted versus Δt in Fig. 2.18a, b.

2.2.4 Gradual Programming Versus Stochastic Programming

As all the pulse schemes are illustrated in the previous section, the most important requirement for emulating synaptic behavior in a large-scale neuromorphic system is the conductance (or resistance) modulation of the PCM device. This is because the conductance of the synapses connecting a pre- and post-neuron in any synaptic network undergoes modification on the basis of specific learning rules, forms of synaptic plasticity, and stimuli [83]. Several previous works have demonstrated gradual programming of phase-change synaptic devices allowing phase-change cell resistance is achieved through 100 steps for both the set and reset transitions [8, 58, 60, 61]. Cross-sectional TEM analysis and simulations also have revealed that modulation of the volume of the amorphous region leads to the formation of numerous intermediate resistance states [8]. Thus, implementing gradual programming to realize the fine control of the conductance of PCM synaptic device has become a common solution. Although the resistance drift mentioned in Sect. 2.1.4 may become a hurdle for gradual programming, plasticity can be implemented in low-resistance range where resistive drift is not very detrimental. In addition, recent work (see Sect. 2.3.1) has shown novel ways to stabilize the resistance, which

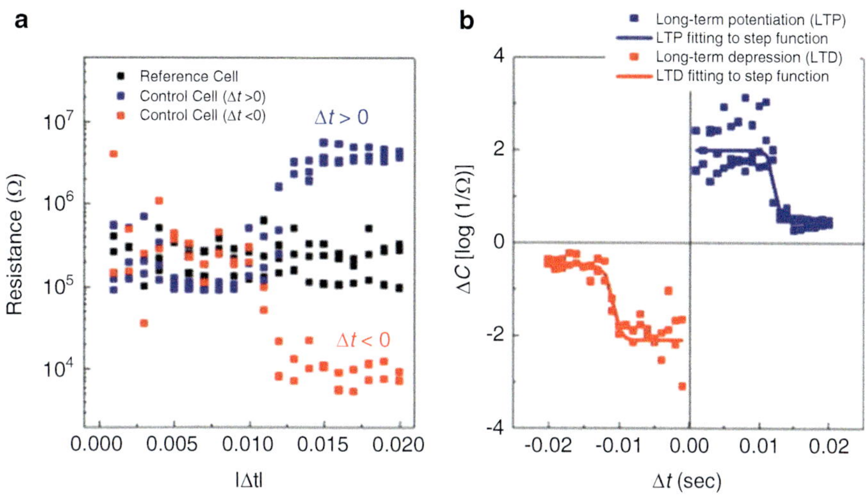

Fig. 2.18 Steplike LTP and LTD behaviors are realized in 2-PCM cells using time-modulated voltage (TMV) scheme (Reprinted with permission from Kang et al. [63])

provides better understanding of resistance drift phenomenon and foundation in realizing more advanced gradual programming. Furthermore, since synaptic updates based on gradual programming of PCM relies on partial crystallization or amorphization of the phase-change material, endurance characteristics are expected to be different from the case of digital memory applications where the phase-change material is programmed to highest- and lowest-resistivity states.

In addition to gradual programming, stochastic phase-change neurons were proposed by *Tuma* et al. [87]. Stochastic phase-change neurons utilized inherent stochastic dynamics of PCM switching. The thickness of the amorphous region created via the melt-quench process and the internal atomic configuration which changes after every reset operation exhibit inherent stochastivity. Furthermore, the stochasticity in a single phase-change neuron over multiple integrate-and-fire cycles leads to a distribution of interspike intervals. It was suggested that the capability of emulating neuronal noise and stochastic neuronal dynamics directly at the device level could enable the creation of dense neural populations for robust representation of signals and neuronal states and foster the development of neuromorphic computers with highly collocated memory and processing units. Inherent stochasticity of PCM devices could also be exploited for population coding to increase robustness against hardware failures in neuronal populations [87].

2.3 Future Outlook

PCM development has made multiple key improvements in the past several years, which improve its viability for synaptic device applications [10, 88]. The following improvements and their ramifications for neuromorphic devices will need to be considered for future PCM synaptic device: firstly, several works have shown improvement in the stability of resistance states. Specifically, solutions to mitigate resistance drift and techniques to increase the endurance have been developed. Secondly, new material exploration has led to doped GeSbTe, new phase-change material systems, and superlattice-based PCM, which can be leveraged for lower switching power, increased endurance, and faster switching time. Finally, the use of phase-change materials as an ovonic threshold switch (OTS) selector device in crossbar structures, a key requirement for designing high-density nonvolatile memories, has shown notable progress. All these advances originally targeted for memory applications will also be useful for synaptic devices.

2.3.1 Resistance State Stability

A fundamental requirement for any nonvolatile memory to be used as a neuromorphic device is the ability to program a wide number of resistance states with predictable read resistances. Inherently, the goals of enabling multilevel cells in nonvolatile memory applications with discrete resistance states overlaps with the goals of enabling multiple probabilistic states to stably exist in neuromorphic devices. Improvements addressing resistance drift and endurance demonstrated by nonvolatile memory community will be discussed here in the context of neuromorphic devices.

Resistance drift for different programmed resistance states for PCM is shown in Fig. 2.19a. The issue with this resistance drift is that the amount of resistance change has variability both between writes and between differing cells, making the expected resistance hard to predict and unstable. Thus, the expected state has a randomness associated with time, which can impact the read resistance. One proposed solution to this is to perform "refresh" operations, similar to DRAM, but this will likely reduce the device endurance as the cell is constantly reprogrammed [89, 90]. More likely solutions involve either using novel structures [91], grain-size engineering [92], or drift-invariant read techniques [93–95]. Using a metal surfactant layer around the GST layer has shown a 6× reduction in drift coefficient and provides an alternate read current path to improve resistance stability. Figure 2.19b shows the measured resistances for a device using a surfactant layer. Using such a surfactant layer significantly reduces the drift [91]. GST with smaller grain sizes (17–10 nm) has shown to have a 2× reduction in drift coefficient as smaller grains have higher surface to volume ratios and thus allow the atoms to efficiently relax without causing large changes in resistance [92]. Finally, the use of different metrics instead of read voltage/resistance can yield to useful parameters. In particular, two works identified the M parameter, the time required to reach a threshold current during a linear voltage ramp, which are weakly dependent on the drift activation energy and instead only depend on the effective amorphous thickness. Measuring M showed a linear time dependence with drift activation energy instead of exponential time dependence, as is the case with measuring resistance of the PCM device [93, 94]. All of these techniques can be used together to reduce the resistance drift and have time-invariant measurable states.

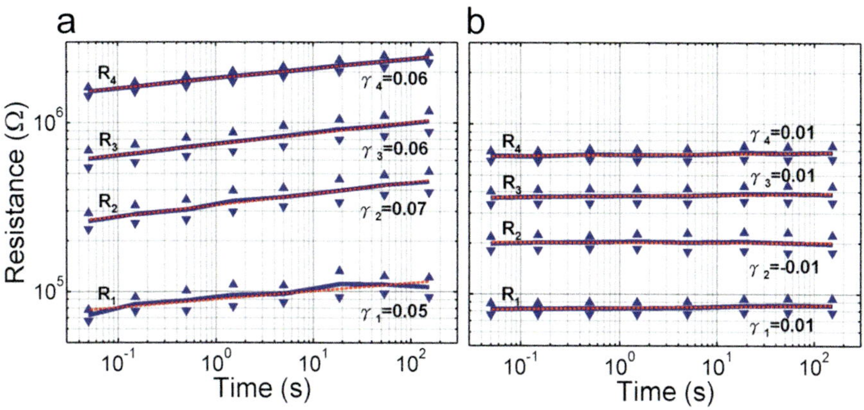

Fig. 2.19 Resistance Drift of PCM. Measured resistance as a function of time for various programmed resistances. For a sample (**a**) without surfactant and (**b**) with surfactant. The effect of resistance drift is much lower for samples with a surfactant (Reprinted with permission from Kim et al. [91])

Additionally, the available endurance of PCM limits how long the device can be used for neuromorphic applications and the application spaces available. There are two key failure mechanisms for GST-based phase-change materials. The first is to void formation between the bottom electrode and the phase-change layer contact. Essentially, each cycle changes the mass density and thus volume required for the phase-change layer. Over prolonged cycling, a void can eventually form and cause the device to be pinned in a high-resistance state [96–98]. The second failure mechanism is the phase segregation of the Sb atoms toward the site of most heat generation, the bottom electrode and phase-change layer contact. This causes the switching region to become Sb rich, which will have lower crystallization temperatures, and thus pins the PCM in the low-resistance state since the Sb-rich region always crystallizes [13, 96, 97]. Several works using PVD or ALD phase-change layers showed endurances longer than 10^{10}–10^{12} cycles [99–102]. PVD-based samples have less atomic movement since switching occurs at the grain interfaces. ALD samples leverage small confined ALD phase-change regions where the entire volume is melted, and thus, the atoms remix after each cycle [100, 103]. Another novel solution regarding endurance is the use of an In situ-Self-Annealing (ISSA) pulse, which involves using long >100 μs pulses to anneal and rejuvenate cycled cells [104]. Improving the endurance of PCM will open up several new opportunities to operate at DRAM levels of frequency.

2.3.2 *New Materials Exploration*

A wide variety of different materials have been developed which show strong improvements to phase-change properties. There are many different areas that are being explored; however, the most promising areas are doped GST; new materials using some of the Ge, Sb, and Te atoms; PVD-based materials; and finally superlattice PCM. Table 2.2 compares the performance of each of these materials. All of the following comparisons are made relative to sputtered $Ge_2Sb_2Te_5$. N-doped GST shows a reduction in reset current by ~4–8× and a significant improvement in the data retention (up to 10 years at 120 °C), but with higher set speeds as a trade-off ~ (80–200 ns) [105]. The result of using GeBiTe (<15% Bi) is a 10× improvement in resistance window, 10^9 cycles, and fast reset and set speeds of ~20–25 ns [106]. A Ga-Sb-Ge material was studied and showed significantly improved 10-year data retentions up to at 220 °C while still exhibiting strong 80 ns set speeds and 10^9 cycles [107]. Another very promising work using nanocrystalline grains <= 5 nm formed by PVD $Ge_2Sb_2Te_5$ with doping has exhibited very low write currents of 20 μA for a 35 nm diameter mushroom cell with 10^{11} endurance cycles, with the trade-off being a lower on/off ratio of 10. The authors claim that the main reason for the different performance compared to sputtered $Ge_2Sb_2Te_5$ was that the switching was happening between grains instead of in bulk [102].

Superlattice PCM also shows very exciting new effects as the inherent physics behind switching derives from the layers in the superlattice and can be different from the conventional phase-change materials such as GST. Currently, only an improvement in power has been observed, but there is substantial room for further exploration and improvement. One of the simpler proposed superlattice structures uses $SnSb_4/SiO_2$. With this structure, the effective volume being heated is halved, and the SiO_2 layer confines the heat in the phase-change layers, thereby reducing the power and improving the retention of the device. Using this structure, the write voltage is reduced compared to GST (from 3.6 V (1000 ns) to 2.8 V (500 ns)) and then measured with 10-year retentions at temperatures >85 °C [108]. Another superlattice structure with $GeTe/Sb_2Te_3$ showed inherently different physics. The switching does not occur in the volume of either superlattice layer. Rather, Ge bonds at the superlattice layer interfaces break and form, causing the change in resistance. Because of the limited atomic movement, the $GeTe/Sb_2Te_3$ superlattice structures showed significant improvements in reset current as well as set speeds. With 50 nm diameter devices, extremely low 70 μA reset currents and ultrafast 10 ns set speeds were measured [109]. Additionally, using 180 nm feature-sized devices, a 50% reduction in current and 40% reduction in reset voltage were measured compared to GST devices [110]. Superlattice structures allow new opportunities for novel physics, which can inherently change the performance of phase-change devices. There remains a wealth of different options and materials that must be explored; there is strong evidence that materials engineering will continue to show significant performance improvements.

2.3.3 *Ovonic Threshold Switch*

The OTS phenomenon was first measured by Stanford Ovshinsky in the late 1960s where a device was found to switch to a highly conductive state after a threshold

Table 2.2 Material performances

Material	RESET current (μA)	Feature size (nm) (D = diameter)	RESET speed (ns)	SET speed (ns)	Endurance (cycles)	Retention (10 years at X°C)	HRS/LRS ratio
N-doped GST [105]	50–150	30	N/A	80	2×10^7	120	6
GeBiTe [106]	250	50	20	25	10^9	106	40
GaSbGe [107]	N/A	$D = 20 \sim 40$	80	80	10^9	220	20
PVD GST [102]	20	$D = 35$	N/A	80	10^{11}	8 y at 70 °C	10
$SnSb_4/SiO_2$ [108]	V = 2 V	180	500	500	N/A	85	100
$GeTe/Sb_2Te_3$ [109, 110]	55	50	40	75	10^8	70	200

voltage was met, and the device would return to the highly resistive state after the threshold voltage was removed [2]. Renewed interest on OTS resurfaces as the requirements for selector devices involved higher drive currents, lower leakage, and repeatable and faster switching, all of which must be created at low-processing temperatures (<400 °C), a challenge which could not be met with Si diodes [111, 112]. Also, in comparison to PCM devices, OTS devices require fast threshold switching without fast crystallization, so material systems different from those used for memory are required.

A few key works have been able to show good performance as well as integration of OTS devices with other memories. In particular, *Kau* et al. from Intel created an OTS + PCM device and showed that performance of the combined structure [43]; *Kim* et al. from Samsung have created AsTeGeSiN-based selector devices with strong performance including high current density (10 MA/cm^2), fast switching (~5 ns), and reasonable endurance (10^8 cycles) for 30 nm feature sizes [113, 114]. A-STAR researchers have recently reported an OTS-based device with excellent performance at the 200 nm feature sizes (10^9 cycles, 0.2 threshold voltage, 10^{-11} leakage current, 10^7 on/off ratio, and <10 ns switching speeds) [115]. Compared to other potential selector devices and the needs of the nonvolatile memories they will be used for, further improvements in switching endurance (to about 10^{12}), while maintaining all of the characteristics described in [114], are required [111, 115]. Overall, OTS devices show strong promise to solve the selector device requirements in larger arrays and show that chalcogenides will likely be used.

Acknowledgments DK acknowledges the Office of Naval Research Young Investigator Award. SWF and HSPW are supported in part by the member companies of the Stanford Non-Volatile Memory Technology Research Initiative (NMTRI), member companies of the Stanford SystemX Alliance, and the National Science Foundation (Grant #DGE-4747 and Expeditions in Computing [Award # 1317560]).

References

1. D. Kuzum, S. Yu, H.S.P. Wong, Synaptic electronics: materials, devices and applications. Nanotechnology. **24**(38), 382001 (2013)
2. S.R. Ovshinsky, Reversible electrical switching phenomena in disordered structures. Phys. Rev. Lett. **21**(20), 1450–1453 (1968)
3. R. Bez, Chalcogenide PCM: a memory technology for next decade, in *IEEE International Electron Devices Meeting (IEDM)* (2009)
4. S. Lai, Current status of the phase change memory and its future, in *IEEE International Electron Devices Meeting (IEDM)* (2003)
5. M.H. Lankhorst, B.W. Ketelaars, R. Wolters, Low-cost and nanoscale non-volatile memory concept for future silicon chips. Nat. Mater. **4**(4), 347–352 (2005)
6. J. Oh et al., Full integration of highly manufacturable 512 Mb PRAM based on 90 nm technology, in *IEEE International Electron Devices Meeting (IEDM)* (2006)
7. G. Servalli, A 45 nm generation phase change memory technology, in *IEEE International Electron Devices Meeting (IEDM)* (2009)

8. D. Kuzum, R.G. Jeyasingh, B. Lee, H.-S.P. Wong, Nanoelectronic programmable synapses based on phase change materials for brain-inspired computing. Nano Lett. **12**(5), 2179–2186 (2011)
9. D.O. Hebb, *The organization of behavior: a neuropsychological theory*. (Psychology Press, 2005)
10. H.-S.P. Wong et al., Phase change memory. Proc. IEEE. **98**(12), 2201–2227 (2010)
11. S. Raoux, Phase change materials. Annu. Rev. Mat. Res. **39**(1), 25–48 (2009)
12. S. Raoux et al., Direct observation of amorphous to crystalline phase transitions in nanoparticle arrays of phase change materials. J. Appl. Phys. **102**(9), 94305–94305 (2007)
13. S. Raoux, F. Xiong, M. Wuttig, E. Pop, Phase change materials and phase change memory. MRS Bull. **39**(08), 703–710 (2014)
14. D. Ielmini, Threshold switching mechanism by high-field energy gain in the hopping transport of chalcogenide glasses. Phys. Rev. B **78**(3), 035308 (2008)
15. A. Owen, J. Robertson, C. Main, The threshold characteristics of chalcogenide-glass memory switches. J. Non Cryst. Solids. **32**(1), 29–52 (1979)
16. A. Pirovano, A.L. Lacaita, A. Benvenuti, F. Pellizzer, R. Bez, Electronic switching in phase-change memories. IEEE Trans. Electron Devices. **51**(3), 452–459 (2004)
17. D. Adler, M. Shur, M. Silver, S. Ovshinsky, Threshold switching in chalcogenide-glass thin films. J. Appl. Phys. **51**(6), 3289–3309 (1980)
18. D. Adler, M. Shur, M. Silver, S. Ovshinsky, Reply to comment on 'Threshold switching in chalcogenide-glass thin films'. J. Appl. Phys. **56**(2), 579–580 (1984)
19. V. Karpov, Y. Kryukov, S. Savransky, I. Karpov, Nucleation switching in phase change memory. Appl. Phys. Lett. **90**(12), 123504 (2007)
20. A. Pirovano, A.L. Lacaita, F. Pellizzer, S.A. Kostylev, A. Benvenuti, R. Bez, Low-field amorphous state resistance and threshold voltage drift in chalcogenide materials. IEEE Trans. Electron Devices. **51**(5), 714–719 (2004)
21. A. Sebastian, D. Krebs, M. Le Gallo, H. Pozidis, E. Eleftheriou, A collective relaxation model for resistance drift in phase change memory cells, in *IEEE International Reliability Physics Symposium (IRPS)* (2015)
22. M. Boniardi, D. Ielmini, Physical origin of the resistance drift exponent in amorphous phase change materials. Appl. Phys. Lett. **98**(24), 243506 (2011)
23. D. Ielmini, S. Lavizzari, D. Sharma, A.L. Lacaita, Physical interpretation, modeling and impact on phase change memory (PCM) reliability of resistance drift due to chalcogenide structural relaxation, in *IEEE International Electron Devices Meeting (IEDM)* (2007)
24. M. Boniardi, A. Redaelli, A. Pirovano, I. Tortorelli, D. Ielmini, F. Pellizzer, A physics-based model of electrical conduction decrease with time in amorphous Ge2Sb2Te5. J. Appl. Phys. **105**(8), 084506 (2009)
25. D. Ielmini, S. Lavizzari, D. Sharma, A. Lacaita, Temperature acceleration of structural relaxation in amorphous Ge2Sb2Te5. Appl. Phys. Lett. **92**(19), 193511 (2008)
26. I. Karpov, M. Mitra, D. Kau, G. Spadini, Y. Kryukov, V. Karpov, Fundamental drift of parameters in chalcogenide phase change memory. J. Appl. Phys. **102**(12), 124503–124503 (2007)
27. A. Chen, J. Hutchby, V. Zhirnov, G. Bourianoff, *Emerging Nanoelectronic Devices* (John Wiley & Sons, Wiley, 2014)
28. M. Mitra, Y. Jung, D.S. Gianola, R. Agarwal, Extremely low drift of resistance and threshold voltage in amorphous phase change nanowire devices. Appl. Phys. Lett. **96**(22), 222111 (2010)
29. E.R. Kandel, J.H. Schwartz, T.M. Jessell, S.A. Siegelbaum, A. Hudspeth, *Principles of Neural Science* (McGraw-Hill, New York, 2000)
30. J. Liang, R.G.D. Jeyasingh, H.-Y. Chen, H.-S. P. Wong, A 1.4 μA reset current phase change memory cell with integrated carbon nanotube electrodes for cross-point memory application, in *Symposium on VLSI Technology* (2011), pp. 100–101

31. F. Xiong, A.D. Liao, D. Estrada, E. Pop, Low-power switching of phase-change materials with carbon nanotube electrodes. Sci. **332**(6029), 568–570 (2011)
32. M. Kang et al., PRAM cell technology and characterization in 20 nm node size, in *IEEE International Electron Devices Meeting (IEDM)* (2011)
33. M.A. Zidan, H.A.H. Fahmy, M.M. Hussain, K.N. Salama, Memristor-based memory: the sneak paths problem and solutions. Microchem. J. **44**(2), 176–183 (2013)
34. P. Narayanan, G.W. Burr, R. Shenoy, K. Virwani, B. Kurdi, Circuit-level benchmarking of access devices for resistive nonvolatile memory arrays, in *IEEE International Electron Devices Meeting (IEDM)* (2009)
35. L. Zhang, S. Cosemans, D.J. Wouters, G. Groeseneken, M. Jurczak, B. Govoreanu, Selector design considerations and requirements for 1 SIR RRAM crossbar array, in *IEEE 6th International Memory Workshop (IMW)* (2014)
36. L. Zhang, S. Cosemans, D.J. Wouters, G. Groeseneken, M. Jurczak, B. Govoreanu, On the optimal ON/OFF resistance ratio for resistive switching element in one-selector one-resistor crosspoint arrays. IEEE Electron Device Lett. **36**(6), 570–572 (2015)
37. Y. Sasago et al., Cross-point phase change memory with 4F 2 cell size driven by low-contact-resistivity poly-Si diode, in *Symposium on VLSI Technology* (2009) pp. 24–25
38. B.S. Kang et al., High-current-density CuO x/InZnOx thin-film diodes for cross-point memory applications. Adv. Mater. **20**(16), 3066–3069 (2008)
39. I. Baek et al., Multi-layer cross-point binary oxide resistive memory (OxRRAM) for post-NAND storage application, in *IEEE International Electron Devices Meeting (IEDM)* (2005)
40. E. Katsia et al., Poly (3-hexylthiophene)/ZnO hybrid pn junctions for microelectronics applications. Appl. Phys. Lett. **94**(14), 143501 (2009)
41. Y.C. Shin et al., (in, Sn) 2O3/TiO2/Pt Schottky-type diode switch for the TiO2 resistive switching memory array. Appl. Phys. Lett. **92**(16) (2008)
42. M.J. Lee et al., Two series oxide resistors applicable to high speed and high density nonvolatile memory. Adv. Mater. **19**(22), 3919–3923 (2007)
43. D. Kau et al., A stackable cross point phase change memory, in *IEEE International Electron Devices Meeting (IEDM)* (2009)
44. H.-S.P. Wong, C. Ahn, J. Cao, H.-Y. Chen, S.B. Eryilmaz, S.W. Fong, J.A. Incorvia, Z. Jiang, H. Li, C. Neumann, K. Okabe, S. Qin, J. Sohn, Y. Wu, S. Yu, X. Zheng, Stanford memory trends, https://nano.stanford.edu/stanford-memory-trends, Accessed 26 Dec 2016.
45. P. Lennie, The cost of cortical computation. Curr. Biol. **13**(6), 493–497 (2003)
46. D. Kuzum, R.G. Jeyasingh, H.-S.P. Wong, Energy efficient programming of nanoelectronic synaptic devices for large-scale implementation of associative and temporal sequence learning, in *IEEE International Electron Devices Meeting (IEDM)* (2011)
47. J. Liang, R.G.D. Jeyasingh, H.-Y. Chen, H.-S.P. Wong, An ultra-low reset current cross-point phase change memory with carbon nanotube electrodes. IEEE Trans. Electron Devices. **59** (4), 1155–1163 (2012)
48. A. Pirovano, A. Lacaita, A. Benvenuti, F. Pellizzer, S. Hudgens, R. Bez, Scaling analysis of phase-change memory technology, in *IEEE International Electron Devices Meeting (IEDM)* (2003)
49. D. Im et al., A unified 7.5 nm dash-type confined cell for high performance PRAM device, in *IEEE International Electron Devices Meeting (IEDM)* (2008)
50. F. Xiong et al., Self-aligned nanotube–nanowire phase change memory. Nano Lett. **13**(2), 464–469 (2013)
51. S.M. Sadeghipour, L. Pileggi, M. Asheghi, Phase change random access memory, thermal analysis, in *Proceedings 10th Intersociety Conference on Thermal and Thermomechanical Phenomena in Electronics Systems* (2006), pp. 660–665
52. S. Fong et al., Thermal conductivity measurement of amorphous dielectric multilayers for phase-change memory power reduction. J. Appl. Phys. **120**(1), 015103 (2016)
53. J. Wu et al., A low power phase change memory using thermally confined TaN/TiN bottom electrode, in *IEEE International Electron Devices Meeting (IEDM)* (2011)

54. C. Kim et al., Fullerene thermal insulation for phase change memory. Appl. Phys. Lett. **92**(1), 013109 (2008)
55. C. Ahn et al., Energy-efficient phase-change memory with graphene as a thermal barrier. Nano Lett. **15**(10), 6809–6814 (2015)
56. J.P. Reifenberg, D.L. Kencke, K.E. Goodson, The impact of thermal boundary resistance in phase-change memory devices. IEEE Elect. Device Lett. **29**(10), 1112–1114 (2008)
57. J.P. Reifenberg et al., Thermal boundary resistance measurements for phase-change memory devices. IEEE Elect. Device Lett. **31**(1), 56–58 (2010)
58. C.D. Wright, Y. Liu, K.I. Kohary, M.M. Aziz, R.J. Hicken, Arithmetic and biologically-inspired computing using phase-change materials. Adv. Mater. **23**(30), 3408–3413 (2011)
59. O. Bichler, M. Suri, D. Querlioz, D. Vuillaume, B. DeSalvo, C. Gamrat, Visual pattern extraction using energy-efficient "2-PCM synapse" neuromorphic architecture. IEEE Trans. Electron Devices. **59**(8), 2206–2214 (2012)
60. M. Suri et al., Phase change memory as synapse for ultra-dense neuromorphic systems: Application to complex visual pattern extraction, in *IEEE Electron Devices Meeting (IEDM)* (2011)
61. B.L. Jackson et al., Nanoscale electronic synapses using phase change devices. ACM J. Emerg. Technol. Comput. Syst. (JETC) **9**(2), 12 (2013)
62. M.J. Breitwisch, R.W. Cheek, C.H. Lam, D.S. Modha, B. Rajendran, System for electronic learning synapse with spike-timing dependent plasticity using phase change memory, ed: Google Patents (2013)
63. D.-H. Kang, H.-G. Jun, K.-C. Ryoo, H. Jeong, H. Sohn, Emulation of spike-timing dependent plasticity in nano-scale phase change memory. Neurobiologia. **155**, 153–158 (2015)
64. Y. Zhong, Y. Li, L. Xu, X. Miao, Simple square pulses for implementing spike-timing-dependent plasticity in phase-change memory. Phys. Status Solidi (RRL)-Rapid Res. Lett. **9** (7), 414–419 (2015)
65. S.B. Eryilmaz et al., Experimental demonstration of array-level learning with phase change synaptic devices, in *IEEE International Electron Devices Meeting (IEDM)* (2013)
66. G.W. Burr et al., Experimental demonstration and tolerancing of a large-scale neural network (165 000 synapses) using phase-change memory as the synaptic weight element. IEEE Trans. Electron Devices. **62**(11), 3498–3507 (2015)
67. S. Kim et al., NVM neuromorphic core with 64 k–cell (256-by-256) phase change memory synaptic array with on-chip neuron circuits for continuous in-situ learning, in *IEEE International Electron Devices Meeting (IEDM)* (2015)
68. S. Ambrogio et al., Unsupervised learning by spike timing dependent plasticity in phase change memory (PCM) synapses, *Frontiers in Neuroscience*, vol. 10, (2016)
69. J. An, C. Choi, Y. Shindo, Y. Sutou, H. Jeong, Y. Song, Investigation of an erasing method for synaptic behaviour in a phase change device using Ge1Cu2Te3 (GCT). Electron. Lett. **52** (18), 1514–1516 (2016)
70. S.B. Eryilmaz et al., Brain-like associative learning using a nanoscale non-volatile phase change synaptic device array, *Frontiers in Neuroscience,* vol. 8, no. 205, (2014)
71. G. Close et al., Device, circuit and system-level analysis of noise in multi-bit phase-change memory, in *IEEE International Electron Devices Meeting (IEDM)* (2010)
72. F. Bedeschi et al., A bipolar-selected phase change memory featuring multi-level cell storage. IEEE J. Solid State Circuits. **44**(1), 217–227 (2009)
73. K. Gerrow, A. Triller, Synaptic stability and plasticity in a floating world. Curr. Opin. Neurobiol. **20**(5), 631–639 (2010)
74. D. Hebb, *Organization of Behavior* (Science Editions, New York, 1961)
75. H. Markram, W. Gerstner, P.J. Sjöström, A history of spike-timing-dependent plasticity. Spike-Timing Dependent Plasticity, 11 (2011)
76. H.Z. Shouval, S. Samuel, G.M. Wittenberg, Spike timing dependent plasticity: a consequence of more fundamental learning rules. Spike-Timing Dependent Plasticity, 60 (2010)

77. G.M. Wittenberg, S.S.-H. Wang, Malleability of spike-timing-dependent plasticity at the CA3–CA1 synapse. J. Neurosci. **26**(24), 6610–6617 (2006)
78. R.C. Froemke, J.J. Letzkus, B.M. Kampa, G.B. Hang, G.J. Stuart, Dendritic synapse location and neocortical spike-timing-dependent plasticity. Spike-Timing Dependent Plasticity, 118 (2010)
79. N. Caporale, Y. Dan, Spike timing-dependent plasticity: a Hebbian learning rule. Annu. Rev. Neurosci. **31**, 25–46 (2008)
80. P.J. Sjöström, E.A. Rancz, A. Roth, M. Häusser, Dendritic excitability and synaptic plasticity. Physiol. Rev. **88**(2), 769–840 (2008)
81. G. Bi, M. Poo, Synaptic modification by correlated activity: Hebb's postulate revisited. Annu. Rev. Neurosci. **24**(1), 139–166 (2001)
82. G.-q. Bi, M.-m. Poo, Synaptic modifications in cultured hippocampal neurons: dependence on spike timing, synaptic strength, and postsynaptic cell type. J. Neurosci. **18**(24), 10464–10472 (1998)
83. M. Suri, B. DeSalvo, Phase change memory and chalcogenide materials for neuromorphic applications: emphasis on synaptic plasticity, in *Advances in Neuromorphic Memristor Science and Applications*. (Springer, 2012), pp. 155–178
84. M. Suri, V. Sousa, L. Perniola, D. Vuillaume, and B. DeSalvo, Phase change memory for synaptic plasticity application in neuromorphic systems, in *International Joint Conference on Neural Networks (IJCNN)* (2011), pp. 619–624
85. D. Kuzum, R.G.D. Jeyasingh, S. Yu, H.-S.P. Wong, Low-energy robust neuromorphic computation using synaptic devices. IEEE Trans. on Electron Devices **59**(12), 3489–3494 (2012)
86. M.A. Woodin, K. Ganguly, M.-M. Poo, Coincident pre- and postsynaptic activity modifies GABAergic synapses by postsynaptic changes in Cl − transporter activity. Neuron **39**(5), 807–820 (2003)
87. T. Tuma, A. Pantazi, M. Le Gallo, A. Sebastian, E. Eleftheriou, Stochastic phase-change neurons, *Nature Nanotechnology* (2016)
88. G.W. Burr et al., Recent progress in phase-change memory technology. IEEE J. Emerging Sel. Top. Circuits Syst. **6**(2), 146–162 (2016)
89. P. Junsangsri, J. Han, F. Lombardi, A system-level scheme for resistance drift tolerance of a multilevel phase change memory, *2014 I.E. International Symposium on Defect and Fault Tolerance in VLSI and Nanotechnology Systems* (2014), pp. 63–68
90. W. Xu, T. Zhang, A time-aware fault tolerance scheme to improve reliability of multilevel phase-change memory in the presence of significant resistance drift. IEEE Trans. Very Large Scale Integr. (VLSI) Syst. **19**(8), 1357–1367 (2011)
91. S. Kim et al., A phase change memory cell with metallic surfactant layer as a resistance drift stabilizer, in *IEEE International Electron Devices Meeting (IEDM)* (2013)
92. W.J. Wang et al., Engineering grains of Ge2Sb2Te5 for realizing fast-speed, low-power, and low-drift phase-change memories with further multilevel capabilities, in *IEEE International Electron Devices Meeting (IEDM)* (2012)
93. A. Sebastian, N. Papandreou, A. Pantazi, H. Pozidis, E. Eleftheriou, Non-resistance-based cell-state metric for phase-change memory. J. Appl. Phys., 084505 (2011)
94. N. Papandreou et al., Drift-resilient cell-state metric for multilevel phase-change memory, in *IEEE International Electron Devices Meeting (IEDM)* (2011)
95. M. Stanisavljevic, A. Athmanathan, N. Papandreou, H. Pozidis, and E. Eleftheriou, Phase-change memory: feasibility of reliable multilevel-cell storage and retention at elevated temperatures, in *IEEE International Reliability Physics Symposium (IRPS)* (2015)
96. A. Padilla et al., Voltage polarity effects in $Ge^2Sb^2Te^5$-based phase change memory devices. J. Appl. Phys. **110**, 054501 (2011)
97. G. Novielli, A. Ghetti, E. Varesi, A. Mauri, R. Sacco, Atomic migration in phase change materials, in *IEEE International Electron Devices Meeting (IEDM)* (2013)

98. K. Do, D. Lee, D.-H. Ko, H. Sohn, M.-H. Cho, TEM study on volume changes and void formation in Ge2Sb2Te5 films, with repeated phase changes. Electrochem. Solid St. **13**(8), H284–H286 (2010)
99. D.H. Im et al., A unified 7.5 nm dash-type confined cell for high performance PRAM device, in *IEEE International Electron Devices Meeting (IEDM)* (2008)
100. I.S. Kim et al., High performance PRAM cell scalable to sub-20 nm technology with below 4F2 cell size, extendable to DRAM applications, in *Symposium on VLSI Technology* (2010), pp. 203–204
101. M.B. Sky et al., Crystalline-as-deposited ALD phase change material confined PCM cell for high density storage class memory, in *IEEE International Electron Devices Meeting (IEDM)* (2015)
102. H.L. Lung et al., Novel low power phase change memory using inter-granular switching, in *Symposium on VLSI Technology* (2016), pp. 128–129
103. D.H. Kang et al., Considerations on highly scalable and easily stackable phase change memory cell array for low-cost and high-performance applications, in *Non-Volatile Memory Technology Symposium (NVMTS)* (2014)
104. W.S. Khwa et al., A novel inspection and annealing procedure to rejuvenate phase change memory from cycling-induced degradations for storage class memory applications, in *IEEE International Electron Devices Meeting (IEDM)* (2014)
105. H. Cheng et al., A thermally robust phase change memory by engineering the Ge/N concentration in $(Ge,N)_xSb_yTe_z$ phase change material, in *IEEE International Electron Devices Meeting (IEDM)* (2012)
106. J. Lee et al., Scalable high-performance phase-change memory employing cvd gebite. IEEE Electron. Device Lett. **32**(8), 1113–1115 (2011)
107. H. Cheng et al., Novel fast-switching and high-data retention phase-change memory based on new Ga-Sb-Ge material, in *IEEE International Electron Devices Meeting (IEDM)* (2015)
108. Y. Lu et al., SnSb 4/SiO 2 multilayered thin films for phase change memory applications, J. Non Cryst. Solids. **432**, 505–509 (2016)
109. T. Ohyanagi et al., Charge-injection phase change memory with high-quality GeTe/Sb2Te3 superlattice featuring 70-μA RESET, 10-ns SET and 100 M endurance cycles operations, in *IEEE International Electron Devices Meeting (IEDM)* (2013)
110. M. Tai et al., 1T-1R pillar-type topological-switching random access memory (TRAM) and data retention of GeTe/Sb2Te3 super-lattice films, in *Symposium on VLSI Technology* (2014)
111. G.W. Burr et al., Access devices for 3D crosspoint memorya. J. Vac. Sci. Technol. B **32**(4), 040802 (2014)
112. W. Czubatyj, S.J. Hudgens, Invited paper: Thin-film ovonic threshold switch: Its operation and application in modern integrated circuits. Electron. Mater. Lett. **8**(2), 157–167 (2012)
113. S. Kim et al., Performance of threshold switching in chalcogenide glass for 3D stackable selector, *Symposium on VLSI Technology* (2013)
114. M.-J. Lee et al., Highly-scalable threshold switching select device based on chaclogenide glasses for 3D nanoscaled memory arrays, in *IEEE International Electron Devices Meeting (IEDM)* (2012)
115. H. Yang et al., Novel selector for high density non-volatile memory with ultra-low holding voltage and 107 on/off ratio, in *Symposium on VLSI Technology* (2015), pp. 130–131

Chapter 3
$Pr_{0.7}Ca_{0.3}MnO_3$ (PCMO)-Based Synaptic Devices

Daeseok Lee and Hyunsang Hwang

3.1 Nanoscale PCMO-Based Synaptic Device

The RRAM has been widely investigated as one of the promising candidates for scale limits of typical electron storage-based memories [1–4]. From its switching mechanisms, the RRAM can be simply categorized as two types: filamentary-type and interfacial-type [3, 5]. Because the filamentary-type RRAM can be operated in a local region, it has fater switching speed and higher scalability than the interfacial-type RRAM. However, to satisfy the requirements of the synaptic device in Chap. 1, the interfacial-type RRAM is more suitable than the filamentary-type RRAM [6, 7]. The interfacial-type RRAM has extremely stable and analog resistance change because it comes from electrochemical reactions between reactive metal layer and resistive switching layer. Figure 3.1 shows a brief operation mechanism of the PCMO-based RRAM. Figure 3.1a shows that the PCMO-based RRAM consists of a reactive top electrode (TE), insulating PCMO layer, and inert bottom electrode (BE). The stable and analog resistance change can be obtained by a redox reaction of TE, as shown in Fig. 3.1b–c. When a positive bias is applied on TE, mobile oxygen ions in PCMO layer can be attracted to TE. Then, those attracted oxygen ions can react with TE at the interface between TE and PCMO layer and form an induced oxide, which can determine a total resistance of the PCMO-based RRAM. This reaction, oxidation of TE and formation of induced

D. Lee (✉)
Department of Electronic Materials Engineering, Kwangwoon University, 20 Kwangwoonro, Nowon-GuSeoul 01897, South Korea
e-mail: leeds@kw.ac.kr

H. Hwang
Department of Materials Science and Engineering, Pohang University of Science and Technology, 77 Cheongamro, Nam-gu Pohang, Gyeongbuk 37673, South Korea
e-mail: hwanghs@postech.ac.kr

S. Yu (ed.), *Neuro-inspired Computing Using Resistive Synaptic Devices*,
DOI 10.1007/978-3-319-54313-0_3

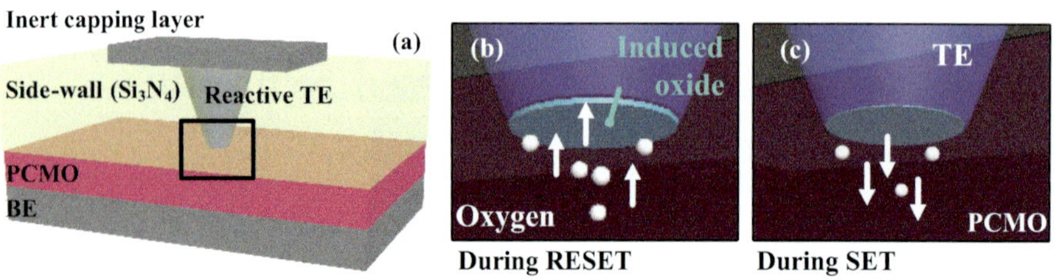

Fig. 3.1 Simple illustrations for operational mechanism of the PCMO-based RRAM. (**a**) Structure of the PCMO-based RRAM. (**b**) Formation and (**c**) dissolution of the induced oxide under applied bias during RESET and SET, respectively

oxide, can increase the total resistance of the PCMO-based RRAM. Thus, the total resistance will change to high-resistance state (HRS), which is defined as RESET for the NVM application. In contrast to RESET, under applied negative bias, oxygen ions of the induced oxide can move back into PCMO layer, which can change the total resistance from HRS to low-resistance state (LRS). This process can be defined as SET for the memory operation [6–11]. Considering the hardware implementation of the neuro-inspired computing system by high-density and power-efficient resistive synaptic devices, the redox reaction between TE and PCMO layer can be the main key factor for improved synaptic characteristics.

As discussed in Chap. 1, the resistive synaptic devices need to be evaluated in different ways with the NVMs. For the hardware implementation of the neuro-inspired computing system, millions of resistive synaptic devices are necessarily integrated with other components. In this respect, the scalability is one of the most important factors for the hardware implementation. In the scalability respect, the PCMO-based synaptic device (PCMO synapse) has advantages in its simple metal–insulator–metal (MIM) structure that can be highly and efficiently integrated by the crossbar array architecture. Figure 3.2 shows a simple fabrication process for the crossbar array architectural PCMO synapse. On the thermal-grown SiO_2 layer (8 in wafer), 50 nm thick Pt layer was deposited and patterned by typical electron beam evaporation and ion etching, respectively. Sequentially, 30 nm thick PCMO layer was deposited by RF magnetron sputtering system at 600 °C to obtain polycrystalline film. For low-temperature process, this thermal budget can be reduced by post-thermal treatment like a high-pressure annealing [12]. As an isolation layer, 80 nm thick SiN_x layer was deposited by plasma-enhanced chemical vapor deposition. On the SiN_x layer, typical lithography and ion etching were conducted to achieve via-hole structure with various diameters from 150 nm to 1 μm. Surface defects which would be generated during the ion etching process were cured by oxygen annealing at 500 °C for 30 min. As TE, 7 nm thick Al layer was utilized, and then 90 nm thick W layer was deposited on TE as a capping layer. Microscopic images of the fabricated PCMO synapse which is nanoscale 1 kilobit (kb) crossbar array architecture are shown in Fig. 3.3a–b. Figure 3.3c shows the cross-sectional transmission electron microscopy (TEM) image of the PCMO synapse. Not only 1 kb but also 8 kb crossbar arrays were fabricated to evaluate the PCMO synapse in

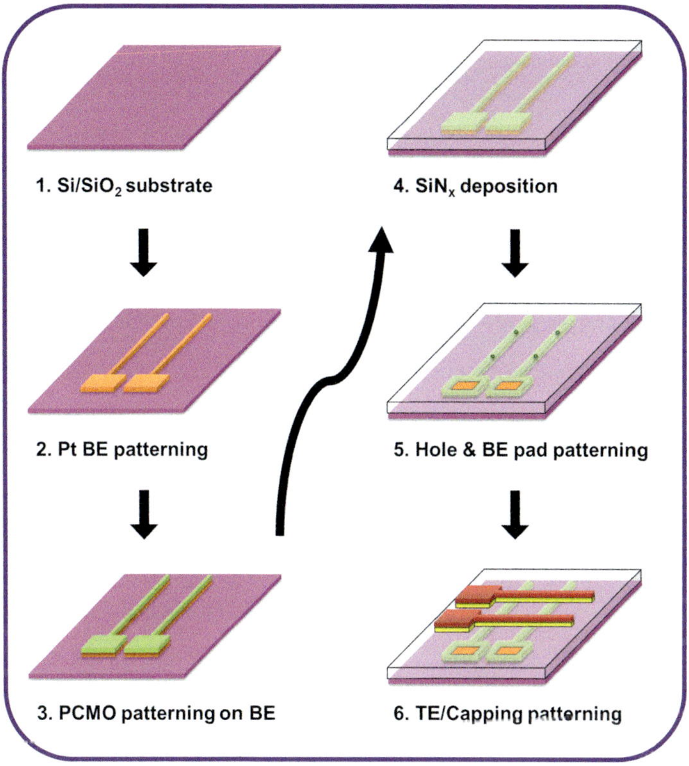

Fig. 3.2 Fabrication flow for PCMO-synapse array

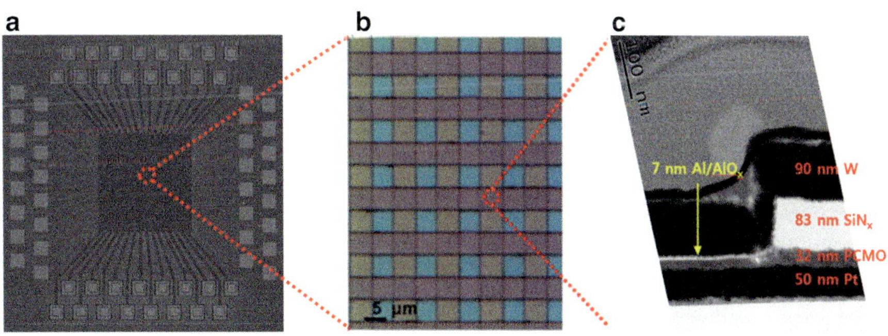

Fig. 3.3 (**a**, **b**) Optical and (**c**) cross-sectional transmission electron microscope (TEM) images of the fabricated PCMO synapse

various conditions, as shown in Fig. 3.4. For electrical characterization of the PCMO synapse, typical semiconductor parameter analyzers (Agilent 4155A and 81104A) were utilized.

Figure 3.5 shows resistance distributions of the fabricated 1 and 8 kb crossbar arrays with PCMO synapses. Randomly selected 20 pristine devices at each three

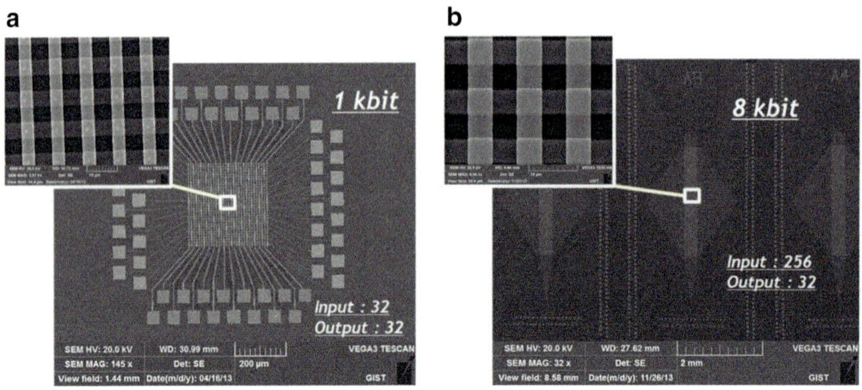

Fig. 3.4 Scanning electron microscope images for (**a**) 1 kb and (**b**) 8 kb crossbar arrays with PCMO synapses

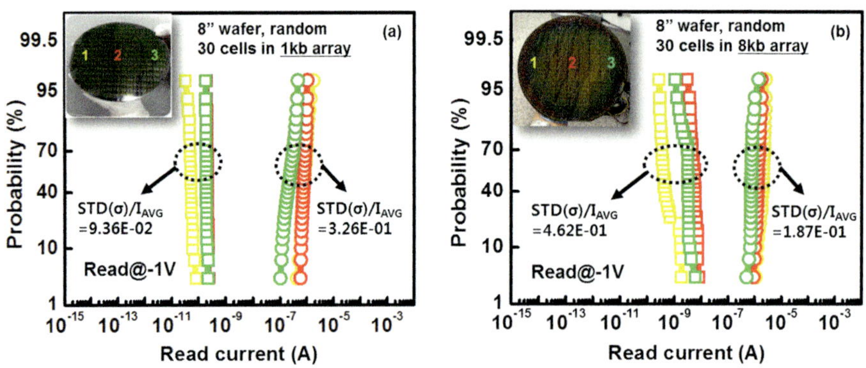

Fig. 3.5 Read current (current at read voltage) distributions of (**a**) fabricated 1 and 8 kb PCMO-synapse arrays

different parts (left, center, and right) of the wafer were evaluated. In the whole wafer region, tight resistance distribution which indicates uniform resistive switching was observed. The similar on/off ratios (resistance ratio between low- and high-resistance states) of 100 bits in the 1 kb crossbar array also obviously support the feasibility of large-scale fabrication of the PCMO synapse, as shown in Fig. 3.6.

Basic electrical characteristics of the fabricated PCMO synapse are shown in Figs. 3.7 and 3.8. Figure 3.7 shows current versus voltage (I–V) curves exhibiting gradual resistance changes from pristine state. When the absolute values of applied bias were increased, the total resistance was gradually increased and decreased during RESET and SET, respectively. According to the previously explained resistive switching mechanism, the gradual resistance changes resulted from the formation and dissolution of the induced oxide layer (in this case, AlO_x) between TE (in this case, Al) and PCMO layer [13]. During RESET, under higher bias, more oxygen ions can be attracted and reacted, which leads to gradually changed total

Top electrode (W)

Bottom electrode (Pt)	1	2	3	4	5	6	7	8	9	10
1	5.5E+3	1.9E+3	247.1E+0	6.1E+3	1.6E+3	160.3E+0	273.2E+0	4.3E+3	3.6E+3	5.3E+3
2	3.0E+3	252.4E+0	315.9E+0	2.5E+3	1.2E+3	6.0E+3	4.2E+3	2.6E+3	3.8E+3	5.1E+3
3	2.5E+3	15.78231	850.5E+0	32.2E+3	2.4E+3	5.5E+3	5.6E+3	119.1E+0	160.2E+0	4.9E+3
4	2.3E+3	42.2E+3	1.4E+3	2.2E+3	47.20497	1.7E+3	5.8E+3	77.05882	1.70E+3	6.6E+3
5	2.4E+3	12.2E+3	1.2E+3	4.6E+3	145.5E+0	1.1E+3	3.3E+3	134.0E+0	9.2E+3	5.5E+3
6	3.0E+3	11.6E+3	16.75462	4.0E+3	1.4E+3	1.2E+3	6.6E+3	5.2E+3	2.9E+3	5.2E+3
7	9.8E+3	5.5E+3	872.9E+0	12.2E+3	1.5E+3	1.9E+3	10.6E+3	2.0E+3	9.2E+3	4.8E+3
8	9.4E+3	4.2E+3	11.2E+3	294.3E+0	455.9E+0	6.0E+3	5.6E+3	284.6E+0	780.4E+0	10.3E+3
9	3.1E+3	3.5E+3	5.2E+3	2.1E+3	978.6E+0	12.1E+3	3.8E+3	2.8E+3	4.7E+3	7.3E+3
10	389.4E+0	16.2E+3	4.8E+3	638.9E+0	305.9E+0	7.7E+3	3.7E+3	5.0E+3	251.5E+3	313.9E+0

On/Off ratio: $10^1 \sim 10^2$ | $10^2 \sim 10^3$ | $10^3 \sim 10^4$ | $10^4 \sim 10^5$

Fig. 3.6 Similar on/off ratio of 100 bits in the 1 kb PCMO-synapse array

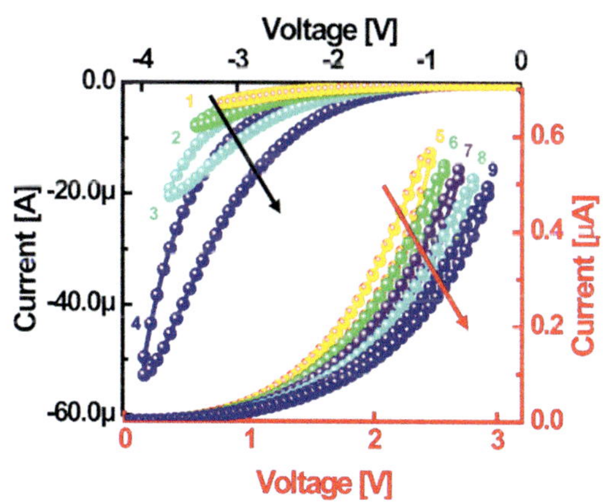

Fig. 3.7 Gradual resistance changes with incremental negative and positive bias

resistance. Higher positive bias can attract more oxygen ions and can form thicker induced oxide layer, which results in higher resistance. In contrast to RESET, during SET, smaller negative bias can move more oxygen ions from the induced oxide layer to the PCMO layer, which leads to thinner induced oxide layer (lower resistance). Because the redox reaction of TE can occur gradually, the analog resistance changes can be observed.

The gradual resistance change that can be considered as conductance change of the PCMO synapse was clearly demonstrated by the incremental biasing method (Fig. 3.7). However, for the neuro-inspired computing system, more practical

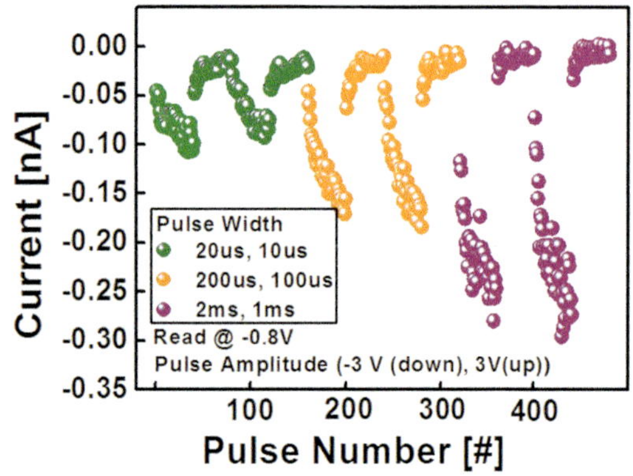

Fig. 3.8 Continuous conductance changes under identical pulse bias with different conditions

methods are necessary because the incremental biasing method can lead to more complex circuit design [14–16]. In this respect, the conductance changes by the number of applied pulse bias are more proper for the hardware implementation of the neuro-inspired computing system, as shown in Fig. 3.8. Under applied identical pulse bias having the same pulse amplitude and width, the current of the PCMO synapse was gradually changed. For weight increase and decrease of the resistive synaptic device, negative pulse bias and positive pulse bias were respectively applied. The PCMO ssynapse exhibited higher current changes under higher pulse amplitude and wider pulse width. These gradual current changes per applied pulse bias can be utilized as the analog conductance change that is a basic property of the resistive synaptic device.

3.2 Reactive Electrode Dependence of PCMO-Based Synaptic Device

As mentioned before, the redox reaction between TE and PCMO layers (TE redox reaction) is a crucial factor for gradual conductance change. Therefore, in this section, the TE redox reaction will be discussed in more detail for synaptic device applications. Figure 3.9a shows the simple fabrication processes of the samples with different TE materials. Firstly, Pt layer is deposited as the BE on thermal-grown SiO_2 layer. Sequentially, 20 nm thick PCMO layer was sputtered, and then Si_3N_4 layer was deposited and patterned to make a via-hole structure. Lastly, various materials such as Ni, Ta, TiN, and Al were deposited as various TEs. To prevent natural oxidation of TE, Pt layer was deposited on all TEs as a capping layer. Consequently, Pt/PCMO/Ni/Pt (Ni/PCMO), Pt/PCMO/Ta/Pt (Ta/PCMO), Pt/PCMO/TiN/Pt (TiN/PCMO), and Pt/PCMO/Al/Pt (Al/PCMO) samples were fabricated. Cross-sectional TEM image of the fabricated TiN/PCMO sample is shown in Fig. 3.9b. For optimization, specific thickness of TE and bias conditions were employed for each sample [6–9, 17]. During the deposition of TE, TE can

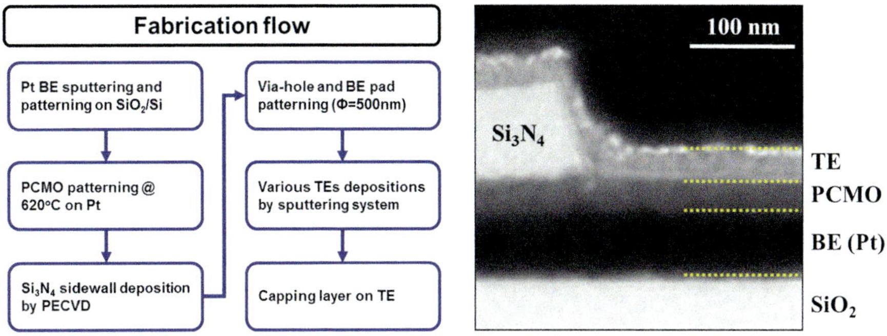

Fig. 3.9 (**a**) Fabrication flow of the evaluated samples with various top electrode (TE) materials. (**b**) Cross-sectional TEM image of Pt/TiN/PCMO/Pt (TiN/PCMO) sample

Fig. 3.10 I–V curves of evaluated samples: (**a**) Ni/PCMO, (**b**) Ta/PCMO, (**c**) TiN/PCMO, and (**d**) Al/PCMO

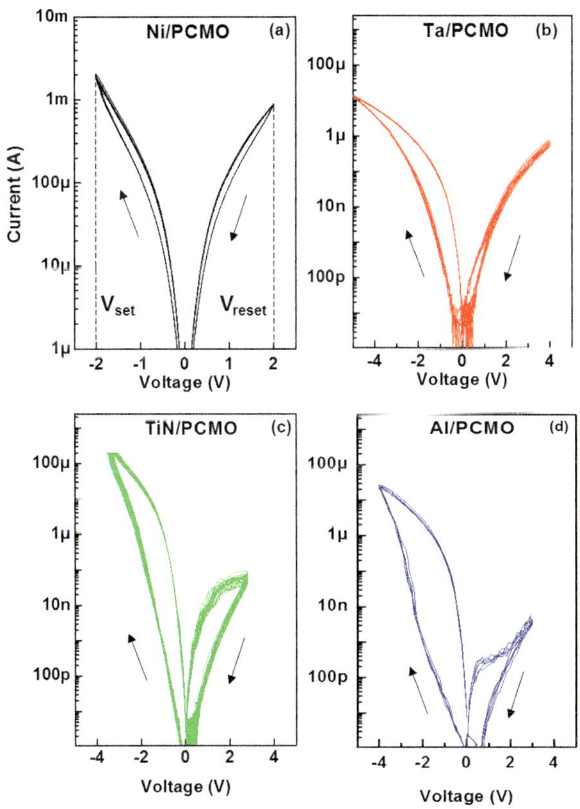

absorb oxygen from the PCMO layer, which results in naturally induced oxide layer.

Figure 3.10 shows the resistance changes of the fabricated samples for five cycles. All the samples exhibited clockwise SET and RESET process, which

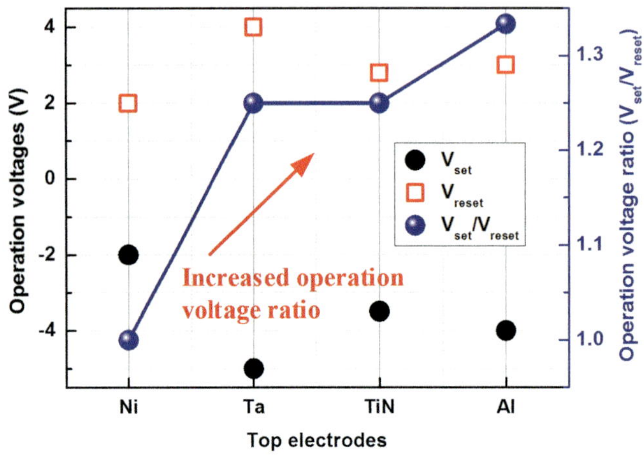

Fig. 3.11 Operation voltages (V_{set} and V_{reset}) and operation voltage ratio ($= V_{\mathrm{set}}/V_{\mathrm{reset}}$) of evaluated samples

can be correlated to the previously explained TE redox reaction. Among those samples, Al/PCMO exhibited wide on/off ratio and low current values, while Ni/PCMO exhibited narrow on/off ratio and high current values. In addition, the samples exhibited different operation voltages: set voltage (V_{set}) and reset voltage (V_{reset}), as shown in Fig. 3.11. To compare the differences between V_{set} and V_{reset} among those samples, we defined operation voltage ratio ($= V_{\mathrm{set}}/V_{\mathrm{reset}}$). When TE was changed from Ni to Ta to TiN to Al, the operation voltage ratio was increased. That means that optimized bias condition is necessary for each sample to obtain proper conductance changes. On the basis of the result in Fig. 3.11, the conductance changes are evaluated with optimized bias condition, as shown in Fig. 3.12. When the conductance is increased and decreased by applied identical pulse bias having the same pulse amplitude and pulse width, it can be considered as potentiation and depression processes, respectively. In Fig. 3.12, the potentiation and depression processes are plotted with the same number of pulse bias. For the potentiation, the number of pulse bias is expressed from 0 to 50, while the depression is plotted from 50 to 0. On the basis of this bilateral symmetric superposition, conductance difference between the potentiation and depression can be defined as asymmetry ratio. To compare the values of conductance among the samples, normalized values (divided by maximum conductance value) are utilized.

Considering that the neuro-inspired computing system needs to be able to update the synaptic weights, conductance of synaptic device needs to be changed with the same incremental value at any conductance value. In other words, at any certain conductance value, the potentiation and depression can be processed randomly [18, 19]. Thus, the incremental values of conductance change (ΔG) need to be the same during the potentiation and depression. For instance, if the potentiation is processed for ten times at certain conductance value, and sequentially the depression process is processed for ten times, then the final conductance value needs to be

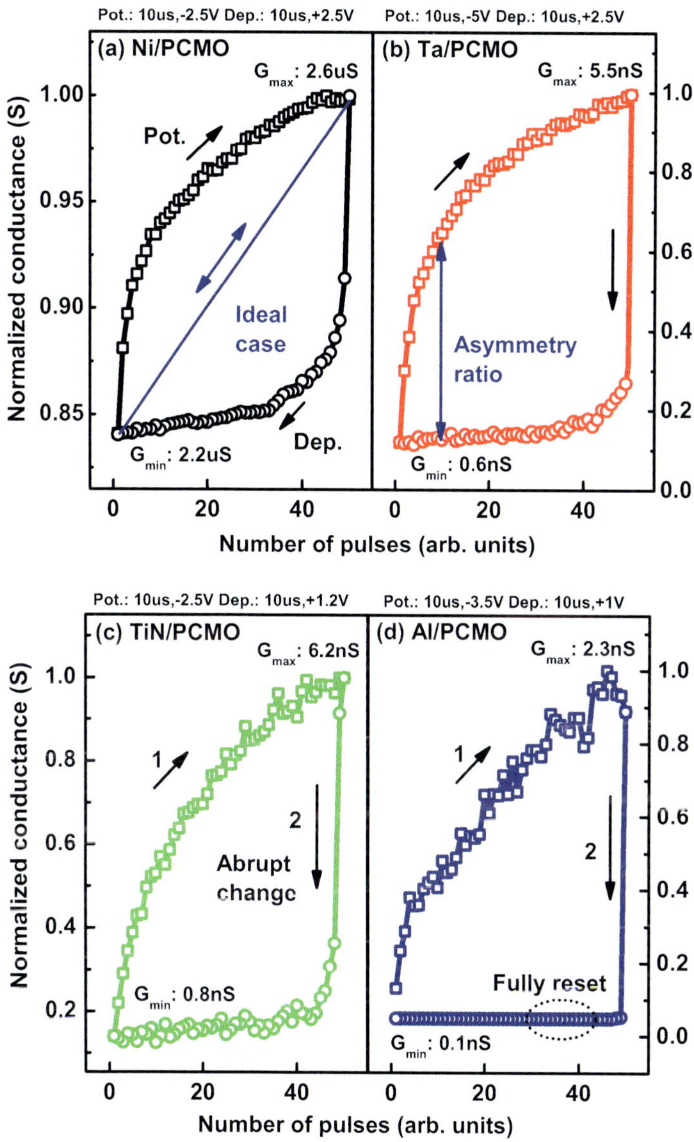

Fig. 3.12 Potentiation (from 0 to 50) and depression (from 50 to 0) processes under identical pulse bias for each samples: (**a**) Ni/PCMO, (**b**) Ta/PCMO, (**c**) TiN/PCMO, and (**d**) Al/PCMO

similar with the initial conductance value. Otherwise, we need more complicated circuits and processes to make a target conductance of the synaptic device. Moreover, the unreliable ΔG can lead to learning accuracy degradation in the computation of neuro-inspired computing system [18, 19]. Therefore, similarity of the ΔG between the potentiation and depression is one of the important factors for designing the synaptic device. Here, to compare fabricated samples, we defined the asymmetry ratio which can show the linearity in weight, indirectly. Based on the maximum and minimum G values ($G_{\max}$ and $G_{\min}$), we assumed that G is changed from $G_{\min}$ to $G_{\max}$ during the potentiation and from $G_{\max}$ to $G_{\min}$ during the depression. Thus, at certain pulse number (N_x), the asymmetry ratio can be defined as $\frac{G_{pot}\ at\ N_x}{G_{dep}\ at\ (N_{\max}-N_x)}$ when G_{pot} is G value during the potentiation (from $G_{\min}$ to $G_{\max}$), G_{dep} is G value during the depression (from $G_{\max}$ to $G_{\min}$), and $N_{\max}$ is total pulse number between $G_{\min}$ and $G_{\max}$. Even though this asymmetry ratio cannot be a direct indicator for the linearity in weight, it can show a difference of G between the potentiation and depression after certain number of pulse. As an example, in an ideal case, ΔG is the same at any conductance value. Therefore, for the N_x, G values will be the same during both potentiation and depression; the asymmetry ratio of the ideal case will be "1," as shown in Fig. 3.12a.

The TE dependencies such as on/off ratio and asymmetry ratio are summarized in Fig. 3.13. The Al/PCMO sample exhibited higher on/off ratio than others, which means that more conductance states can be obtained in the Al/PCMO sample [18, 19]. However, the highest asymmetry ratio was obtained in the Al/PCMO sample, which can degrade the performance of neuro-inspired computing system. Considering the tendencies of on/off ratio and asymmetry ratio in all evaluated samples, these results mean that there is a trade-off between the on/off ratio and asymmetry ratio. This trade-off can be explained by a metal–oxide free energy (Gibbs free energy of the induced oxide) [20–22]. In the Al/PCMO sample, AlO_x layer having a smaller metal–oxide free energy than that of PCMO is induced, as shown in Fig. 3.14. It means that the induced AlO_x layer is more stable than the PCMO layer. Therefore, higher activation energy is required for reduction of the Al TE (SET or potentiation process) than that for oxidation of the Al TE (RESET or depression process) [21, 22]. On the basis of the different metal–oxide free energies, in the Al/PCMO sample, the abrupt conductance change of depression which is a dominant origin of the high asymmetry ratio can be explained. During the depression, the induced oxide can be formed, which can reduce effective bias drop

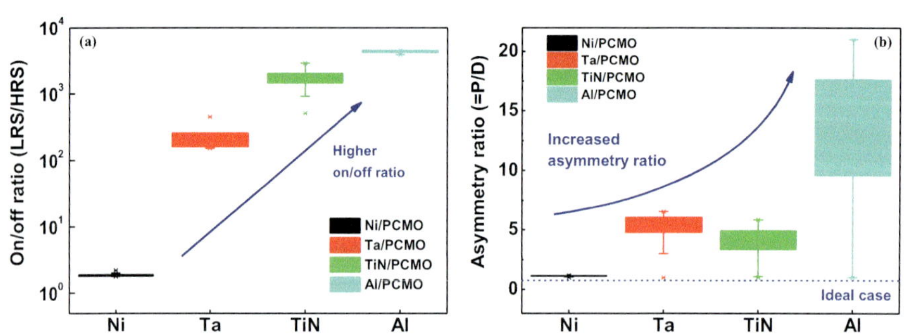

Fig. 3.13 Each sample exhibited different (**a**) on/off ratio and (**b**) asymmetry ratio

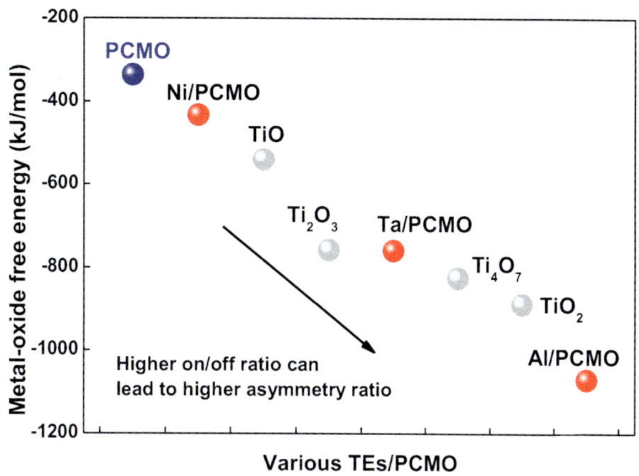

Fig. 3.14 Trade-off between the dynamic range and linearity in weight

on the PCMO layer. Thus, for enough multilevel states, high amplitude bias is necessary, which can lead to the abrupt conductance change [21, 22].

In Fig. 3.13, similar tendencies between the TiN/PCMO and Ta/PCMO samples can be explained by the various titanium oxide phases such as TiO, Ti_2O_3, Ti_4O_7, and TiO_2. In the Ti/PCMO sample, the Ti TE can induce various oxides having different metal–oxide free energy. Comparing to the metal–oxide free energy of the Ta/PCMO sample, Ti_2O_3 and Ti_4O_7 have more similar metal–oxide free energy than that of TiO and TiO_2. Thus, the TiN/PCMO and Ta/PCMO samples exhibited similar values in the on/off ratio and asymmetry ratio.

On the basis of the metal–oxide free energy differences, these TE dependencies clearly show the trade-off between the on/off ratio and asymmetry ratio. It can be also considered as a trade-off between the number of multilevel states and linearity in weight of the synaptic device. For the PCMO synapse, the metal–oxide free energy of TE should be carefully considered; high metal–oxide free energy of TE can result in smaller asymmetry ratio but narrower on/off ratio.

3.3 Nitrogen Treatment for Tunable Synaptic Characteristics

Considering various requirements, multilevel states, dynamic range, linearity in weight, etc., among previously discussed structures (Ni/PCMO, Ta/PCMO, TiN/PCMO, and Al/PCMO), the TiN/PCMO is more proper for the synaptic device application. However, for hardware implementation of the neuro-inspired computing system, synaptic characteristics of TiN/PCMO need to be improved. Thus, here, as one of the approaches for the improved synaptic characteristics, nitrogen treatment will be discussed.

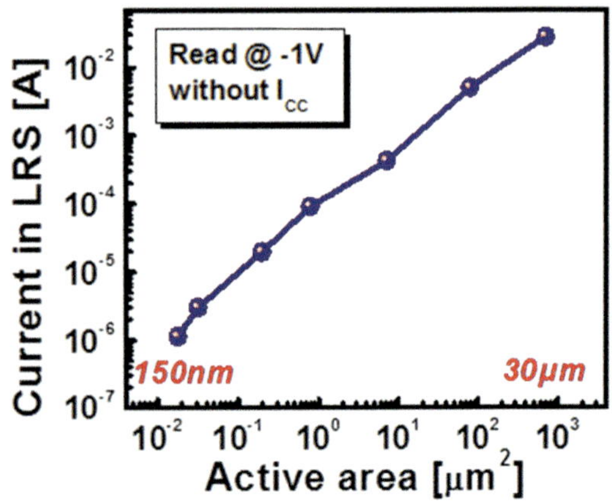

Fig. 3.15 Cell size dependence of maximum current in TiN/PCMO-based synaptic device

Even though the interfacial-type TiN/PCMO has lots of advantages such as low programming energy consumption, dynamic range, multilevel states, and uniformity for the synaptic device application, it is necessary to match the dynamic range and operational bias range (operational ranges) of synaptic device to the CMOS-based neuron parts for the hardware implementation. The simplest way to obtain the proper operational ranges for the interfacial-type TiN/PCMO-based synaptic device (TiN/PCMO synapse) is to optimize the cell size, as shown in Fig. 3.15. However, in this approach, the total size of the hardware-implemented neuro-inspired computing system will be dependent on the size of the synaptic device. Thus, tunable operational ranges need to be achieved in the nanoscale TiN/PCMO synapse.

Considering the work functions of the TiN TE (4.2 eV) and PCMO layers (4.9 eV), Schottky-like contact which can affect resistance changes can be formed at the interface [11, 23, 24]. For easier understanding, simplified illustrations of SET and RESET are shown in Fig. 3.16. Under positive bias, oxygen ions can be attracted to the TE, which results in induced oxide. Then, the induced oxide can lead to the metal–oxide–semiconductor (MOS) contact at the interface. In contrast to the RESET, during the SET, oxygen ions in the induced oxide can move back to the PCMO layer. Thus, metal–semiconductor (M–S) contact, which results in Schottky diode-like behavior, can be obtained [25]. Figure 3.17 shows in situ TEM analyses during the SET and RESET, which obviously exhibits the formation of induced oxide layer. When applied positive bias is increased, thicker- and brighter-induced oxide was observed. In addition, in HRS, clearly observed oxygen component at the induced oxide layer indicates that TiON layer is formed by applied bias as the induced oxide layer.

On the basis of this, tunable operational ranges can be obtained by modifying work function of the TiN TE. Figure 3.18 shows I–V curves with various TiN TE deposition conditions with nitrogen treatment. During deposition of the TiN TE, various rates of nitrogen gas which can be defined as $R_N = \frac{N_2}{(Ar+N_2)}$ were employed. When the R_N increased from 0% to 100%, higher current values of LRS and smaller operational voltage were obtained. Moreover, higher on/off ratio (at $V_{read} = -1$ V)

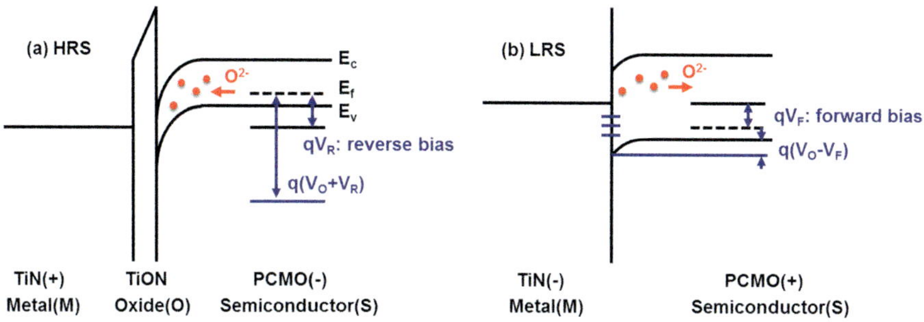

Fig. 3.16 Simple energy band diagram for (**a**) RESET and (**b**) SET operations

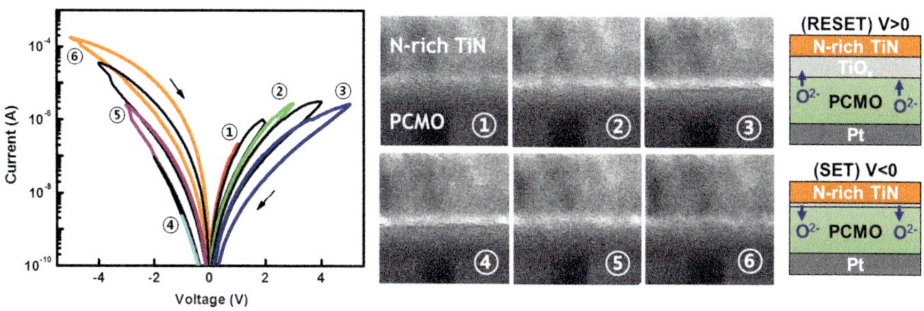

Fig. 3.17 Real-time TEM images (from ① to ⑥) during analog resistance changes

Fig. 3.18 I–V curves with various nitrogen gas rates from 0% to 100%

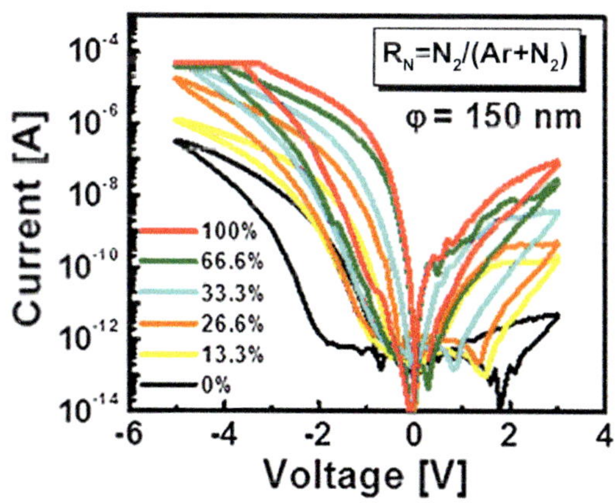

was obtained, as shown in Fig. 3.19. Compared to 0% condition, in 100% condition, the on/off ratio and maximum current were increased about 1.4×10^5 and 1×10^3 times, respectively. On the basis of the increased maximum current level, in Fig. 3.20, modified work function of the TiN TE can be estimated by Schottky diode equation:

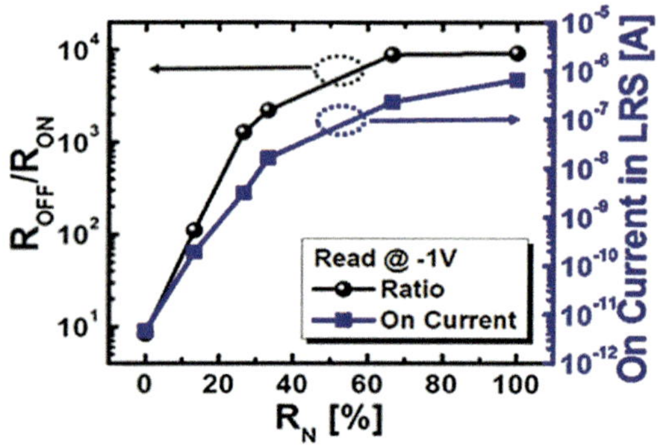

Fig. 3.19 LRS current values (on current) and on/off ratio with various nitrogen gas rates

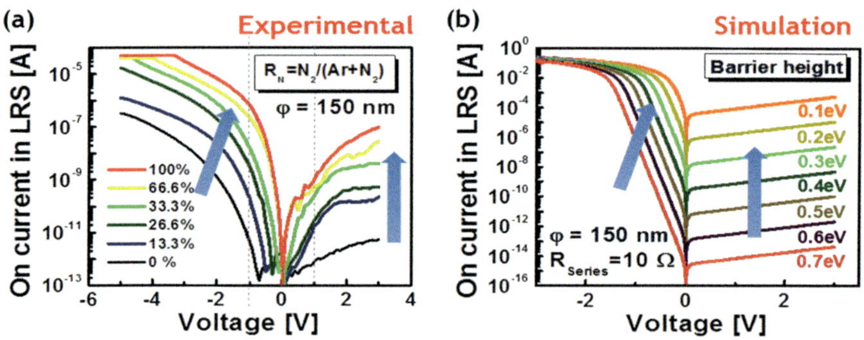

Fig. 3.20 (**a**) Experimental results with various nitrogen gas rate. (**b**) Schottky diode model with various barrier heights to explain the influence of nitrogen treatment

$$I = S \cdot A^{**} \cdot T^2 \exp\left(-\frac{q\phi_{\mathrm{B}}}{k_{\mathrm{B}}T}\right)\left[\exp\left(\frac{qV}{nk_{\mathrm{B}}T} - 1\right)\right] = I_{\mathrm{s}}\left[\exp\left(\frac{qV}{nk_{\mathrm{B}}T} - 1\right)\right]$$

where S is device area, A^{**} is effective Richardson constant, k_B is Boltzmann constant, Φ_B is Schottky barrier height, T is absolute temperature, n is ideality factor, and I_{S} is saturated current [13]. Figure 3.21 shows R_{N} dependence of the estimated work functions, which results in lowering of the barrier height (Fig. 3.22). These results obviously demonstrate that the tunable operational ranges can be achieved by the nitrogen treatment method in the nanoscale TiN/PCMO synapse [26].

3.4 Approaches to Improve Synaptic Characteristics

Even though the TiN/PCMO can satisfy several requirements such as device dimension, multilevel states, dynamic range, programming energy consumption, retention, endurance, uniformity, variability, and resistive crossbar array

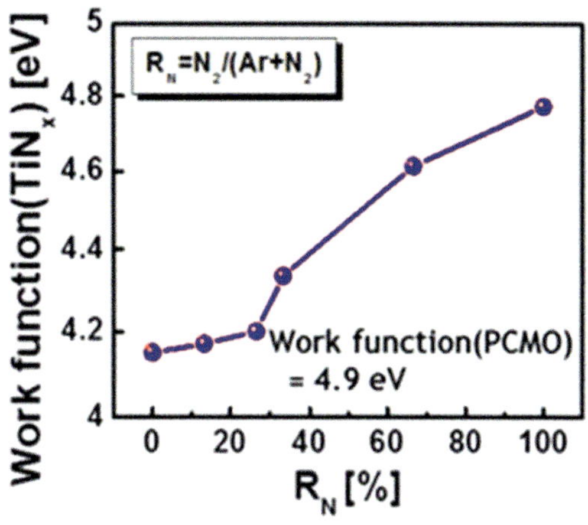

Fig. 3.21 Estimated work function of TiN_x with various nitrogen gas rates

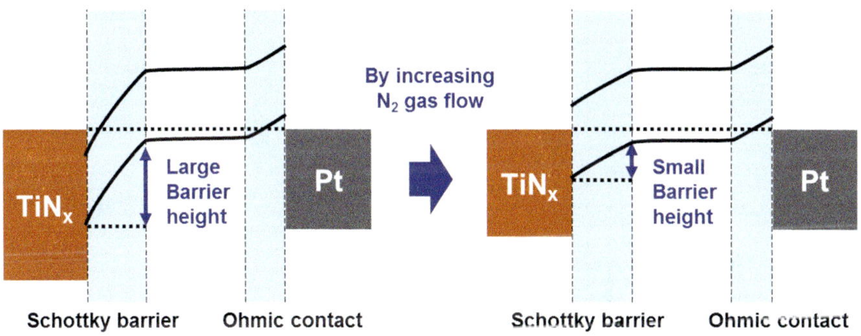

Fig. 3.22 Band diagram showing reduced barrier height with increased nitrogen gas rate

architecture, the linearity in weight update still needs to be intensively investigated. For the PCMO synapse, here, two approaches for improving the linearity in weight update will be discussed.

Figure 3.23 shows effects of the linearity in weight update on recognition rate of a face recognition problem [19]. From Yale Face Database A, randomly selected 88 images of eight individuals (11 images for each person) were used for the face recognition problem [27]. When the asymmetry ratio is the maximum value ($-1 = -5/5$ or $5/-5$ for $P_{\text{potentiation}}/P_{\text{depression}}$), the minimum recognition rate was achieved. Contrast to the maximum asymmetry ratio, the minimum asymmetry ratio ($1 = P_{\text{potentiation}}/P_{\text{depression}}$) shows more than 90% recognition ratio. These mean that not only the linearity in weight update but also the asymmetry ratio can be a critical factor for the high-performance neuro-inspired computing system. For the improved asymmetry ratio, two approaches, modifying external bias condition or device structure, can be considered. For instance, the modified bias condition can be utilized, as shown in Fig. 3.24. The Al/PCMO which exhibited the highest asymmetry ratio was investigated with various bias conditions such as the identical

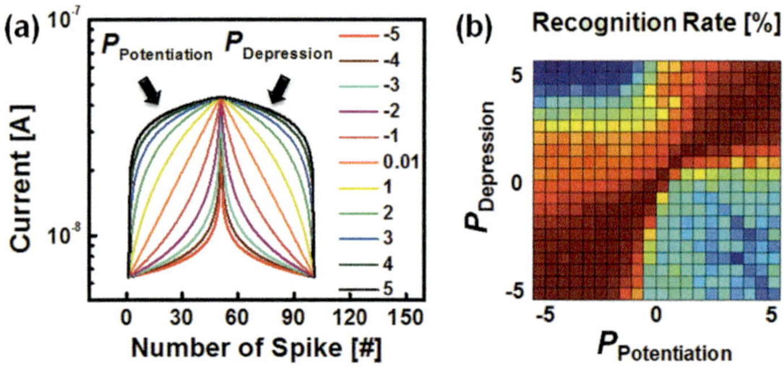

Fig. 3.23 (**a**) Various weight update during the potentiation and depression. (**b**) Asymmetry ratio dependence of recognition rate

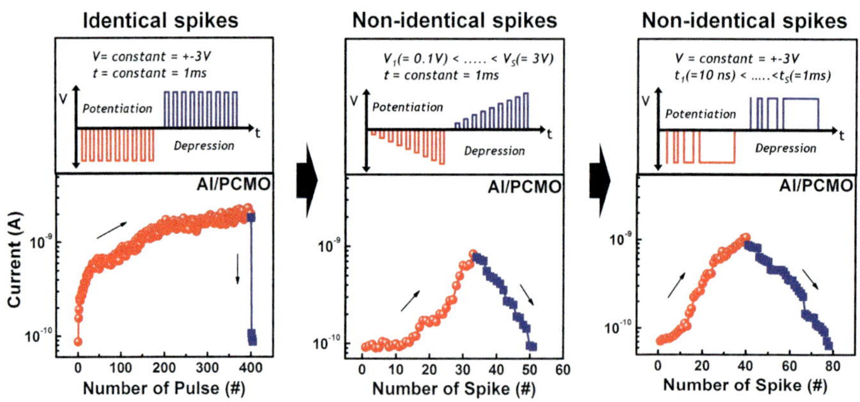

Fig. 3.24 Potentiation and depression with various programming spikes: identical and nonidentical spikes

spikes and nonidentical spikes. When the identical spikes having the same pulse amplitude and duration were applied, the Al/PCMO exhibited gradual potentiation, but abrupt change during the depression. The origin of abrupt change can be explained by the different metal–oxide free energy, as previously discussed. Considering the metal–oxide free energy and induced oxide layer, nonidentical spikes which have different pulse amplitude or duration can be a solution for the high asymmetry ratio, as shown in Fig. 3.24**b**, **c**. However, this method (applying nonidentical spikes) can increase complexity of the peripheral circuits and the neuro-inspired computing system. Moreover, increasing pulse amplitude or duration can lead to higher power dissipation and disturbance in the large-scale synaptic device crossbar array. Thus, for the hardware implementation of the neuro-inspired computing system, the gradual and linear weight changes are necessary with the identical pulse bias.

As the other approach, device structure engineering can be investigated, as shown in Fig. 3.25. On the basis of the metal–oxide free energy, TiN TE can be

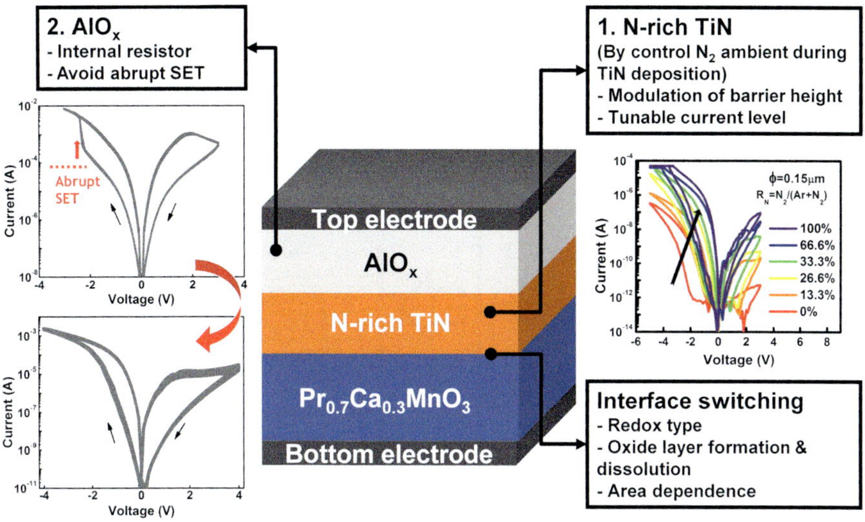

Fig. 3.25 Summary of device structural engineering to obtain improved synaptic characteristics

selected. Then the nitrogen treatment method which is discussed previously can be conducted to achieve the tunable operational ranges: tunable current level and voltage reduction effect. Moreover, on the TiN TE, insulating layer (here, AlO_x) can be adopted as an internal resistor that can prevent abrupt resistance changes and lead to more gradual resistance changes. These approaches can significantly improve the synaptic device, as shown in Fig. 3.26. Compared to the Al/PCMO, the TiN/PCMO having the internal resistor exhibited more gradual and symmetry conductance changes. Considering various respects such as energy consumption, integration size, circuit complexity, etc., the device structure engineering is the more proper solution for the hardware implementation of the neuro-inspired computing system.

3.5 Summary and Outlook

As discussed in this chapter, the PCMO synapse can be one of the promising candidates based on its various advantages: nanoscale device dimension, enough multilevel states, tunable dynamic range, low programming energy consumption, retention, endurance, uniformity, variability, and already demonstrated crossbar array structure. However, linear and symmetric weight update is still necessary and needs further improvement for the hardware implementation. The modification of external bias condition can highly improve the linearity and symmetry in weight update, but it can also increase the complexity of neuro-inspired computing system. Thus, for more efficient neuro-inspired computing system, device structure

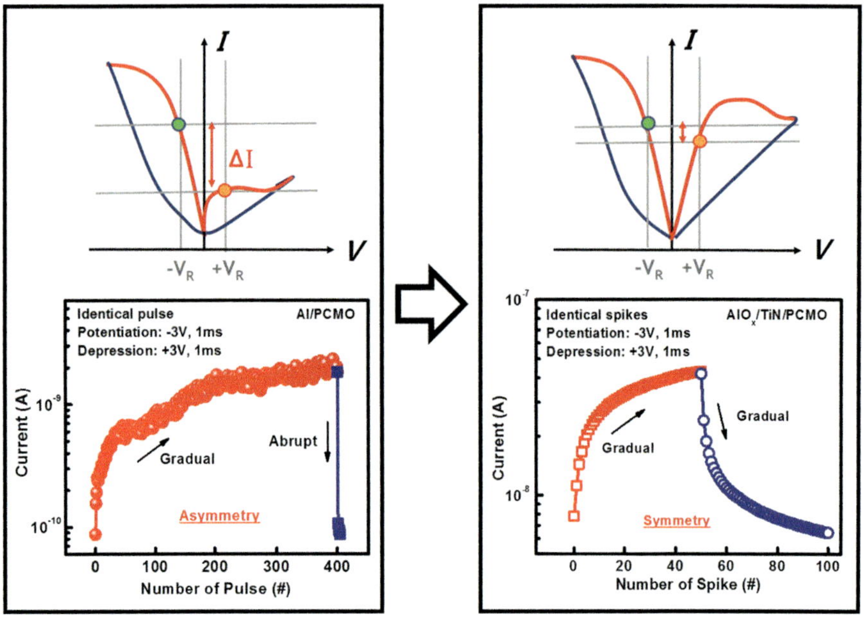

Fig. 3.26 Significantly improved synaptic characteristics based on the structural engineering

engineering is required with careful considerations as discussed in the preliminary results in this chapter.

References

1. R. Waser, M. Aono, Nanoionics-based resistive switching memories. Nat. Mater. **6**(11), 833–840 (2007)
2. K. Szot, W. Speier, G. Bihlmayer, R. Waser, Switching the electrical resistance of individual dislocations in single-crystalline $SrTiO_3$. Nat. Mater. **5**(4), 312–320 (2006)
3. International Technology Roadmap for Semiconductor (ITRS) (2015)
4. H.Y. Chen, S. Yu, B. Gao, P. Huang, J. Kang, H.S.P. Wong, HfO_x based vertical resistive random access memory for cost-effective 3D cross-point architecture without cell selector, in *IEEE International Electron Devices Meeting (IEDM),* (2012)
5. H. Akinaga, H. Shima, Resistive random access memory (ReRAM) based on metal oxides. Proc. IEEE **98**(12), 2237–2251 (2010). doi:10.1109/JPROC.2010.2070830
6. S. Park, H. Kim, M. Choo, J. Noh, A. Sheri, S. Jung, K. Seo, J. Park, S. Kim, W. Lee, J. Shin, D. Lee, G. Choi, J. Woo, E. Cha, J. Jang, C. Park, M. Jeon, B. Lee, B. H. Lee, H. Hwang, RRAM-based synapse for neuromorphic system with pattern recognition function, in *IEEE International Electron Devices Meeting (IEDM),* (2012)
7. S. Park, A. Sheri, J. Kim, J. Noh, J. Jang, M. Jeon, B. Lee, B. R. Lee, B. H. Lee, H. Hwang, Neuromorphic speech systems using advanced RRAM-based synapse, in *IEEE International Electron Devices Meeting (IEDM),* (2013)
8. P. Sangsu, N. Jinwoo, C. Myung-lae, S. Ahmad Muqeem, C. Man, K. Young-Bae, K. Chang Jung, J. Moongu, L. Byung-Geun, L. Byoung Hun, H. Hyunsang, Nanoscale RRAM-based

synaptic electronics: toward a neuromorphic computing device. Nanotechnol. **24**(38), 384009 (2013)

9. P. Sangsu, S. Manzar, N. Jinwoo, L. Daeseuk, M. Kibong, W. Jiyong, L. Byoung Hun, H. Hyunsang, A nitrogen-treated memristive device for tunable electronic synapses. Semicond. Sci. Technol. **29**(10), 104006 (2014)
10. Z.L. Liao, Z.Z. Wang, Y. Meng, Z.Y. Liu, P. Gao, J.L. Gang, H.W. Zhao, X.J. Liang, X.D. Bai, D.M. Chen, Categorization of resistive switching of metal-$Pr_{0.7}Ca_{0.3}MnO_3$-metal devices. Appl. Phys. Lett. **94**(25), 253503 (2009)
11. T. Nakamura, K. Homma, K. Tachibana, Thin film deposition of metal oxides in resistance switching devices: electrode material dependence of resistance switching in manganite films. Nanoscale Res. Lett. **8**(1), 76 (2013)
12. W. Lee, M. Jo, J. Park, J. Lee, S. Park, S. Kim, S. Jung, J. Shin, D. Lee, M. Siddik, H. Hwang, Improved switching characteristics of perovskite oxide-based resistance random access memory by high-pressure oxygen annealing at low temperature. Phys. Status Solidi A **208**(1), 202–205 (2011)
13. M. Jo, D.J. Seong, S. Kim, J. Lee, W. Lee, J.B. Park, S. Park, S. Jung, J. Shin, D. Lee, H. Hwang, Novel cross-point resistive switching memory with self-formed schottky barrier, in *Symposium on VLSI Technology*, (2010)
14. K. Duygu, Y. Shimeng, H.S.P. Wong, Synaptic electronics: materials, devices and applications. Nanotechnol. **24**(38), 382001 (2013)
15. S. Park, M. Chu, J. Kim, J. Noh, M. Jeon, B. Hun Lee, H. Hwang, B. Lee, B.-G. Lee, Electronic system with memristive synapses for pattern recognition. Sci. Rep. **5**, 10123 (2015), 05/05/ online 2015. doi:10.1038/srep10123
16. J.W. Jang, S. Park, G.W. Burr, H. Hwang, Y.H. Jeong, Optimization of conductance change in $Pr_{1-X}Ca_xMnO_3$-based synaptic devices for neuromorphic systems. IEEE Electron Device Lett. **36**(5), 457–459 (2015). doi:10.1109/LED.2015.2418342
17. D.J. Seong, M. Hassan, H. Choi, J. Lee, J. Yoon, J.B. Park, W. Lee, M.S. Oh, H. Hwang, Resistive-switching characteristics of Al/$Pr_{0.7}Ca_{0.3}$/MnO_3 for nonvolatile memory applications. IEEE Electron Device Lett. **30**(9), 919–921 (2009). doi:10.1109/LED.2009.2025896
18. G.W. Burr, R.M. Shelby, C.D. Nolfo, J.W. Jang, R.S. Shenoy, P. Narayanan, K. Virwani, E.U. Giacometti, B. Kurdi, H. Hwang, Experimental demonstration and tolerancing of a large-scale neural network (165,000 synapses), using phase-change memory as the synaptic weight element, in *IEEE International Electron Devices Meeting (IEDM)*, (2014)
19. J.W. Jang, S. Park, Y.H. Jeong, H. Hwang, ReRAM-based synaptic device for neuromorphic computing, in *2014 I.E. International Symposium on Circuits and Systems (ISCAS)*, (2014)
20. H.-G. Lee, *Chemical Thermodynamics for Metals and Materials* (Imperial College Press, London, 1999)
21. G.V. Samsonov, *The Oxide Handbook*, 1st edn. (IFI/Plenum, New York, 1973)
22. S. Shirasaki, H. Yamamura, H. Haneda, K. Kakegawa, J. Moori, Defect structure and oxygen diffusion in undoped and La-doped polycrystalline barium titanate. J. Chem. Phys. **73**(9), 4640–4645 (1980)
23. D.W. Reagor, S.Y. Lee, Y. Li, Q.X. Jia, Work function of the mixed-valent manganese perovskites. J. Appl. Phys. **95**(12), 7971–7975 (2004)
24. H.B. Michaelson, The work function of the elements and its periodicity. J. Appl. Phys. **48**(11), 4729–4733 (1977)
25. S.M. Sze, *Physics of Semiconductor Devices* (Wiley, New York, 1981)
26. S. Park, S. Jung, M. Siddik, M. Jo, J. Park, S. Kim, W. Lee, J. Shin, D. Lee, G. Choi, J. Woo, E. Cha, B.H. Lee, H. Hwang, Self-formed schottky barrier induced selector-less Rram for cross-point memory applications. Phys. Status. Solidi. Rapid. Res. Lett. **6**(11), 454–456 (2012)
27. P.N. Belhumeur, J.P. Hespanha, D.J. Kriegman, Eigenfaces vs. Fisherfaces: recognition using class specific linear projection. IEEE. Trans. Pattern. Anal. Mach. Intell. **19**(7), 711–720 (1997)

Chapter 4
TaO_x-/TiO_2-Based Synaptic Devices

I-Ting Wang and Tuo-Hung Hou

4.1 Device Fabrication Method

In the last few years, numerous intensive studies have investigated the characteristics of Ta/TaO_x/TiO_2/Ti RRAM devices for memory applications [1–3]. Here, we described the typical fabrication flow of Ta/TaO_x/TiO_2/Ti RRAM devices. A shadow-mask process was adopted initially for basic characteristic and switching mechanism investigation in 2D Ta/TaO_x/TiO_2/Ti synaptic device. 70 nm of TiO_2 and 20 nm of TaO_x were sequentially deposited on 100 nm-thick Ti bottom electrode (BE), followed by a deposition of 100-nm-thick Ta top electrode (TE) patterned by shadow-mask with the area of 10^4 μm^2. The Ti and Ta electrodes were fabricated by magnetron DC sputtering, while TiO_2 and TaO_x were deposited by reactive magnetron DC sputtering using the Ti (99.99%) and Ta (99.95%) targets accordingly. And the ratios of Ar:O_2 gas flow were 2:1 for TiO_2 and 20:3 for TaO_x. The entire fabrication was done at room temperature.

In addition, 3D Ta/TaO_x/TiO_2/Ti synaptic device was adopted for further investigation. In order to realize 3D synaptic network, a four-mask process flow was utilized for fabricating a 3D 4 × 4 Ta/TaO_x/TiO_2/Ti synaptic cross-point array. A multilayer stack of SiO_2/Ti/SiO_2/Ti/SiO_2 was deposited with each layer of 100 nm thick. SiO_2 here is used as an isolation layer deposited by plasma-enhanced chemical vapor deposition. Ti as BE was deposited by DC magnetron sputtering using a Ti (99.99%) target. And the multilayer stack was patterned and defined by photolithography and dry etching (HBr and CCl_4), respectively, to form a pillar structure. After the pillar formation, Ti layers on the sidewall were exposed to serve as two separate horizontal BEs for top cell (TC) and bottom cell (BC) cladded with SiO_2 layers.

I.-T. Wang • T.-H. Hou (✉)
Department of Electronics Engineering and Institute of Electronics, National Chiao Tung University, 1001 University Road, Hsinchu 300, Taiwan
e-mail: thhou@mail.nctu.edu.tw

S. Yu (ed.), *Neuro-inspired Computing Using Resistive Synaptic Devices*,
DOI 10.1007/978-3-319-54313-0_4

Next, similar to the process of 2D device, TiO_2 layer and TaO_x layer were deposited sequentially by reactive DC magnetron sputtering using Ti (99.99%) and Ta (99.95%) targets, respectively. A 250-nm-thick vertical Ta TE was patterned by photolithography and deposited by DC magnetron sputtering using a Ta (99.95%) target and followed by liftoff process. Finally, the contacts of Ti BEs for TCs and BCs were patterned and etched using two additional masks. The effective device area was 20 μm^2 defined by the cross-point region of horizontal Ti BE and vertical Ta TE at the pillar sidewall. The entire fabrication was also done at room temperature.

The electrical characterizations were measured by HP 4156B and Agilent B1500A. And the synaptic behaviors were performed using Agilent B1530A waveform generator/fast measurement unit. All the measurements were realized at room temperature with biased Ta TEs and grounded Ti BEs.

4.2 Basic Device Characteristics

Figure 4.1a, b illustrate the typical current–voltage (*I–V*) characteristics of the device; a multistate capability in low resistance states (LRS) and high resistance states (HRS) is observed in the negative voltage region (−2 V) by using SET-controlled (Fig. 4.1a) and RESET-controlled (Fig. 4.1b) operation modes. Moreover, the device exhibits numerous desirable features such as forming-free, self-compliance, self-rectification, and extremely high endurance [1–3]; the forming-free and self-compliance properties simplify peripheral circuit design, while the self-rectifying characteristic suppresses unwanted sneak current path [4] and allows a larger attainable cross-point array size without requiring selection devices, such as unipolar diodes [5, 6] and bipolar selectors [7, 8]. With a record-high endurance of more than 10^{12} cycles [1, 3] and a fast switching speed of 50 ns,

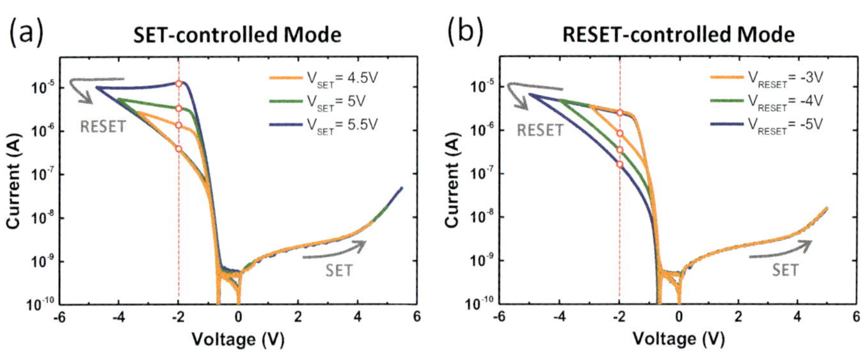

Fig. 4.1 Typical *I–V* characteristics of Ta/TaO_x/TiO_2/Ti RRAM under (**a**) SET-controlled and (**b**) RESET-controlled operation modes. A low current is obtained when a positive bias is applied to the device because of the self-rectifying characteristic. Consequently, multiple resistance states are only recognized using negative read voltage

Ta/TaO_x/TiO_2/Ti RRAM is favorable to fit the time gap between hard disk (>10 ms) and DRAM (<100 ns) in the memory hierarchy [9, 10], making it a suitable candidate for storage-class memory applications [1, 3].

4.3 Switching Mechanism

In this section, we discuss the switching mechanism for a Ta/TaO_x/TiO_2/Ti RRAM. The experimental findings are first demonstrated. Unlike traditional filamentary RRAMs where the resistive switching is determined by the connecting or rupturing of the conducting filaments [11], the switching mechanism of a Ta/TaO_x/TiO_2/Ti RRAM can be elucidated using oxygen ion migration, which causes homogeneous barrier modulation (HBM) [1–3]. A physical simulation based on these non-filamentary characteristics is then introduced, demonstrating its accuracy in a theoretical study of the device.

4.3.1 Experimental Findings

Strong evidence for the homogeneous current conduction is shown in Fig. 4.2a, where area-dependent current conduction is achieved [2, 3, 12]. A decrease in current in both LRS and HRS is associated with a decrease in device area,

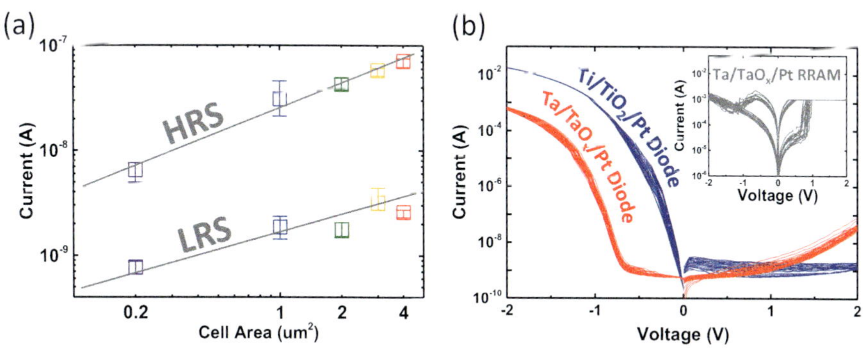

Fig. 4.2 (**a**) Area-dependent current conduction with the current proportional to the device area in both LRS and HRS, obtained in Ta/TaO_x/TiO_2/Ti RRAM, providing evidence for the proposed HBM model. (**b**) The *I–V* characteristics of the Ta/TaO_x/Pt and Ti/TiO_2/Pt diodes, which were made specifically for investigating the roles of the critical interfaces Ta/TaO_x and Ti/TiO_2 in the switching mechanism. Both diodes exhibit self-rectifying behavior; the Ti/TiO_2 diode provides a more ohmic-like barrier with a higher current. The inset displays the *I–V* characteristic after the forming process; only the Ta/TaO_x/Pt diode indicates stable resistive switching, while the Ti/TiO_2/Pt diode indicates a complete breakdown of behavior (data not shown), suggesting TaO_x plays a dominant role in the Ta/TaO_x/TiO_2/Ti resistive switching process

indicating that the homogeneous interface under the device area is involved during resistive switching instead of a localized region such as conducting filaments.

To further investigate the switching mechanism, $Ta/TaO_x/Pt$ and $Ti/TiO_2/Pt$ devices were fabricated for exploring the role of Ta/TaO_x and Ti/TiO_2, the two critical interfaces in a $Ta/TaO_x/TiO_2/Ti$ device stack. An inert metal (Pt) was not involved in the switching process; hence, the *I–V* characteristic was strictly influenced by the other interface. Both $Ta/TaO_x/Pt$ and $Ti/TiO_2/Pt$ devices demonstrated basic diode characteristics within ± 2 V sweeps, as illustrated in Fig. 4.2b, where voltage was applied to Ta and Ti TEs, whereas the Pt BEs were grounded. The rectifying characteristic was investigated in both devices because of the asymmetric Schottky barriers at the Ta/TaO_x and TaO_x/Pt interfaces for $Ta/TaO_x/Pt$ and at the Ti/TiO_2 and TiO_2/Pt interfaces for $Ti/TiO_2/Pt$. Moreover, a more ohmic-like transparent barrier formed at the Ti/TiO_2 interface, yielding a higher current at a negative bias than at the Ta/TaO_x interface. After the electrical forming process at a positive bias ($\sim$ 5 V) with a current compliance of 1 μA, only the $Ta/TaO_x/Pt$ device demonstrated stable bipolar resistive switching, as displayed in the inset of Fig. 4.2b. This investigation indicated that the Ta/TaO_x interface may play a predominant role in the switching behavior of the $Ta/TaO_x/TiO_2/Ti$ device. Further experimental findings included an oxygen-deficient TaO_x layer found through energy-dispersive X-ray spectroscopy (EDX) analysis [1–3] and a more critical influence of TaO_x thickness than TiO_2 on $Ta/TaO_x/TiO_2/Ti$ resistive switching behavior [3]. The oxygen-deficient TaO_x layer explains the strong oxygen scavenging capability of Ta, while the influence of TaO_x thickness suggests that TaO_x is the main resistive switching layer [3].

From the experimental results, we conclude that the resistive switching in the HBM mechanism was triggered by the uniform migration of oxygen ions in TaO_x close to the Ta interface. By applying a positive or negative bias on the Ta TE, oxygen ions migrate according to the applied electric field and modulate the electron tunnel barrier at the Ta/TaO_x interface, thereby affecting device conductance. A more detailed theoretical analysis on the switching mechanism is discussed in the next section.

4.3.2 *Physical Modeling*

A one-dimensional simulation has been demonstrated for the proposed HBM model [13]. An oxygen-vacancy-rich (V_O^{2+}-rich) layer is located at the Ta/TaO_x interface, which is supported by experimental findings [1–3]. Here, we assume the non-lattice oxygen ions (O^{2-}) with negative charges in this V_O^{2+}-rich region migrate easily under a local electric field, while V_O^{2+} serve as immobile shallow donor-like dopants with a positive charge. The O^{2-} migration is thus obtained by the diffusion and drift components of O^{2-} hopping through local potential wells [14, 15]. The O^{2-} density (N_O) is described by the continuity equation [13]:

$$\frac{\partial N_O}{\partial t} = D\frac{\partial^2 N_O}{\partial x^2} - v\frac{\partial N_O}{\partial x} \tag{4.1}$$

where the diffusion coefficient D is obtained by [13]

$$D = \frac{1}{2}a^2 f \cdot \exp\left(-\frac{E_a}{kT}\right) \tag{4.2}$$

and the drift velocity v is obtained by [13]

$$v = af \cdot \exp\left(-\frac{E_a}{kT}\right) \cdot \sinh\left(-\frac{q\gamma aF}{kT}\right) \tag{4.3}$$

The effective hopping distance a is 0.05 nm, which is used as a mesh size in numerical calculation [3, 16]; f is the attempt-to-escape frequency [3, 16] set to 10^{13} Hz; E_a is the activation energy for O^{2-} migration; kT is the thermal energy at $T = 300$ K; q is the elementary charge; γ is a field-dependent fitting parameter; F is the electric field. The local electric field with an initial O^{2-} profile is calculated by applying voltage and the Poisson equation [13]:

$$\frac{\partial F}{\partial x} = \frac{2q}{\varepsilon_r \varepsilon_0}(-N_O + N_{V+}) \tag{4.4}$$

where ε_r is the dielectric constant of TaO_x or TiO_2 and $\varepsilon_0$is the permittivity of vacuum. N_{V+} describes the immobile V_O^{2+} density, which is assumed to have an exponential function decaying from the Ta/TaO_x interface; this assumption corroborates the experimental observation of the gradient sub-stoichiometric TaO_x composition [3]. By considering the effect of both N_O and N_{V+}, the Poisson equation is solved, and the entire Ta/TaO_x/TiO_2/Ti potential profile can be obtained. The Wentzel–Kramers–Brillouin (WKB) approximation is used to calculate the tunneling current through the TaO_x barrier with the obtained potential profile by considering the field-dependent barrier lowering and the thermionic emission component.

Figure 4.3a displays schematic diagrams of the Ta/TaO_x/TiO_2/Ti device, illustrating the HBM-based switching mechanism with oxygen ions migrating in the TaO_x region during SET and RESET operations. When applying $-V_{RESET}$ on the Ta TE (RESET operation), O^{2-} migrates toward the bulk of TaO_x and increases the effective barrier width for electron tunneling from Ta to TaO_x, thus decreasing device conductance. By contrast, when applying $+V_{SET}$ on the Ta TE (SET operation), O^{2-} migrates toward the Ta/TaO_x interface, thus decreasing the effective barrier width for electron tunneling from Ta to TaO_x and increasing device conductance. Figure 4.3a, c demonstrate the simulated band diagrams of the device under read operation when applying −2 V and +2 V read voltage at the Ta TE, respectively. The O^{2-} substantially affects the tunnel barrier at the Ta/TaO_x

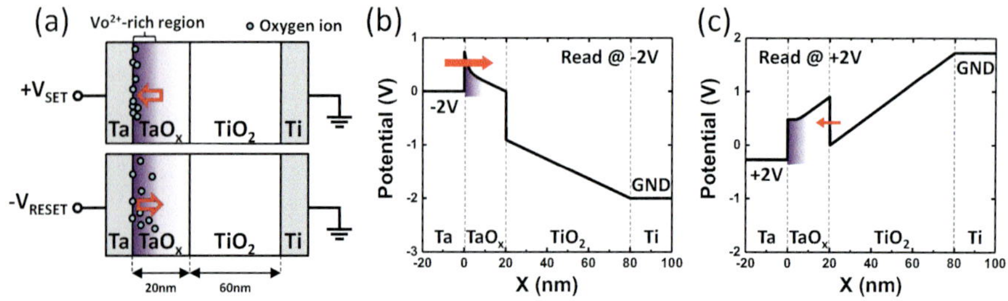

Fig. 4.3 (**a**) Schematic of Ta/TaO_x/TiO_2/Ti device under SET and RESET operations by applying positive and negative bias on the Ta TE, respectively. The oxygen ions in the Vo^{2+}-rich region migrate according to the local electric field, thus changing the tunnel barrier at the Ta/TaO_x interface. After SET operation, the calculated band diagrams are demonstrated under read operation when applying (**b**) −2 V and (**c**) +2 V on the Ta TE. At a −2 V read condition, the tunnel barrier at Ta/TaO_x modulated by oxygen ion migration in Vo^{2+}-rich region shows dramatic influence on electron tunneling from the Ta TE. By contrast, at a +2 V read condition, the unaffected tunnel barrier suppresses the electron tunneling from TiO_2 to TaO_x, thus reducing the current

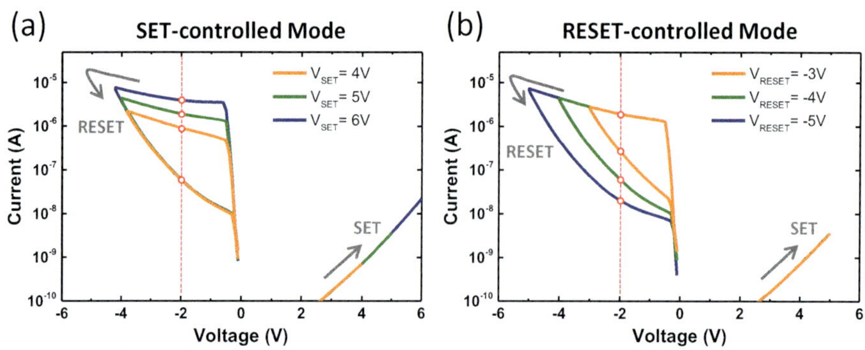

Fig. 4.4 Simulated *I–V* characteristics of Ta/TaO_x/TiO_2/Ti RRAM under (**a**) SET-controlled and (**b**) RESET-controlled operation modes. A low current is obtained when a positive bias is applied to the device because of the self-rectifying characteristic. As the result, multilevel resistance states are only recognized using negative read voltage (−2 V)

interface during SET and RESET operations; however, a high tunnel barrier remains unchanged at the TaO_x/TiO_2 interface because of the relatively thick tunnel barrier at the TaO_x/TiO_2 interface, which is far from the Vo^{2+}-rich region. Hence, the device's conductance change can only be recognized by applying negative read voltage at the Ta TE while the current at a positive bias is substantially lower. Consequently, the device exhibits inherent self-rectifying behavior, which provides a desirable solution to the sneak current problem in the cross-point array without requiring the use of selection devices [17]. Moreover, the measured *I–V* characteristics can be reproduced by the proposed HBM model. Specific features including the SET-control mode (Fig. 4.4a), RESET-control mode (Fig. 4.4b), and self-rectifying characteristic demonstrate reasonable agreement.

4.4 Synaptic Function Realization and Modeling Results

Synapses are indispensable components in a neural network; biologically, they represent the connection strength between neurons by synaptic weight modulation, which is related to learning and memory in human brain. In a neuromorphic computing system, a synaptic device exhibits synaptic weight by device conductance, and the amount of weight modulation is related to the strength and frequency of the input signals [18, 19].

Aside from the favorable memory applications of Ta/TaO_x/TiO_2/Ti RRAM, we observed that the device also provided synaptic characteristics which were suitable for neuromorphic computing [20]. Hence, in this section, we introduce the measured synaptic behaviors obtained in the Ta/TaO_x/TiO_2/Ti synaptic device, and we demonstrate the modeling results, including compact modeling and physical simulation.

4.4.1 Potentiation and Depression

The plasticity of a synaptic device describes the increase (potentiation, P) and the decrease (depression, D) in a synaptic weight state; these phenomena can be achieved in a RRAM device by using SET and RESET training patterns, respectively. By using consecutive P or D pulses, a continual tuning of weight states can be realized. Figure 4.5 displays the synaptic plasticity of the Ta/TaO_x/TiO_2/Ti synaptic device, where alternating ten P pules (+9 V/50 μs) and ten D pulses

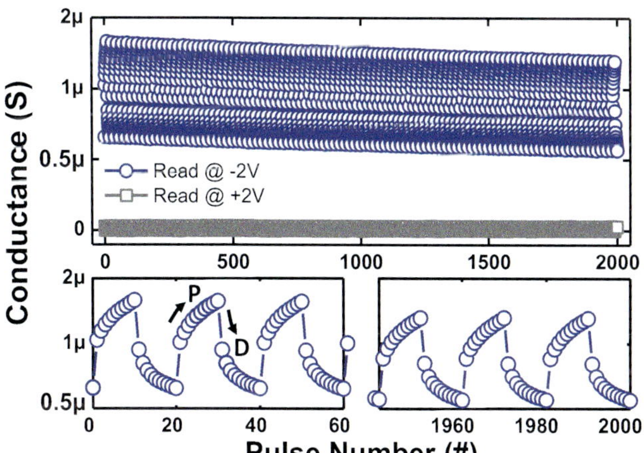

Fig. 4.5 Synaptic plasticity of 2D Ta/TaO_x/TiO_2/Ti device showing alternative potentiation (*P*) and depression (*D*) characteristics for 2000 training pulses (*P*, +9 V/50 μs; *D*, −8 V/50 μs). The device exhibits noticeable plasticity by using −2 V/4 ms read pulses, while it exhibits no plasticity by using +2 V/ 4 ms read pulses. The responses of the first and last 60 pulses are enlarged

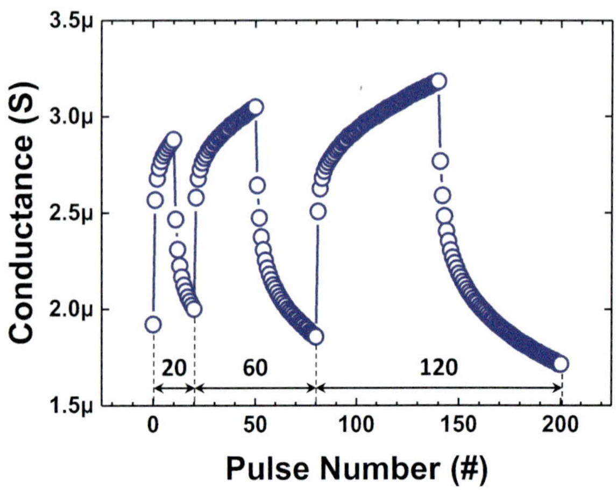

Fig. 4.6 A pulse number-dependent weight window of the device is achieved by using various numbers of training pulses (*P*, +9 V/50 μs; *D*, −8 V/50 μs). Higher and lower weight states can be achieved by increasing the P and D pulse numbers, respectively, and a larger weight window achieved by a larger pulse number successfully mimics biological memory enhancement through increased stimulation

(−8 V/50 μs) were used as training patterns. The weight state was then read by the read pulse (−2 V/4 ms) after each training pulse. The device exhibited a stable and repeatable synaptic plasticity without obvious variation over 2000 training pulses, which demonstrates its stability and robustness. Furthermore, the homogeneous current conduction increased the device's immunity against localized random ion migration. The plasticity could only be read by −2 V; it could not be read by +2 V because of its self-rectifying characteristic. The device exhibited a nonlinear P/D response toward training pulses (i.e., the amount of weight change caused by one pulse was not the same along the consecutive P or D processes). Moreover, by fixing the training pulse strength (i.e., pulse width or pulse amplitude) while increasing the consecutive number of P/D pulses, higher and lower weight states could be achieved, and a wider weight window was observed, as shown in Fig. 4.6. This is similar to a biological brain, wherein memory is enhanced by increasing stimulation. It is noteworthy that the weight states of the device could be continually tuned, even after 800 pulses without saturation (Fig. 4.7). This analog-like behavior with multilevel states is also desirable for improving the capacity and robustness of a neuromorphic computing system [21].

Generally, the behaviors of memristive device can be described using the following equations [22]:

$$I = G(\mathrm{g}) \cdot V \tag{4.5}$$

$$\frac{dg}{dt} = f(\mathrm{g}, V) \tag{4.6}$$

The conductance G of the memristive device is determined according to its internal state variable g, and a change in rate of this internal state variable is related to its present state and the applied voltage V. Based on the measurement results, a compact model is generated using analytical equations in physical simulation and can be expressed as [13]

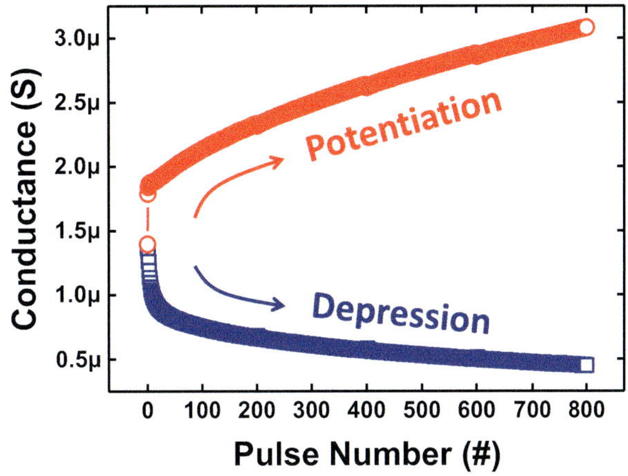

Fig. 4.7 Potentiating and depressing characteristics of the device (*P*, +9 V/50 μs; *D*, −8 V/50 μs). The device shows continual weight tuning for over 800 training pulses without saturation

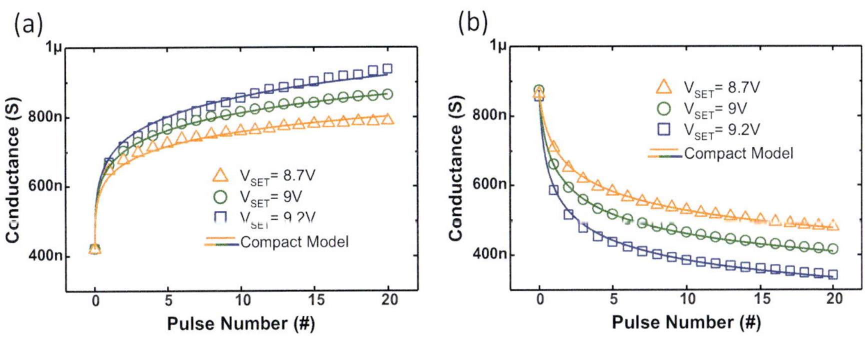

Fig. 4.8 Synaptic plasticity of (**a**) *P* and (**b**) *D* characteristics is achieved by using various voltage amplitudes while fixing the same width of 50 μs. The device exhibits different magnitudes of *P* and *D* demonstrated by experimental results (*indicated by symbols*), and the compact model calculation (*indicated by lines*) indicates strong agreement

$$I = I_0 \cdot \exp\left(-\frac{g}{g_0}\right) \cdot \sinh\left(\frac{V_R}{V_0}\right) \tag{4.7}$$

$$\frac{dg}{dt} = \exp\left(-\frac{E_A}{kT}\right) \cdot \sinh\left[C(g) - \gamma(g) \cdot \frac{V_W}{V_0}\right] \tag{4.8}$$

In Eq. (4.7), the tunnel current through TaO_x during read operation with read voltage of V_R is described, with a constant current I_0 and an effective tunneling gap g. And g_0 and V_0 here are the normalized constants. Equation (4.8) depicts the change in g due to O^{2-} migration through drift and diffusion processes during P/D operations. C(g) is the state-dependent function and γ(g) is the field-acceleration function. These two equations are used to describe the state-dependent velocity of O^{2-} migration. As a result of different pulse amplitudes for P/D measurements (Fig. 4.8a, b), the device demonstrates a voltage-dependent weight update.

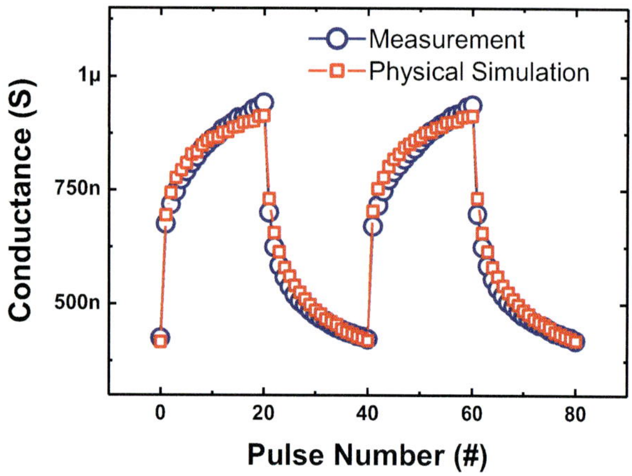

Fig. 4.9 Potentiating and depressing characteristics simulated by the physical model, indicating excellent agreement with the measurement results

The compact model fitting result (indicated by line) agrees strongly with the experimental data (indicated by symbols), and the model may facilitate the development of a large-scale system-level simulation by considering the device's characteristics.

The measured synaptic plasticity was also simulated by the physical model. Because of the voltage and time dependence in ion migration [23], the progressively increasing number of pulses applied to the device was equivalent to the higher voltage operation, which caused a more severe barrier change. Hence, the physical model (red square symbols) indicates good agreement for both the P and D processes with the measurement data (blue circle symbols) in Fig. 4.9. Moreover, the physical model successfully depicted and reasonably explained the nonlinearity (*NL*) of the device. During the first P pulse, when a positive pulse was applied at the Ta TE, O^{2-} migrated rapidly from the bulk of the TaO_x toward the Ta/TaO_x interface, which dramatically reduced the tunnel barrier and increased the current. However, the accumulated O^{2-} at the Ta/TaO_x interface reduced the internal electric field and suppressed further O^{2-} migration to the interface, which gradually saturated conductance change. By contrast, a similar phenomenon occurred when the first D pulse was applied, wherein O^{2-} migration from the Ta/TaO_x interface toward the bulk of the TaO_x drastically increased the tunnel barrier and decreased the current. The decreasing current gradually saturated when the accumulated O^{2-} at the bulk of the TaO_x weakened the internal electric field, and further migration of O^{2-} was diminished.

4.4.2 Spike-Timing-Dependent Plasticity

In a biological nervous system, synaptic weight is modulated by spike-timing-dependent plasticity (STDP), where synaptic plasticity is triggered by the relative timing of presynaptic and postsynaptic spikes [24]. STDP is a crucial function related to memory and learning; it is related to the Hebbian learning rule [25]. Here, instead of square pulses, the action potential-like waveform is adopted for the STDP measurement, and the simulation of the Ta/TaO_x/TiO_2/Ti synaptic device, as shown in Fig. 4.10a, is generated as follows [13]:

$$V = \begin{cases} A \cdot \exp\left(\frac{-t}{\tau}\right), & t > 0 \\ -A \cdot \exp\left(\frac{-t}{\tau}\right), & t < 0 \end{cases} \tag{4.9}$$

where A determines the spike amplitude and τ is the decay time constant. The pre-spike is sent from the Ti BE, and the post-spike is sent from the Ta TE. The net voltage drop across the entire device gravitates toward P when the post-spike

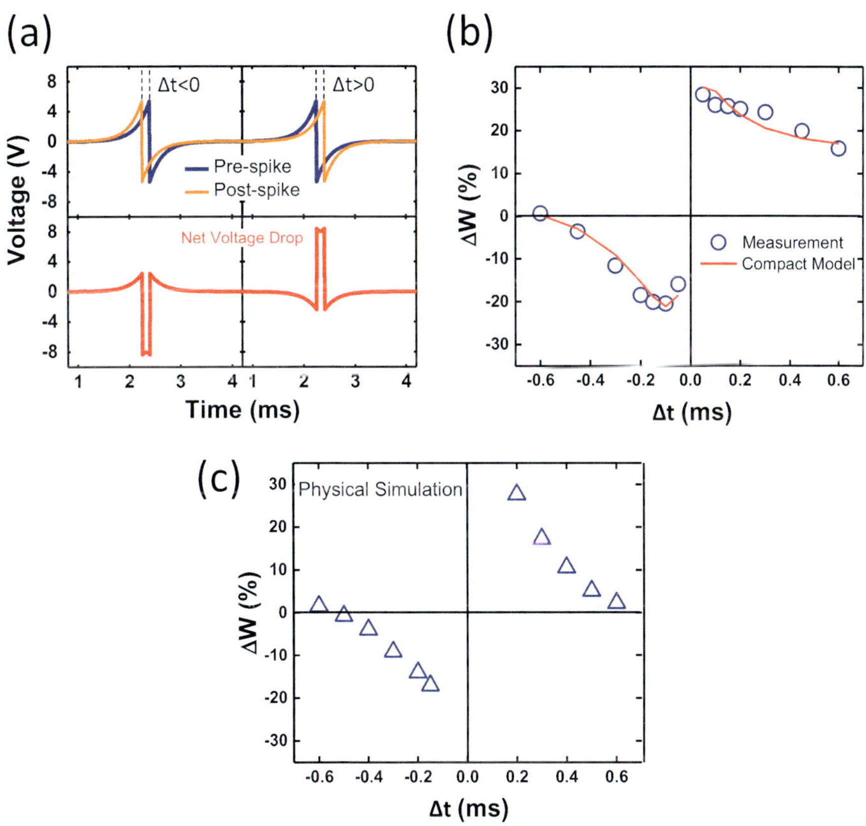

Fig. 4.10 (**a**) Biomorphic action potential-like waveforms are used for STDP measurement when the pre-spike is applied to the Ti BE and the post-spike is applied to the Ta TE. The potentiation and depression of the device is controlled by the net voltage drop, determined by the relative timing of pre-spikes and post-spikes (Δt). (b) Measured STDP characteristics and the fitting using a compact model calculation. (c) Simulation of STDP characteristics based on the physic-based HBM model

arrives earlier than the pre-spike (relative timing $\Delta t > 0$), while it moves toward D when $\Delta t < 0$. The STDP measurement is displayed in Fig. 4.10b (symbol). The synaptic weight change (ΔW) is defined by

$$\Delta W = \frac{G' - G_0}{G_0} \times 100\% \tag{4.10}$$

where G_0 and G' represent the device conductance obtained before and after the spikes, respectively. The Hebbian learning rule states that cells that are wired together fire together [25]. The maximum synaptic weight change occurs at Δt, close to zero. Moreover, similar behavior is fitted by the compact model and is reproduced by the physical model, as shown in Fig. 4.10b (line) and Fig. 4.10c, respectively; this indicates strong agreement with the measurement data.

4.4.3 Paired-Pulse Facilitation

Paired-pulse facilitation (PPF) involves temporal summation of input stimuli in biological neural networks. For neuromorphic systems, PPF provides flexibility to process information in the frequency domain, which determines the enhancement of synaptic weight by modifying the time interval between two sequential pulses generated from the neurons. Figure 4.11a displays a PPF measurement result in the $Ta/TaO_x/TiO_2/Ti$ synaptic device. The paired-pulse as a training configuration is also indicated in the inset. The PPF is successfully described by [13]

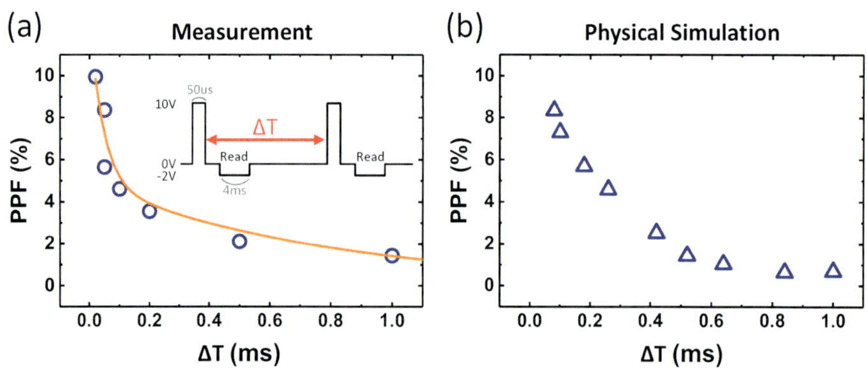

Fig. 4.11 (**a**) PPF experimental results (*symbols*) of the device. The inset displays the paired-pulse waveform for PPF measurement and simulation. By increasing the time interval (ΔT) between pulses, the measurement result exhibits exponential decay with increasing ΔT, fitted by (4.11) where τ_1 (45 ms) and τ_1 (800 ms) are fitting time constants used to represent fast- and slow-decaying terms, respectively. The fitting time constants agree with the time scales of biological synapses [26]. (**b**) Simulated PPF results demonstrate similar characteristics to the measurement results

$$\mathrm{PPF} = \frac{G_2 - G_1}{G_1} \times 100\% = C_1 \cdot \exp\left(\frac{-\Delta T}{\tau_1}\right) + C_2 \cdot \exp\left(\frac{-\Delta T}{\tau_2}\right) \quad (4.11)$$

where G_1 and G_2 describe the conductance value obtained by the first and second pulses, respectively, in the paired-pulse configuration. $C_1$and C_2 are the fitting parameters; τ_1(45 ms) and τ_2 (800 ms) used as fitting time constants represent the fast- and slow-decaying terms that are comparable to biological synapses [26]; ΔT is the time interval between the paired-pulses. Maximum synaptic plasticity occurs at the smallest ΔT and indicates exponential decay behavior. Moreover, PPF behavior can also be simulated by the physical model wherein O^{2-} diffuses from the Ta/TaO_x interface toward the bulk TaO_x during the time interval between pulses. If the time interval is long enough for O^{2-} to diffuse away from the Ta/TaO_x interface, the effect of barrier lowering is diminished. Figure 4.11b displays simulated PPF behavior in a physical model. The physical simulation results exhibit similar behavior to the measurement data.

4.5 3D Synaptic Network

Our brains process information efficiently and intelligently, which allows us to be trainable and capable of solving complicated tasks. The brain contains more than 10 billion neurons and 100 trillion synapses yet consumes only 10 W in a compactly layered structure [27]. To date, computer scientists have successfully developed several adaptive, massively parallel and fault-tolerant artificial neural networks in supercomputers such as IBM's Watson [28], Facebook's DeepFace [29], and Google's DeepMind [30], which are the most advanced examples of software-based neuromorphic computing system. However, limitations for these neural networks include the scalability of power consumption, form factor, and cost [21], which hinder their future development. Therefore, to facilitate future hardware implementation of neural networks on chip, both 3D and low-power synaptic devices are preferable to deal with current obstacles in neuromorphic computing.

Hence, a 3D synaptic network based on the Ta/TaO_x/TiO_2/Ti synaptic device has been realized. Firstly, its synaptic characteristics are demonstrated and discussed. Finally, weight update nonlinearity (*NL*), which is a critical element in experimental synaptic device is comprehensively studied.

4.5.1 3D Synaptic Network Realization

Figure 4.12a displays an optical microscopic image of the fabricated 3D 4 × 4 Ta/TaO_x/TiO_2/Ti synaptic cross-point array, and its schematic is presented in Fig. 4.12b. The device uses two Ti layers, isolated by SiO_2 buffer layers and

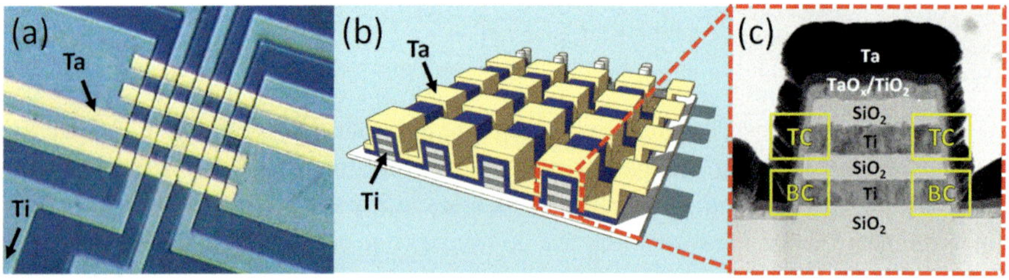

Fig. 4.12 (**a**) Optical microscopic image of fabricated 3D 4 × 4 Ta/TaO_x/TiO_2/Ti synaptic cross-point array. (**b**) Schematic of detailed structure of the two-layer 3D 4 × 4 Ta/TaO_x/TiO_2/Ti array; the zoomed-in cross-sectional TEM image is displayed in (**c**). The two Ti horizontal electrodes are cladded by SiO_2 layers and form the TC and BC on the pillar sidewalls. Each cell consists of two sub-cells at both sides of the pillar sidewalls also indicated in (**c**). The total effective device area is 20 μm^2

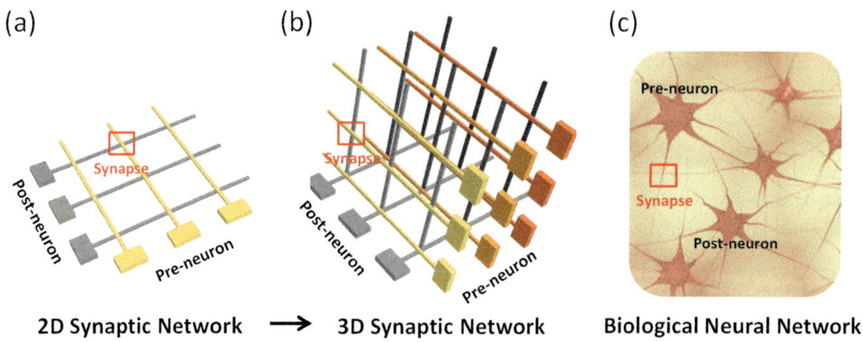

Fig. 4.13 Evolution of synaptic network from (**a**) 2D to (**b**) 3D architecture indicating the significance in increasing integration density. The 3D synaptic network mimics the intricate neural network in a human brain, illustrated in (**c**). The pre-neuron, post-neuron, and synapse are indicated

contains vertically stacked TCs and BCs, as displayed in the zoomed-in cross-sectional transmission electron microscope (TEM) image in Fig. 4.12c. Each cell comprises two sub-cells at both sides of the Ti electrode, and the total effective device area is 20 μm^2. The 3D implementation of the synaptic cross-point array significantly increases the integration density compared with a traditional 2D array, as illustrated in Fig. 4.13a, b. The 3D array successfully depicts a 3D neural network similar to the human brain, where a massive neural connection is intricately connected by countless neurons and synapses (Fig. 4.13c).

Figure 4.14a is the synaptic plasticity measurement result for the 3D device obtained using alternate 50 P pulses (+3 V/5 ms) and 50 D pulses (−3 V/5 ms) as training patterns; a stable and repeatable cycling endurance was achieved within 5000 training pulses, indicating minimal degradation. Moreover, the device exhibited similar behavior in the TC and the BC, as displayed in Fig. 4.14b, c, respectively, suggesting that the array density could be increased by increasing the

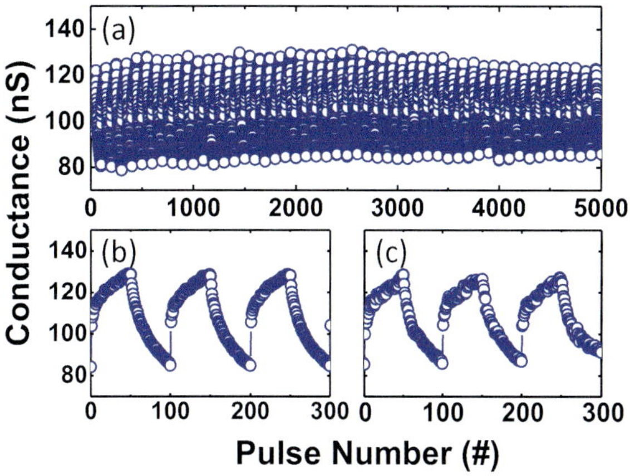

Fig. 4.14 (**a**) Synaptic plasticity of the 3D $Ta/TaO_x/TiO_2/Ti$ device exhibiting excellent endurance over 5000 training pulses (*P*, +3 V/5 ms; *D*, −3 V/5 ms; read, −1.5 V/1 ms). The enlarged plasticity characteristics in (**b**) TC and (**c**) BC indicate uniformity among cells

stacking number of Ti layers in a vertical direction, similar to the industrial integration of 3D NAND flash [31]. Notably, the effective cell size was 20 μm^2 in the 3D device, which was much smaller than that of 2D device, with 10^4 μm^2. Consequently, the conductance value in the 3D device was much smaller because of the area-dependent current conduction mechanism in the HBM model [2, 3, 12]. This phenomenon suggests that the higher density synaptic arrays with lower power consumption can be achievable by further reducing the device area. The reported synaptic behaviors operate at a relatively low frequency (5 ms), which is comparable with that of biological neural networks. This relatively lower frequency of 3D device compare to 2D one (50 μs) is mainly due to the smaller applied voltage.

There are crucial influential factors such as training pulse number and pulse strength that affect the effectiveness of plasticity in synaptic devices. In Fig. 4.6, a larger weight window could be achieved by increasing the training pulse number with a fixed pulse width and amplitude. By contrast, the enhancement of both P and D demonstrated in Fig. 4.8a, b, respectively, could be realized by increasing pulse amplitude while maintaining the same pulse width and pulse numbers. The pulse number-dependent plasticity (Fig. 4.6) and voltage-dependent plasticity (Fig. 4.8) have been demonstrated. However, the influences of both pulse width and pulse amplitude are yet to be discussed. To investigate the relation between pulse width and pulse amplitude, another measurement was adopted. By varying training pulse widths and amplitudes while maintaining a similar weight window (Fig. 4.15), the device exhibited an exponential *V–t* dependence when it was operated not only with low voltage amplitude at a low frequency (±1.5 V/100 ms) but with high voltage amplitude at a high frequency (±9 V/50 μs). This phenomenon is termed the *voltage–time dilemma* [23, 32], and the exponential *V–t* relation can be explained as O^{2-} ions migrating over the local potential well at TaO_x [13, 23]. Furthermore, an extremely low energy consumption (<10 fJ per spike) similar in scale to biological synapses was experimentally achieved by using −18 V/10 ns D pulses [20]. This implementation of extremely low energy consumption was also agreed with the prediction of the exponential *V–t* relation.

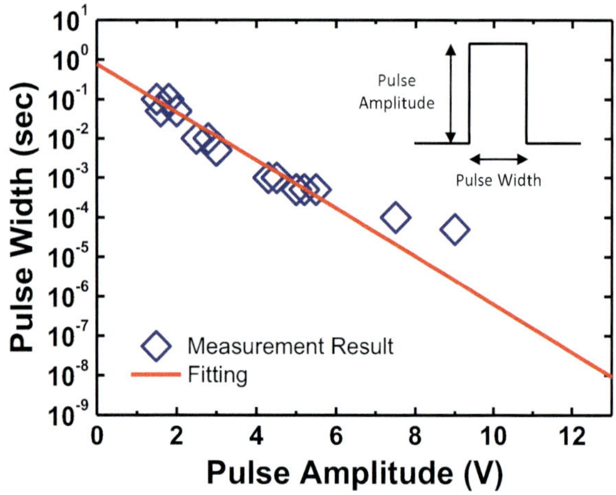

Fig. 4.15 Exponential behavior of the V–t dependence in the 3D $Ta/TaO_x/TiO_2/Ti$ device. By varying the pulse width and pulse amplitude indicated in the inset while maintaining similar synaptic plasticity, the device provides a nonlinear V–t dependence wherein the hardware neural network system can be operated in various domains depending on the application

4.5.2 Linearity Tuning

A nonlinear weight update is frequently observed in reported synaptic devices [18, 33, 34] and has become a critical non-ideal factor in hardware neural network development [35–37]. Figure 4.16a displays a simple identical training pulse configuration commonly used as a traditional training scheme (UP-scheme), and all the pulses are in unipolar polarity for P or D. A normalized P/D measurement performed in the $Ta/TaO_x/TiO_2/Ti$ synaptic cross-point array by using 50 consecutive P/D pulses, illustrated in Fig. 4.16b. However, a severe *NL* of 0.78 is obtained by [38]

$$\mathrm{NL} = \mathrm{Max}|G_P(n) - G_D(n)| \text{ for } n = 1 \text{ to } N \tag{4.12}$$

The *NL* value is normalized to a total plasticity range from zero to one, and $G_P(n)$ and $G_D(n)$ describe the conductance value after *n*th P pulse and D pulse, respectively; *N* defines the total consecutive P/D pulse number. For a perfectly linear weight update, as displayed by the linear dashed line, *NL* is zero, which suggests an equal conductance change was obtained by every training pulse. However, because of the intrinsic property which is supported by the simulation of O^{2-} migration [13], the device with a nonlinear weight update does not easily follow simple addition and subtraction rules assumed in software-based neural networks.

To improve the nonlinear weight update in a synaptic device, many solutions have been proposed [39]. Among them, the state-dependent training scheme (SD-scheme) is commonly adopted [39]. Figure 4.16c illustrates an example of a non-identical training pulse configuration in a SD-scheme. By exponentially increasing the pulse width while maintaining the same amplitude with an increasing pulse number, the training pulse sequence has the following relation:

$$\mathrm{PW}(n+1) = \mathrm{PW}(0) \cdot \exp\left(\frac{\alpha n}{N}\right) \tag{4.13}$$

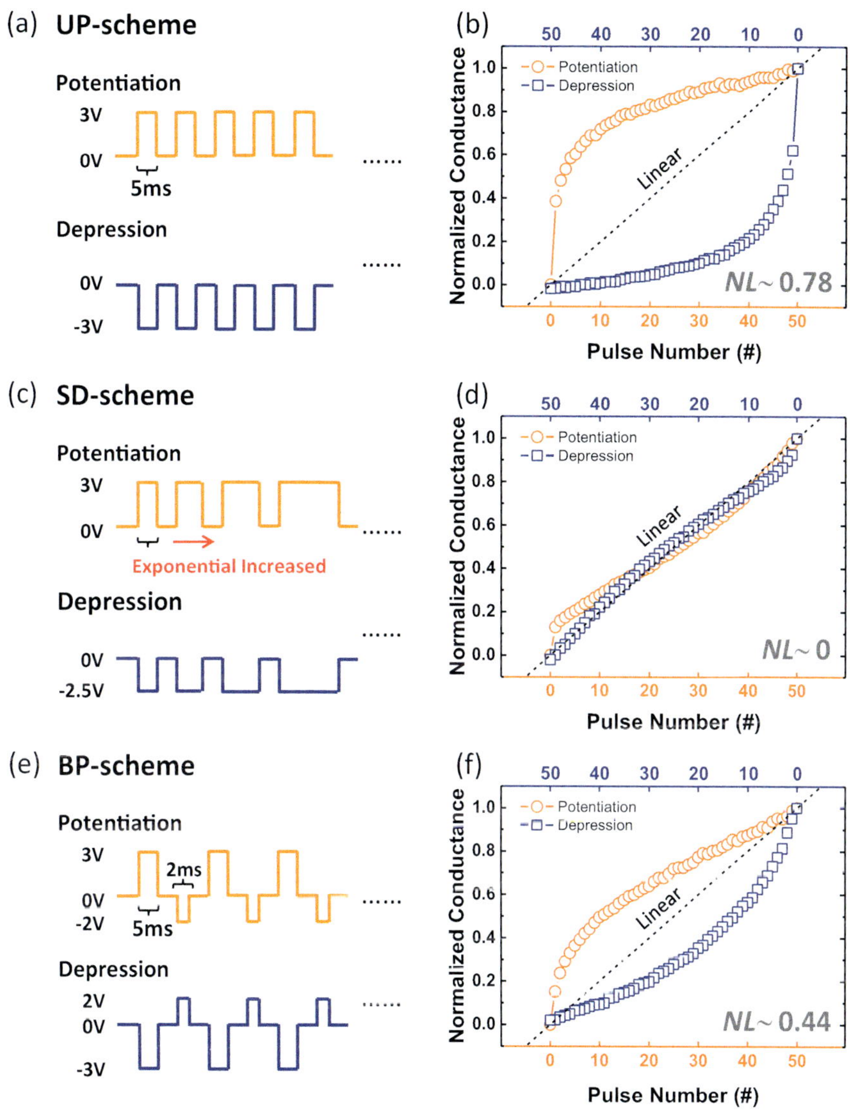

Fig. 4.16 (**a**) The training pulse configuration of the UP-scheme wherein identical square pulses are used for both potentiating (P) and depressing (D) operations, and the measurement results of the 3D Ta/TaO_x/TiO_2/Ti device under the UP-scheme are provided in (**b**), with a severe nonlinear weight update (NL ~ 0.78). (**c**) The training pulse configuration of the SD-scheme, wherein the non-identical square pulses demonstrated exponential increases of pulse width for both P and D, and the measurement results of the device are provided in (**d**), with an approximately linear weight update (NL ~ 0). (**e**) The training pulse configuration of the BP-scheme wherein identical bipolar paired-pulses are used, and the measurement results of the device are provided in (**f**), with the improved weight update (NL ~ 0.44)

where PW($n+1$) represents the training pulse width at the ($n+1$)th spike, N is the total training pulse number, and α is a tuning parameter that determines the different degrees of increasing widths. In the P process, $\alpha = 6$ and PW(0) = 0.1 ms, while in the D process, $\alpha = 4$ and PW(0) = 1 ms. The first few weaker pulses successfully suppress the drastic ion migration that mitigates the rapid increase in conductance, while the last few stronger pulses effectively enhance the remaining ions to migrate further, which overcomes the gradually saturated conductance. The SD-scheme works perfectly; the normalized P/D curve is almost linear (NL $\sim$ 0) in the measurement result, as illustrated in Fig. 4.16d. However, the non-identical training pulses differ according to the current weight state of the device, which increases the complexity of peripheral circuit design because the system requires memory units to record the current state and needs verification before every training pulse. Therefore, an identical training scheme (BP-scheme) is proposed [38] wherein each pulse consists of bipolar pulses for P or D, as illustrated in Fig. 4.16e. For the P process, each identical pulse consists of a 3 V/5 ms P pulse, followed by a −2 V/2 ms offset pulse. By contrast, a −3 V/5 ms D pulse is followed by a 2 V/2 ms offset pulse in the D operation. The measurement result under the BP-scheme is illustrated in Fig. 4.16f, and the device indicates an improved *NL* of 0.44, attributed to the suppression of rapid conductance change through its additional offset pulses. Although identical pulses are used in both UP- and BP-schemes, the proposed BP-scheme exhibits a substantially superior *NL*, which yields more accurate addition and subtraction results for hardware neural networks. Moreover, statistic results of a 3D 4 × 4 Ta/TaO_x/TiO_2/Ti synaptic cross-point array have higher uniformity of *NL* (0.42–0.54) under the BP-scheme than under the UP-scheme (0.6–0.81) [38], indicating the superior effectiveness of the proposed scheme.

To investigate the influence of the offset pulses in the BP-scheme, measured P processes with both the UP- and BP-schemes are illustrated in Fig. 4.17a, and the synaptic curves are expressed by [38]

$$G(n) = G_{\text{initial}} + G_1 \cdot \left[1 - \exp\left(\frac{-n}{\tau_1}\right)\right] + G_2 \cdot \left[1 - \exp\left(\frac{-n}{\tau_2}\right)\right] + G_{\text{offset}} \quad (4.14)$$

where G_{initial} is set to 70 nS indicates the device conductance state before training; G_1, τ_1, G_2, and τ_2 are empirical fitting parameters; G_{offset} is the offset value induced by the BP-scheme related to the strength of offset pulse. The measured P process under UP-scheme is fitted with $G_{\text{offset}} = 0$; by slightly increasing G_{offset} while keeping the other parameters the same, the fitting curve moves downward and successfully fit the measured P process under BP-scheme until $G_{\text{offset}} = 27$ nS. Note that the engineered offset pulses shift the synaptic curve nearly in parallel. Figure 4.17b displays the fitting results of P and D by varying the G_{offset}; symmetric P and D processes are assumed. As a result, an improved *NL* is obtained by increasing the offset value. Although the detailed mechanism is still under investigation, the fitting result provides further evidence for the BP-scheme's effectiveness, wherein the *NL* is successfully improved by the offset pulses.

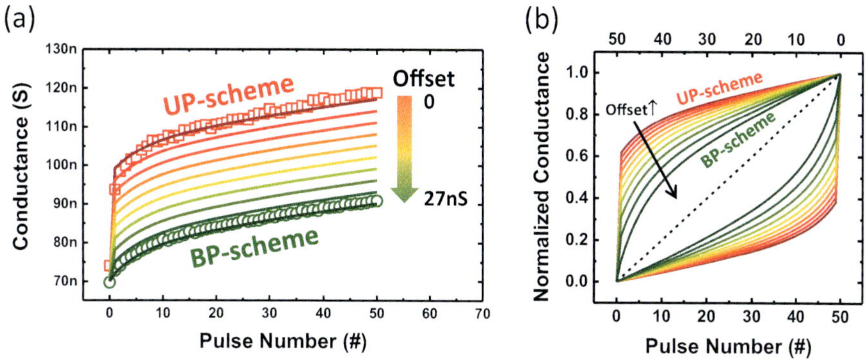

Fig. 4.17 (**a**) Measured potentiating characteristics of the 3D $Ta/TaO_x/TiO_2/Ti$ device under the UP-scheme (*open square*), the BP-scheme (*open circle*), and the fitting results (*color lines*) obtained using (14). By gradually increasing the offset value from 0 to 27 nS, the fitting curve moves downward and in parallel. (**b**) Normalized potentiation and depression characteristics using different offset values. An improved *NL* is obtained with an increasing offset. Here, symmetric potentiation and depression behavior is assumed for simplicity

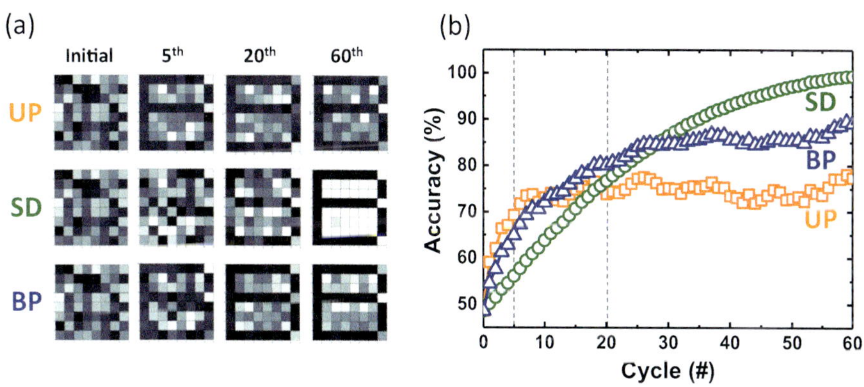

Fig. 4.18 (**a**) Simulated training weight maps of an 8 × 8 synaptic array trained by a binary alphabetic "B" pattern at the initial, 5th, 20th, and 60th training cycles under the UP-, SD-, and BP-schemes. The darker color represents higher conductance values, while the paler color represents lower conductance values. (**b**) Simulated training accuracy is calculated by Eq. (4.15), recorded in each cycle

To investigate the influence of nonlinear weight updates on training accuracy in hardware neural networks, an 8 × 8 binary alphabetic "B" [40] was adopted as the training pattern for simulating the training evolution. The device characteristics of the three different schemes (UP-, SD-, and BP-schemes) were used, as displayed in Fig. 4.16. The measurement result of the SD-scheme was assumed to be linear (NL = 0) for simplicity. Fig. 4.18a displays the weight maps from the pattern training process at the 5th, 20th, and 60th cycles. Starting from the random initial conductance values among the 8 × 8 synapses, the synapses were potentiated or depressed according to the following training pattern: 1 (black) or 0 (white). In each cycle, the 8 × 8 training pattern was intentionally set to have 10% error bits. The trained weight maps under the three schemes thus demonstrated different degrees of resemblance to the "B" pattern. For comparison, the training accuracy of each cycle is calculated by [38]:

$$\text{Accuracy} = \left[1 - \sqrt{\sum_{i=1}^{m} \frac{\left(G_{\text{target}}(i) - G_{\text{trained}}(i)\right)^2}{m}}\right] \times 100\% \tag{4.15}$$

where $G_{\text{target}}(i)$ and $G_{\text{trained}}(i)$ represent the target and trained conductance at the ith synapse in a hardware neural network system with the total m synapses, and the total conductance values are normalized to a plasticity range from zero to one. The calculated training accuracy is plotted in Fig. 4.18b, recorded from each training cycle. A 100% training accuracy was achieved with the SD-scheme, which is reasonable because the linear weight update is less influenced by training noise in this scheme. During the first few training cycles, the UP-scheme demonstrated relatively high initial training accuracy because of the nonlinear conductance change of the device. However, the UP-scheme was strongly affected by training noise and exhibited an unstable fluctuation of training accuracy, which was also caused by the severe *NL*. The learning on the frequent training data was more easily saturated and then overwritten by the less frequent noise. Because of the lower *NL* of the BP-scheme, the training accuracy indicated superior improvement (90%) than the UP-scheme's (75%).

4.6 Summary and Outlook

In this chapter, we reviewed the development of a $Ta/TaO_x/TiO_2/Ti$ device. Numerous desirable features suggest its suitability for 3D cross-point array applications. Through thorough experimental findings and physical simulations, device current conduction and switching mechanisms were investigated. Moreover, the device with analog-like synaptic plasticity provided extensive applications for a synaptic device, indicating adequate synaptic weight change in accordance with stimuli induced by neurons. Finally, a 3D synaptic network was realized, indicating the potential for mimicking the low-power, high-density, and high-connectivity of the human brain.

In the future, intricate and powerful hardware neural networks, which provide efficient and precise functionalities similar to the human brain, will continue to be developed and eventually transform the landscape of information processing technology in the future. However, unlike the software-based approach where the parameters of synaptic behavior are ideal and accurate through arithmetic calculation, the non-ideal effects existing in an experimental synaptic device remain a great challenge for hardware neural network implementations. Hence, a deeper understanding of the switching mechanism for the synaptic device must be achieved, and the non-ideal characteristics must be carefully engineered at the device level and evaluated in the hardware neural network system. Furthermore, a substantial effort must be made in tailored learning algorithms that may be inherently less susceptible to these non-ideal effects or even take advantage of them. We believe that the future development of hardware neural networks will be challenging yet fascinating.

References

1. C.-W. Hsu, I.-T. Wang, C.-L. Lo, M.-C. Chiang, W.-Y. Jang, C.-H. Lin, T.-H. Hou, in *Symposium on VLSI Technology*. Self-rectifying bipolar TaO_x/TiO_2 RRAM with superior endurance over 10^{12} cycles for 3D high-density storage-class memory (IEEE, Piscataway, 2013)
2. C.-W. Hsu, C.-C. Wan, I.-T. Wang, M.-C. Chen, C.-L. Lo, Y.-J. Lee, W.-Y. Jang, C.-H. Lin, T.-H. Hou, in *IEEE International Electron Devices Meeting (IEDM)*. 3D vertical TaO_x/TiO_2 RRAM with 10^3 self-rectifying ratio and sub-μA operating current (IEEE, Piscataway, 2013)
3. C.-W. Hsu, Y.-F. Wang, C.-C. Wan, I.-T. Wang, C.-T. Chou, W.-L. Lai, Y.-J. Lee, T.-H. Hou, Homogeneous barrier modulation of TaO_x/TiO_2 bilayers for ultra-high endurance three-dimensional storage-class memory. Nanotechnology **25**, 165202 (2014)
4. A. Flocke, T.G. Noll, in *IEEE European Solid-State Circuits Conference (ESSCIRC)*. Fundamental analysis of resistive nano-crossbars for the use in hybrid nano/CMOS-memory (IEEE, Piscataway, 2007)
5. M.-J. Lee, Y. Park, B.-S. Kang, S.-E. Ahn, C. Lee, K. Kim, W. Xianyu, G. Stefanovich, J.-H. Lee, S.-J. Chung, Y.-H. Kim, C.-S. Lee, J.-B. Park, I.-G. Baek, I.-K. Yoo, in *IEEE International Electron Devices Meeting (IEDM)*. 2-stack 1D-1R cross-point structure with oxide diodes as switch elements for high density resistance RAM applications (IEEE, Piscataway, 2007)
6. J.-J. Huang, T.-H. Hou, C.-W. Hsu, Y.-M. Tseng, W.-H. Chang, W. Y. Jang, C.-H. Lin, Flexible one diode-one resistor crossbar resistive-switching memory. Jpn. J. Appl. Phys. **51**, 04DD09 (2012)
7. J. Shin, I. Kim, K.P. Biju, M. Jo, J. Park, J. Lee, S. Jung, W. Lee, S. Kim, S. Park, H. Hwang, TiO_2-based metal-insulator-metal selection device for bipolar resistive random access memory cross-point application. J. Appl. Phys. **109**, 033712 (2011)
8. J.-J. Huang, Y.-M. Tseng, W.-C. Luo, C.-W. Hsu, T.-H. Hou, in *IEEE International Electron Devices Meeting (IEDM)*. One selector-one resistor (1S1R) crossbar array for high-density flexible memory applications (IEEE, Piscataway, 2011)
9. R.F. Freitas, W.W. Wilcke, Storage-class memory: The next storage system technology. IBM J. Res. Dev. **52**, 439–447 (2008)
10. G.W. Burr, B.N. Kurdi, J.C. Scott, C.H. Lam, K. Gopalakrishnan, R.S. Shenoy, Overview of candidate device technologies for storage-class memory. IBM J. Res. Dev. **52**, 449–464 (2008)
11. H.-S.P. Wong, H.-Y. Lee, S. Yu, Y.-S. Chen, Y. Wu, P.-S. Chen, B. Lee, F.T. Chen, M.-J. Tsai, Metal-oxide RRAM. Proc. IEEE **100**(6), 1951–1970 (2012)
12. J.-C. Liu, I.-T. Wang, C.-W. Hsu, W.-C. Luo, T.-H. Hou, in *Symposium on VLSI Technology, Systems and Applications (VLSI-TSA)*. Investigating MLC variation of filamentary and non-filamentary RRAM (IEEE, Piscataway, 2014)

13. Y.-F. Wang, Y.-C. Lin, I.-T. Wang, T.-P. Lin, T.-H. Hou, Characterization and modeling of nonfilamentary Ta/TaO_x/TiO_2/Ti analog synaptic device. Sci. Rep. **5**, 10150 (2015)
14. N.F. Mott, R.W. Gurney, *Electronic Processes in Ionic Crystals* (Clarendon Press, Oxford, 1948)
15. S. Yu, H.-S.P. Wong, A phenomenological model for the reset mechanism of metal oxide RRAM. IEEE Electron Device Lett. **31**(21), 1455–1457 (2010)
16. S. Kim, S.-J. Kim, K.M. Kim, S.R. Lee, M. Chang, E. Cho, Y.-B. Kim, C.J. Kim, U.-I. Chung, I.-K. Yoo, Physical electro-thermal model for resistive switching in bi-layered resistance-change memory. Sci. Rep. **3**, 1680 (2013)
17. X.A. Tran, B. Gao, J.F. Kang, X. Wu, L. Wu, Z. Fang, Z.R. Wang, K.L. Pey, Y.C. Yeo, A.Y. Du, M. Liu, B.Y. Nguyen, M.F. Li, H.Y. Yu, in *IEEE International Electron Devices Meeting (IEDM)*. Self-rectifying and forming-free unipolar HfO_x based-high performance RRAM built by fab-available materials (IEEE, Piscataway, 2011)
18. S.H. Jo, T. Chang, I. Ebong, B.B. Bhadviya, P. Mazumder, W. Lu, Nanoscale memristor device as synapse in neuromorphic systems. Nano Lett. **10**, 1297–1301 (2010)
19. T. Ohno, T. Hasegawa, T. Tsuruoka, K. Terabe, J.K. Gimzewski, M. Aono, Short-term plasticity and long-term potentiation mimicked in single inorganic synapses. Nat. Mater. **10**, 591–595 (2011)
20. I.-T. Wang, Y.-C. Lin, Y.-F. Wang, C.-W. Hsu, T.-H. Hou, in *IEEE International Electron Devices Meeting (IEDM)*. 3D synaptic architecture with ultralow sub-10 fJ energy per spike for neuromorphic computation (IEEE, Piscataway, 2014)
21. D. Kuzum, S. Yu, H.-S.P. Wong, Synaptic electronics: Materials, devices, and applications. Nanotechnology **24**, 382001 (2013)
22. L.O. Chua, S.M. Kang, Memristive devices and systems. Proc. IEEE **64**(2), 209–223 (1976)
23. S. Menzel, M. Waters, A. Marchewka, U. Böttger, R. Dittmann, R. Waser, Origin of the ultra-nonlinear switching kinetics in oxide-based resistive switches. Adv. Funct. Mater. **21**, 4487–4492 (2011)
24. C. Zamarreño-Ramos, L.A. Camuñas-Mesa, J.A. Pérez-Carrasco, T. Masquelier, T. Serrano-Gotarredona, B. Linares-Barranco, On spike-timing-dependnet-plasticity, memristive devices, and building a self-learning visual cortex. Front. Neurosci. **5**, 26 (2011)
25. D.O. Hebb, *The Organization of Behavior: A Neuropsychological Study* (Wiley, New York, 1949)
26. R.S. Zucker, W.G. Regehr, Short-term synaptic plasticity. Annu. Rev. Physiol. **64**, 355–405 (2002)
27. C. Mead, Neuromorphic electronic systems. Proc. IEEE **78**(10), 1629–1636 (1990)
28. D. Ferrucci, E. Brown, J. Chu-Carroll, J. Fan, D. Gondek, A.A. Kalyanpur, A. Lally, J.W. Murdock, E. Nyberg, J. Prager, N. Schlaefer, C. Welty, Building Watson: An overview of the DeepQA project. AI Mag. **31**(3), 59–79 (2010)
29. Y. Taigman, M. Yang, M. Ranzato, L. Wolf, in *IEEE Computer Vision and Pattern Recognition (CVPR)*. DeepFace: Closing the gap to human-level performance in face verification (IEEE, Piscataway, 2014)
30. D. Silver, A. Huang, C.J. Maddison, A. Guez, L. Sifre, G. van den Driessche, J. Schrittwieser, I. Antonoglou, V. Panneershelvam, M. Lanctot, S. Dieleman, D. Grewe, J. Nham, N. Kalchbrenner, I. Sutskever, T. Lillicrap, M. Leach, K. Kavukcuoglu, T. Graepel, D. Hassabis, Mastering the game of go with deep neural networks and tree search. Nature **529**, 484–489 (2016)
31. D. Kang, W. Jeong, C. Kim, D.-H. Kim, Y.S. Cho, K.-T. Kang, J. Ryu, K.-M. Kang, S. Lee, W. Kim, H. Lee, J. Yu, N. Choi, D.-S. Jang, J.-D. Ihm, D. Kim, Y.-S. Min, M.-S. Kim, A.-S. Park, J.-I. Son, I.-M. Kim, P. Kwak, B.-K. Jung, D.-S. Lee, H. Kim, H.-J. Yang, D.-S. Byeon, K.-T. Park, K.-H. Kyung, J.-H. Choi, in *IEEE International Solid-State Circuits Conference (ISSCC)*. 256Gb 3b/cell V-NAND flash memory with 48 stacked WL layers (IEEE, Piscataway, 2016)

32. W.-C. Luo, J.-C. Liu, Y.-C. Lin, C.-L. Lo, J.-J. Huang, K.-L. Lin, T.-H. Hou, Statistical model and rapid prediction of RRAM SET speed-disturb dilemma. IEEE Trans. Electron Devices **60** (11), 3760–3766 (2013)
33. S. Park, H. Kim, M. Choo, J. Noh, A. Sheri, S. Jung, K. Seo, J. Park, S. Kim, W. Lee, J. Shin, D. Lee, G. Choi, J. Woo, E. Cha, J. Jang, C. Park, M. Jeon, B. Lee, B.H. Lee, H. Hwang, in *IEEE International Electron Devices Meeting (IEDM)*. RRAM-based synapse for neuromorphic system with pattern recognition function (IEEE, Piscataway, 2012)
34. S. Yu, B. Gao, Z. Fang, H. Yu, J. Kang, H.-S.P. Wong, in *IEEE International Electron Devices Meeting (IEDM)*. A neuromorphic visual system using RRAM synaptic devices with sub-pJ energy and tolerance to variability: Experimental characterization and large-scale modeling (IEEE, Piscataway, 2012)
35. P.-Y. Chen, B. Lin, I.-T. Wang, T.-H. Hou, J. Ye, S. Vrudhula, J.-S. Seo, Y. Cao, S. Yu, in *IEEE/ACM International Conference on Computer-Aided Design (ICCAD)*. Mitigating effects of non-ideal synaptic device characteristics for on-chip learning (IEEE Press books, Piscataway, 2015)
36. S. Yu, P.-Y. Chen, Y. Cao, L. Xia, Y. Wang, H. Wu, in *IEEE International Electron Devices Meeting (IEDM)*. Scaling-up resistive synaptic arrays for Neuro-inspired architecture: Challenges and Prospect (IEEE, Piscataway, 2015)
37. G.W. Burr, R.M. Shelby, C. di Nolfo, J.W. Jang, R.S. Shenoy, P. Narayanan, K. Virwani, E.U. Giacometti, B. Kurdi, H. Hwang, in *IEEE International Electron Devices Meeting (IEDM)*. Experimental demonstration and tolerancing of a large-scale neural network (165,000 synapses), using phase-change memory as the synaptic weight element (IEEE, Piscataway, 2014)
38. I.-T. Wang, C.-C. Chang, L.-W. Chiu, T. Chou, T.-H. Hou, 3D Ta/TaO_x/TiO_2/Ti synaptic array and linearity tuning of weight update for hardware neural network applications. Nanotechnology **27**, 365204 (2016)
39. S. Park, A. Sheri, J. Kim, J. Noh, J. Jang, M. Jeon, B. Lee, B.R. Lee, B.H. Lee, H. Hwang, in *IEEE International Electron Devices Meeting (IEDM)*. Neuromorphic speech system using advanced ReRAM-based synapse (IEEE, Piscataway, 2013)
40. T. Chou, J.-C. Liu, L.-W. Chiu, I.-T. Wang, C.-M. Tsai, T.-H. Hou, in *Symposium on VLSI Technology, Systems and Applications (VLSI-TSA)*. Neuromorphic pattern learning using HBM electronic synapse with excitatory and inhibitory plasticity (IEEE, Piscataway, 2015)

Part II
Array-Level Demonstrations of Resistive Synaptic Devices and Neural Networks

Chapter 5
Training and Inference in Hopfield Network Using 10 × 10 Phase Change Synaptic Array

Sukru Burc Eryilmaz and H.-S. Philip Wong

5.1 Introduction

Resistive nonvolatile memories such as phase change memory (PCM) [1], metal-oxide resistive memory (RRAM) [2–4], conductive bridge memory (CBRAM) [5], and ferroelectric memory (FeRAM) [6] have been shown to exhibit gradual resistance change properties in analog fashion, similar to synaptic weights in biological and artificial neural networks. Some of these demonstrations have been described in previous chapters. These works have been limited to single device measurements. In this chapter, an array-level demonstration is presented using an array of phase change synapses. This section describes unsupervised training of a Hopfield network for associative learning using Hebbian learning rule, a variant of symmetric spike timing-dependent plasticity [7]. The results presented in this chapter are taken from [8, 9].

5.2 Phase Change Memory Array for Synaptic Operation

10 × 10 phase change memory array consists of 100 mushroom-type phase change memory cells. The cells are connected in a crossbar fashion as shown in Fig. 5.1a. Each memory cell consists of a phase change element connected in series with a selection transistor (1T1R structure). The cells can be accessed by selecting 1 of 10 bitlines (BL) and 1 of 10 wordlines (WL) (see Fig. 5.1a). SET programming of a memory cell is achieved by applying a long (from a few hundred ns to few μs)

S.B. Eryilmaz (✉) • H.-S.P. Wong
Department of Electrical Engineering,
Stanford University, 420 Via Palou, Stanford, CA 94305, USA
e-mail: eryilmaz@stanford.edu; hspwong@stanford.edu

S. Yu (ed.), *Neuro-inspired Computing Using Resistive Synaptic Devices*,
DOI 10.1007/978-3-319-54313-0_5

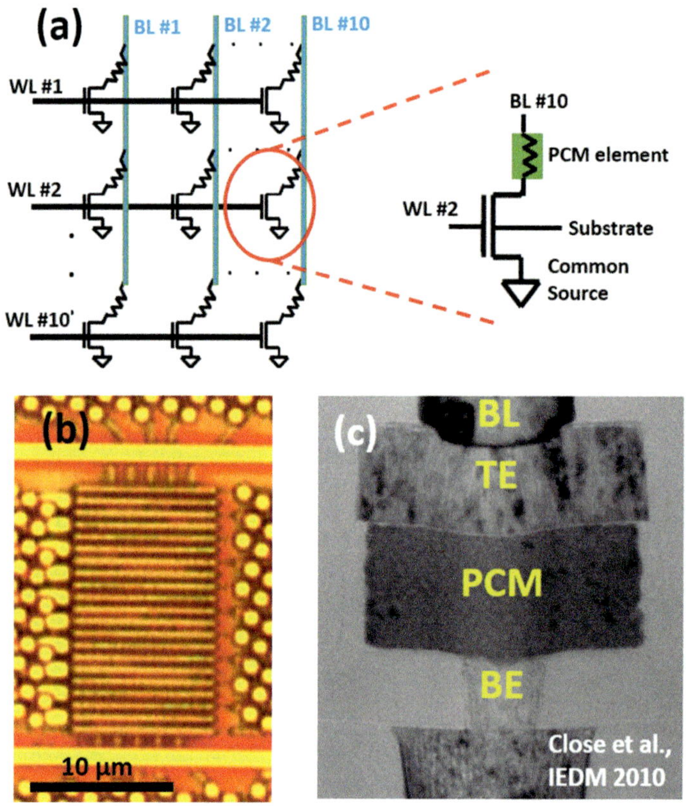

Fig. 5.1 (**a**) 10 × 10 phase change memory (*PCM*) array schematics on left-hand side. PCM elements are shown as resistances connected in series with the selector transistors. The figure on the right-hand side shows the schematic of single memory cell. This particular cell can be accessed by applying biases at WL #2 and BL#10. Substrate and common source terminals are grounded in the experiment. (**b**) Optical microscope image of 10 × 10 memory array is shown on the left. (**c**) TEM image of a single memory cell. Mushroom-type cell structure can be seen by observing the bitline (*BL*), top electrode (*TE*), phase change material, and bottom electrode (*BE*) stack [10]

current pulse through the PCM element to crystallize the phase change material via Joule heating. In a gradual SET programming, depending on the amplitude of the current pulse, resistance of the PCM reduces gradually, rather than abruptly going into the lowest resistance (fully SET) state (see Fig. 5.2d). RESET (high resistance) programming is achieved by amorphizing the phase change material of the memory cell by applying a larger current pulse with a very sharp (2–10 ns fall time) falling edge. A large amplitude of current pulse results in melting of PCM material through Joule heating; the sharp falling edge quenches the cell, without allowing time for the phase change material to go into the more stable crystalline state, leaving it in the amorphous state. Typical DC switching characteristics of a single device arbitrarily chosen from an array are shown in Fig. 5.2a. For DC characterization,

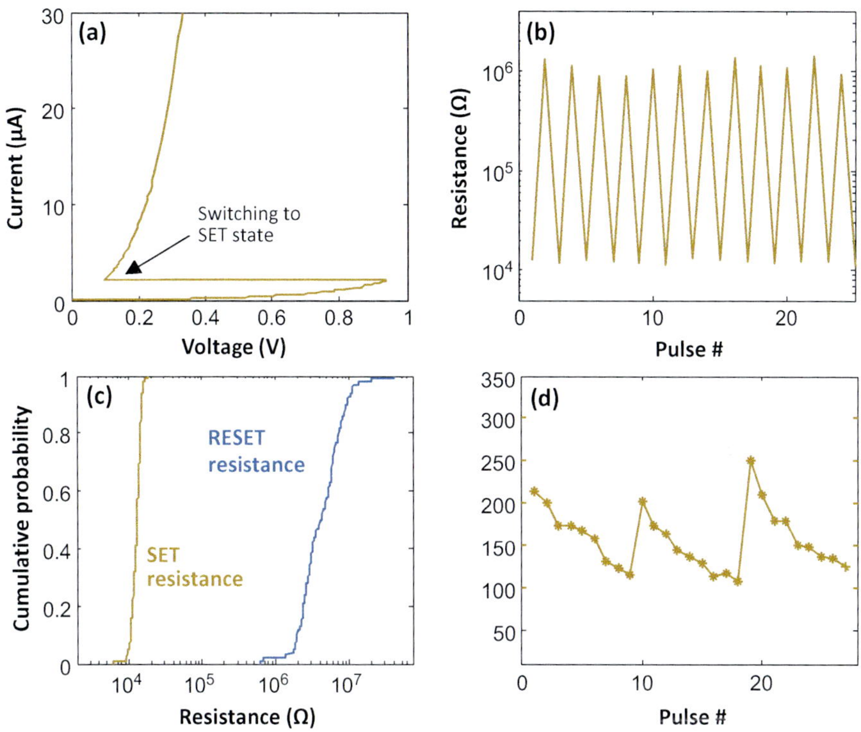

Fig. 5.2 Electrical characterization of memory devices. (**a**) DC switching of a single memory cell arbitrarily selected in the array. Binary switching cycles are shown in (**b**). SET pulse is applied at pulse #1, 3, ... and RESET pulse is applied at pulse #2, 4, ... The plot shows the measured resistance of memory cell right after the programming pulse is applied. Array-level binary resistance distribution is shown in (**c**). Resistance window for binary switching is ~100. Gradual resistance change in a single cell is shown in (**d**). This plot is obtained by applying several gradual SET pulses right after the cell is abruptly programmed to RESET state. The plot shows three gradual programming cycles

3.3 V is applied at the WL of an arbitrarily selected cell, and the BL node is swept from 0 V up to a value slightly higher than switching threshold. Figure 5.2a shows that switching threshold for a cell in the array in a fully RESET state is around 0.8 V, and the current at which switching occurs is 2 μA. Note that these values vary across the memory array due to device-to-device variation. For pulse switching, set and reset pulses with amplitudes 1 and 1.5 V and with (50 ns/300 ns/1 μs) and (20 ns/50 ns/5 ns) rise/width/falltime is applied at WL node, while BL node is held at 3.3 V. Figure 5.2b shows the pulse switching characteristics. This data is obtained by applying SET and RESET pulses alternatingly for binary switching (SET pulses for pulse #1, 3, 5... and RESET pulses for pulse #2, 4, 6....). These SET and RESET pulses are used for characterizing binary resistance distribution across the array, shown in Fig. 5.2c. RESET resistance is distributed around 3 MΩ, and SET resistance is distributed around 10kΩ. Instead of binary characteristics, gradual resistance change characteristics of memory cells are utilized for synaptic operation. Specifically, our system uses gradual SET programming of memory cells. To characterize gradual resistance change from the RESET state to the partially SET state, a 1.1 V RESET pulse and then 9 SET pulses with 0.85 V amplitude are applied. Gradual resistance change characteristics from RESET to SET state for a single cell are shown in Fig. 5.2d for a few cycles of gradual SET characterization. Energy consumption for gradual SET is lower than gradual RESET, but variability is larger for gradual SET since gradual SET is probabilistic in nature [11]. The reason behind this is the intrinsic stochasticity of the nucleation of crystalline clusters during gradual SET operation. The cycle-to-cycle variability is also observed in Fig. 5.2d; the same resistance levels are not accurately repeatable from cycle to cycle. Due to variability in gradual resistance change, multilevel cell (MLC) memory applications use a write-and-verify technique since the data storage applications require deterministic binary resistance levels [12]. However, as we discuss later in the chapter, massively parallel brain-like architectures can tolerate such variations and do not require the use of write-and-verify that is needed to achieve an accurate resistance level.

5.3 Hebbian Learning in Synaptic Array

A fully connected recurrent Hopfield network [7] is implemented for the learning experiments (see Fig. 5.3a). The Hopfield network consists of 10 recurrently connected neurons and 100 synaptic devices, as shown in Fig. 5.3a. Note that in Hopfield network architecture, all neurons serve as both input and output neurons. Integrate-and-fire neurons are implemented by software (that controls the measurement equipment), and memory cells serve as synaptic devices between neurons. Figure 5.3a illustrates how the network is mapped onto the memory cell array. The input terminal of the i-th neuron is connected to BL #i, and output terminal of the j-th neuron is connected to WL #j, where $i, j = 1, 2, \ldots, 10$, i.e., neuron #1 input and output is connected to BL #1 and WL #1, respectively, neuron #2 input and output is connected to BL #2 and WL #2, respectively, etc. To initialize resistances before the experiment, all synapses are programmed to the RESET state. A learning experiment consists of epochs during which synaptic weights are updated depending on neuron firing events. After training, the pattern is presented with an incorrect pixel this time, and the incorrect pixel is expected to be recalled in the recall phase as a result of weight adaptations after training is performed (Fig. 5.3b). A complete pattern is presented during the training phase of an epoch, and an incomplete pattern with an incorrectly OFF pixel is presented during the recall phase. All patterns consist of 10 pixels, and each neuron in Hopfield network architecture is associated with a pixel. The mapping between neurons and pixels for two different patterns considered in this work is shown in Fig. 5.3c. Figure 5.3b shows the pulsing scheme for firing and nonfiring neurons in both update and recall phases at a given iteration. When a pattern is presented during training phase,

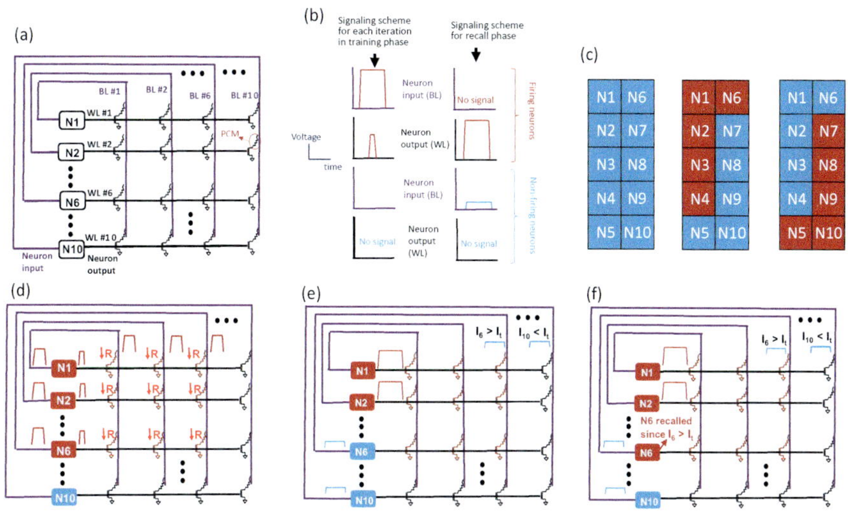

Fig. 5.3 Mapping Hopfield network on memory array. (**a**) shows the schematic for recurrently connected Hopfield network implemented in the learning experiment. Pulsing schemes during training and recall phases are shown in (**b**). Network is trained with two patterns shown in (**c**), where *red* pixels correspond to ON and *blue* pixels correspond to OFF pixels. Numbers in pixels correspond to the neuron number associated with that pixel. During update phase shown in (**d**), resistance of synaptic elements connected to nonfiring neurons do not change, since no pulse is applied at the WL node of nonfiring neurons in update phase. However, the synaptic connections between firing neurons are programmed by the pulses applied at the BL and WL of the corresponding memory cell. The pulse characteristics are predetermined for gradual SET programming of the memory cell; hence the resistance is reduced with an amount, and the connection gets stronger. (**e**) During the read phase, a small amplitude voltage applied at the BL node of nonfiring neurons sense the total current due to the synapses of that neuron connected to firing neurons, since a pulse applied at the output of the firing neurons turns the selection transistor on simultaneously. (**f**) In this example, during the recall phase, N1, N2, N3, and N4 are presented with N6 OFF (not firing), but N6 is recalled since the input current of N6 is larger than the threshold

neurons associated with ON (red pixels in Fig. 5.3c) pixels are stimulated, resulting in firing of these neurons. As shown in Fig. 5.3d, when a neuron spikes during training phase, it applies programming pulses at its input (corresponding BL) and output (corresponding WL). This results in gradual SET programming of the synaptic element (PCM element) between the two firing neurons. For example, when neuron 1 and neuron 2 fire, programming pulses are applied at WL1, WL2, BL1, and BL2, as described in the pulsing scheme in Fig. 5.3b. These pulses result in current going through PCM elements and hence gradual SET programming of memory cells that connect neuron 1 and neuron 2 (see Fig. 5.3d). After training, the recall phase begins. During the recall phase, a pattern with an incorrect pixel (an ON pixel replaced by an OFF pixel) is presented (Fig. 5.3e). Again, the neurons associated with ON pixels during recall phase fire, and appropriate pulses are applied at the input and output of neurons as shown in Fig. 5.3b. Neurons associated with OFF pixels during recall phase do not fire. Note that there is a low-amplitude pulse applied at the input of nonfiring neurons during recall phase. This voltage pulse, together with the large amplitude voltage pulse applied at the firing neurons' output during recall phase, creates an input current feeding into nonfiring neurons. The amplitude of this current through a nonfiring neuron is determined by the resistance values of synaptic connections between that neuron and the firing neurons. This input current of nonfiring neurons during recall phase is analogous to membrane potential of biological neurons. In biological neurons, the postsynaptic current into a neuron accumulates charge on capacitive membrane, forming a membrane potential. Typically, this is modeled by a time constant that is determined by membrane capacitance. In this experiment, neurons fire simultaneously during the recall phase, while at the same time, the input current through the nonfiring neurons is measured. Since the delays and timing properties of the neurons are not included in the neuron model, the membrane capacitance is not included in neurons. Hence, input current through a neuron is actually equivalent to membrane potential in our experiments. In this paper, we will use the terms input current and membrane voltage interchangeably, for the reasons explained above. The input current into a nonfiring neuron during recall phase can be expressed as:

$$I_i = V_{\mathrm{read}} \sum_{j \in F} \frac{1}{R_{ij}} \tag{5.1}$$

In Eq. 5.1, I_i is the input current into the i-th neuron (a nonfiring neuron), F is the set of indices of firing neurons, R_{ij} is the resistance of synaptic element between BL i and WL j, and V_{read} is the read voltage at the input of nonfiring neurons during recall phase (see Fig. 5.3b), which is 0.1 V in our experiments. As shown in Fig. 5.3b, if a neuron is not associated with an OFF pixel at the beginning of recall phase, it fires, and the reading voltage V_{read} at its input is 0, making its input 0.

If the input current through a nonfiring neuron exceeds a threshold during the recall phase, then the neuron associated with the pixel fires (Fig. 5.3f). The membrane potential of neurons is set to 0 at the beginning of each epoch, so that it does not transfer to the next epoch. We define "missing pixel" as the pixel that is ON in the pattern used for training but OFF in the input pattern during recall phase. Note that the pixel missing from the pattern in recall phase still fires in update phase during training, resulting in SET programming of the corresponding memory cells between this neuron and other firing neurons. This results in a decrease in the resistance values between the neuron corresponding to missing pixel and other firing neurons (ON pixels) as shown in Fig. 5.3d, increasing the input current of the neuron corresponding to missing pixel during the recall phase (Fig. 5.3f). Hence, recall is expected to occur after few epochs, at which point the membrane potential exceeds some threshold. This learning scheme is a form of Hebbian learning rule, since the synaptic connections between coactive neurons during training phase get stronger, due to reduced resistances of these synaptic connections. The time window that defines the firing of two neurons as being coactive is defined by the width of the pulse applied at the input of firing neurons during update phase

(Fig. 5.3b). This time window is 100 μs in this experiment. As an illustration of Hebbian learning process, two simple 10-pixel patterns are chosen to be learnt. These two patterns are shown in Fig. 5.3c. The network is first trained with pattern 1 (on the left-hand side in Fig. 5.3c) and then pattern 2 (on the right-hand side in Fig. 5.3c).

During training with pattern 1, until the pattern is recalled, the complete pattern is presented; and the pattern with pixel 6 missing is presented during recall phase. After pattern 1 is recalled, the same procedure is performed for pattern 2, this time with pixel 5 missing in the recall phases of epochs. This experiment is performed for four cases, each corresponding to cases with different initial device-to-device resistance variations across the 10 × 10 array. Initial device-to-device variation here refers to the variation of device resistances across the array after all cells are programmed to RESET before learning experiment begins. Different initial variation values are obtained by individually programming the memory cells in different arrays. The evolution of synaptic weights is shown in Fig. 5.4 for the case where initial variation is 60%. Note that the synaptic weight map in Fig. 5.4 shows the normalized synaptic weight of each synaptic device. Each data point in the heatmap shows the resistance of the synaptic device after the corresponding epoch that is divided by the initial RESET resistance (right before the experiment when all devices are RESET programmed as explained above) of that device. Hence, the

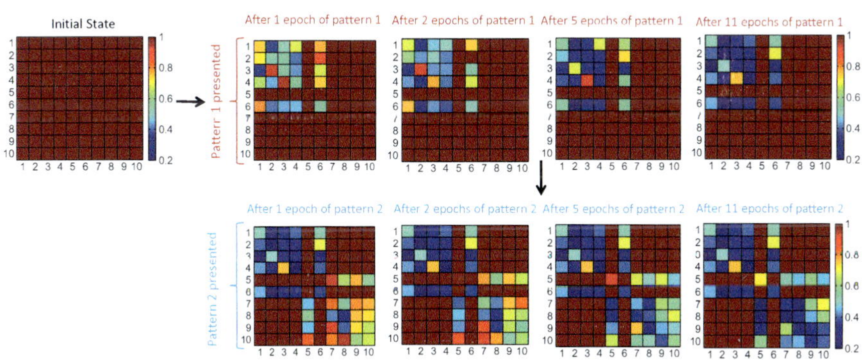

Fig. 5.4 Evolution of normalized resistance of synaptic devices is shown for the 60% initial variation case. All normalized resistances are initially 1 since the normalized resistance map shows the current resistance of synaptic devices divided by their initial resistances. Row and column numbers correspond to BL and WL of the synaptic devices. First, pattern 1 is presented to the network. For pattern 1, ON neurons for the complete pattern during update phase are N1, N2, N3, N4, and N6 and, for the recall phase, N6 is OFF and expected to be recalled (i.e., expected to fire) after training with a certain number of epochs. The gradual decrease in the normalized resistance of synaptic connections between firing neurons during update phase is observed. When recall phase is performed after 11 epochs, OFF pixel #6 (neuron #6) is recalled (meaning neuron #6 fires in recall phase), and then pattern 2 is presented for training. For pattern 2, the complete pattern is represented by N5, N7, N8, N9, and N10 being ON, and N5 is missing (OFF) in the recall phase

map does not include the variations of initial RESET state resistances across the array. The next section describes the effect of variation on learning performance. As shown in Fig. 5.4, after presenting each pattern to the network, synapses between the ON neurons gradually get stronger (resistance of the corresponding cell decreases); after 11 epochs, patterns are recalled. The overall energy consumed in synaptic devices during this experiment is 52.8 nJ. This energy does not include the energy consumed in the neurons and the wires and is the energy consumed by the synaptic devices during training and recalling of pattern 1. Our measurements indicate that roughly 10% of this energy is consumed in phase change element, while around 90% is consumed in selector transistors. Note that the number of epochs and the overall energy consumed strongly depend on the choice of the threshold membrane potential of neurons. If threshold membrane potential is kept low, the number of epochs would be reduced, but a wrong pixel might fire (hence turn on) in the output of recall phase due to variations, hence recalling a wrong pattern. This is explained in the next section.

5.4 Effects of Device-to-Device Variation on Associative Learning Performance

Figure 5.5a shows the absolute resistance map of synaptic connections after 11 epochs for the experiment above, along with the resistance distribution (on the left-hand side in Fig. 5.5) when all the cells are in the RESET state before the experiment. As the synaptic connections evolve during training for two patterns, synapses between coactive neurons get stronger. Resistance maps in Fig. 5.5 also illustrate the resistance variation across the array when all cells are in RESET state before training. In our experiment, neuron firing threshold is the important parameter that can be tuned to tolerate the variation. This threshold value has to be large enough so that a wrong pixel will not turn on in recall phase but low enough to guarantee that the overall energy consumed is minimal and the missing pixel will actually turn on in recall phase, hence recalling the original pattern. To this end, the firing threshold of neurons is selected as:

$$I_{thr} = C.\max_{N,i}\left(V_{\text{read}} \sum_{j \in N} \frac{1}{R_{ij}}\right) \tag{5.2}$$

In Eq. 5.2, N is constrained to be a four-element subset of the set $\{1, 2, 3, ...9, 10\}$, and R_{ij} is the initial RESET resistance of the memory cell defined by bitline i and wordline j, and V_{read} is defined as in Eq. 5.1. This equation means that the threshold current is a constant C times the largest input current that a neuron can possibly have in the recall phase, given the resistance values for each cell. The reason for considering four-element subsets is because four pixels are assumed to be ON in the input

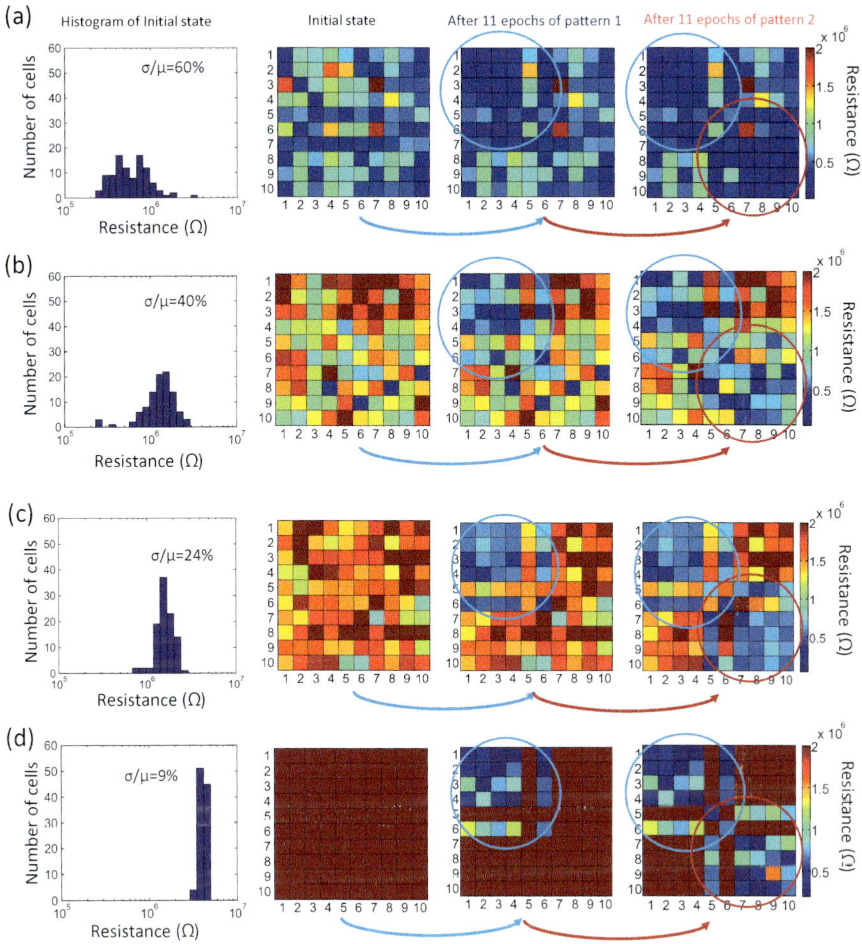

Fig. 5.5 Evolution of absolute resistances of phase change synapses is shown for four different experiments, where initial resistance variation across memory array is (**a**) 60 %, (**b**) 40 %, (**c**) 24 %, and (**d**) 9 % The representation of synaptic devices in these resistance maps is the same as in Fig. 5.4, but here resistance values are not normalized. The variation across the arrays is apparent here for (**a–d**). With decreased initial variation, the difference in resistance values between potentiated synapses and the synapses that remain unchanged becomes more pronounced

during recall phase, and the threshold should be large enough to avoid firing of a neuron during recall phase when it is actually not ON in the true pattern. In its current form, this scheme might not work successfully when different numbers of pixels are missing, for example, when three pixels are ON in recall phase, while five pixels are ON in the actual pattern. This limitation can be mitigated by allowing negative weights, equivalently using 2-PCM synapse suggested in [13] or adaptive threshold method suggested in [7]. The requirement that $C > 1$ guarantees that during the training, the wrong pixel will not be recalled at any epoch. This is because the resistance of the synaptic connections between an arbitrary OFF pixel in the original pattern and other neurons do not decrease, as the OFF pixels do not fire during training. We choose $C = 2$ for our experiments. Choosing $C = 2$ also allows, without requiring negative synaptic weights, to generalize recall to some extent for inputs with incorrectly ON pixels, in addition to incorrectly OFF pixels as given in our example. This idea is similar to adaptive threshold method in [7], where neuron threshold is increased while keeping the weights positive, instead of using negative weights. With larger variation, the low-resistance tail of the initial RESET resistance distribution (leftmost histograms in Fig. 5.5a–d) extends toward lower resistance values. This results in a decrease in minimum resistance values, as shown in histograms in Fig. 5.5a–d. Hence, maximum neuron input current increases for the case where four neurons fire. This increases the max term in Eq. 5.2, resulting in higher number of epochs needed to recall the missing pixel for larger variation. The resistance maps for other variation cases are shown in Fig. 5.5b–d. As initial variation reduces, the same number of epochs yields a more pronounced overall difference between the weights that get stronger and the weights that do not change, as illustrated in Fig. 5.5.

The evolution of the membrane potential with each epoch for different variation cases is shown in Fig. 5.6. While it takes 11 epochs to recall a pattern when there is 60% initial variation, one epoch is sufficient when the initial resistance variation is 9%. It is worth mentioning that the variation in read voltage in the experiment is negligible since the reading of memory cell resistances is performed with electronic equipment. When synaptic grid is integrated with actual CMOS neurons, however, it is expected to have some variation in read voltage, which results in variation in the input current of neurons. This variation in input current might cause variations in the number of epochs needed for training. Figure 5.6 shows that while it takes 3% input current variation (hence read voltage variation) to change the number of epochs needed for 60% variation case (Fig. 5.6a), it takes 40% variation in read voltage to change the number of training epochs required for 9% initial variation case (Fig. 5.6d). This is because as the number of epochs increases, resistances of programmed synapses begin to converge to low-resistance values. To minimize the effect of read voltage variation, synaptic device properties and the proper pulsing scheme during training should be carefully chosen, considering the read voltage variation of CMOS neuron circuit. The increase in the required number of epochs to recall the pattern results in a larger energy consumption. Overall energy consumption for 9% initial resistance variation case is 4.8 nJ, whereas it is 52.8 nJ for 60% initial variation case. Figure 5.7 illustrates energy consumption and number of epochs needed for different initial resistance variation cases. Figure 5.7 shows that overall energy reduces significantly with lower initial resistance variation. These energy values represent only the energy consumed in the synaptic devices for training and recall phases for pattern 1. They do not include the energy consumed in the wires or the neurons. Energy consumption in the wires can be a substantial part of the overall energy consumption for a large array [14]. It is also

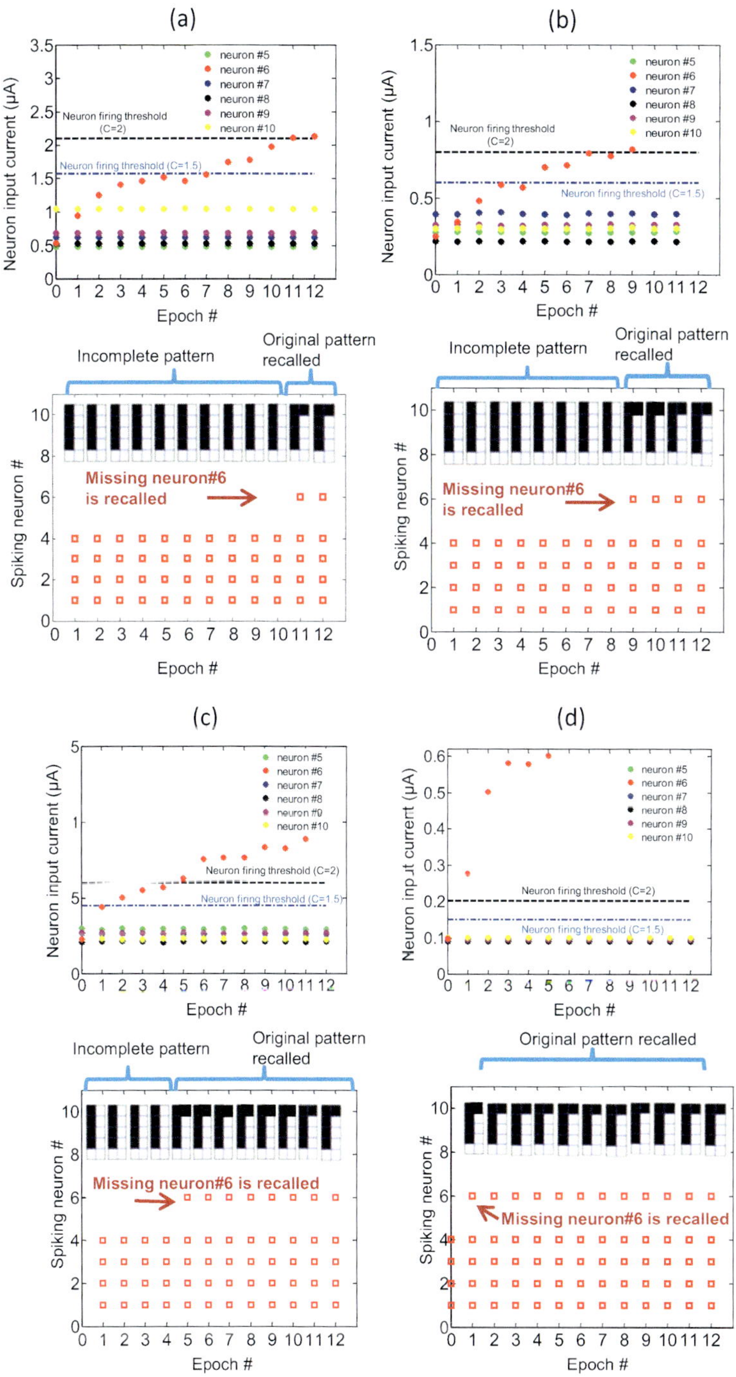

Fig. 5.6 Recall of the missing pixel for training with pattern 1 for different initial device-to-device variation cases, (**a**) 60%, (**b**) 40%, (**c**) 24%, and (**d**) 9%, are illustrated. For each case, figures on the top show what the input current of neurons that do not fire would be if the recall is performed after the

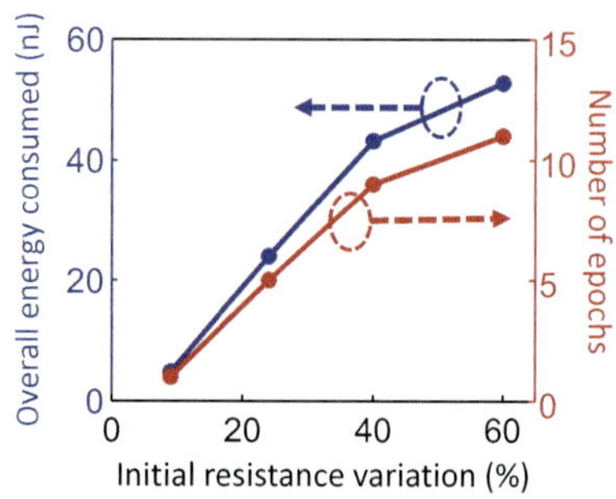

Fig. 5.7 Total number of epochs required for training and overall energy consumed in the synaptic devices during training and recall phases for pattern 1, for each cases of different initial resistance variation. This plot shows the tradeoff between the ability of the system to tolerate variation, and energy consumption as well as speed. As the network needs to tolerate larger variation, larger firing threshold is required for neurons. This increases the number of epochs and energy consumption required for training

worth noting that since the timescale between the epochs in these experiments is on the order of seconds, we did not observe any effects of drift in our measurements, which would require a timescale of μs or ms to observe [15].

5.5 Summary

In this chapter, we report brain-like learning in hardware using a 10×10 array of phase change synaptic devices. We demonstrated in hardware experiments that synaptic network can implement robust pattern recognition through brain-like learning. Test patterns were shown to be stored and recalled associatively via Hebbian plasticity in a manner similar to the biological brain. Increasing the number of training epochs provides a better tolerance for initial resistance variations, at the cost of increased energy consumption. Demonstration of robust brain-inspired learning in a small-scale synaptic array is a significant milestone toward building large-scale computation systems with brain-level computational efficiency.

Fig. 5.6 (continued) corresponding number of epochs, and figures on the bottom show neurons that fire if the recall was performed after the corresponding number of epochs for $C = 2$ (see the text for description of parameter C). Cases for $C = 1.5$ and $C = 2$ that correspond to different threshold levels are shown in the figures on the top. When the input current exceeds the threshold after some epoch, missing pixel N6 fires. For $C = 2$, the number of epochs required to fire N6 in each case is 11 (60% variation), 9 (40% variation), 5 (24% variation), and 1 (9% variation)

References

1. D. Kuzum, R.G.D. Jeyasingh, B. Lee, H.-S.P. Wong, Nanoelectronic programmable synapses based on phase change materials for brain-inspired computing. Nano Lett. **12**, 2179–2186 (2012)
2. S. Yu, Y. Wu, R. Jeyasingh, D. Kuzum, H.-S.P. Wong, An electronic synapse device based on metal oxide resistive switching memory for neuromorphic computation. IEEE Trans. Electron Devices **58**, 2729–2737 (2011)
3. S.H. Jo, T. Chang, I. Ebong, B.B. Bhadviya, P. Mazumder, W. Lu, Nanoscale memristor device as synapse in neuromorphic systems. Nano Lett. **10**, 1297–1301 (2010)
4. I.-T. Wang, Y.-C. Lin, Y.-F. Wang, C.-W. Hsu, T.-H. Hou, 3D synaptic architecture with ultralow sub-10 f. energy per spike for neuromorphic computation, in *IEEE International Electron Devices Meeting (IEDM)* (2014)
5. T. Ohno, T. Hasegawa, T. Tsuruoka, K. Terabe, J.K. Gimzewski, M. Aono, Short-term plasticity and long-term potentiation mimicked in single inorganic synapses. Nat. Mater. **10**, 591–595 (2011)
6. A. Chanthbouala, V. Garcia, R.O. Cherifi, K. Bouzehouane, S. Fusil, X. Moya, S. Xavier, H. Yamada, C. Deranlot, N.D. Mathur, M. Bibes, A. Barthélémy, J. Grollier, A ferroelectric memristor. Nat. Mater. **11**, 860–864 (2012)
7. J. Hertz, A. Krough, R.G. Palmer, *Introduction to the Theory of Neural Computation* (Addison-Wesley Publishing, Redwood City, 1991)
8. S.B. Eryilmaz, D. Kuzum, R.G.D. Jeyasingh, S. Kim, M. BrightSky, C. Lam, H.-S.P. Wong, Experimental demonstration of array-level learning with phase change synaptic devices, in *IEEE International Electron Devices Meeting (IEDM)* (2013)
9. S.B. Eryilmaz, D. Kuzum, R. Jeyasingh, S. Kim, M. BrightSky, C. Lam, H.-S.P. Wong, Brain-like associative learning using a nanoscale non-volatile phase change synaptic device array. Front. Neurosci. **8**, 205 (2014)
10. G.F. Close, U. Frey, M. Breitwisch, H.L. Lung, C. Lam, C. Hagleitner, E. Eleftheriou, Device, circuit and system-level analysis of noise in multi-bit phase-change memory, in *IEEE International Electron Devices Meeting (IEDM)* (2010)
11. S. Braga, A. Cabrini, G. Torelli, Experimental analysis of partial-SET state stability in phase-change memories. IEEE Trans. Electron Devices **58**, 517–522 (2010)
12. D.-H. Kang, J.-H. Lee, J.H. Kong, D. Ha, J. Yu, C.Y. Um, J.H. Park, F. Yeung, J.H. Kim, W.I. Park, Y.J. Jeon, M.K. Lee, Y.J. Song, J.H. Oh, G.T. Jeong, H.S. Jeong, Two-bit cell operation in diode-switch phase change memory cells with 90 nm technology, in *Symposium on VLSI Technology* (2008)
13. O. Bichler, M. Suri, D. Querlioz, D. Vuillaume, B. DeSalvo, C. Gamrat, Visual pattern extraction using energy-efficient "2-PCM synapse" neuromorphic architecture. IEEE Trans. Electron Devices **59**, 2206–2214 (2012)
14. S.B. Eryilmaz, D. Kuzum, S. Yu, H.-S.P. Wong, Device and system level design considerations for analog-non-volatile-memory based neuromorphic architectures, in *IEEE International Electron Devices Meeting (IEDM)* (2015)
15. I.V. Karpov, M. Mitra, D. Kau, G. Spadini, Y.A. Kryukov, V.G. Karpov, Fundamental drift of parameters in chalcogenide phase change memory. J. Appl. Phys. **102**, 124503 (2007)

Chapter 6
Experimental Demonstration of Firing Rate Neural Networks Based on Metal-Oxide Memristive Crossbars

Farnood Merrikh Bayat, Mirko Prezioso, and Bhaswar Chakrabarti

6.1 Introduction

Two main challenges standing on the path toward having a power-efficient and ultrafast neuromorphic system are the compact implementation of high-performance vector-by-matrix multiplication, which in turn requires a compact implementation of synaptic weights, and the effective integration of these multiplication accelerators with neuron circuitries usually constructed using CMOS technology. Although CMOS-based synapses can facilitate the integration process, implementing synapses through CMOS technology is insufficient for meeting these challenges mostly because they are excessively bulky [1–3]. One possible solution to tackle this density problem is to use resistive switching ("memristive") memory arrays as synapses monolithically integrated on top of existing CMOS-based neuron circuitries, an architecture known as CMOL (**CMOS** + **M**olecular scale devices) [4] (see Fig. 6.1).

The key feature of CMOL architecture is the area-distributed interface, implemented by using tilted and segmented crossbars and a double decoding scheme (see Ref. [4, 5]). Such specific interface allows to uniquely address every cross-point device without the need for overlay alignment between the nanoscale memristive devices and the underlying much larger CMOS-based circuits. Recent experiments and theoretical estimations based on realistic simulations have shown that such hybrid circuits, if scaled down to a realistic 15-nm pitch, may provide a 1000× increase in speed and energy efficiency in comparison with similar networks implemented using conventional circuits.

F.M. Bayat (✉) • M. Prezioso • B. Chakrabarti
Department of Electrical and Computer Engineering, University of California Santa Barbara, Santa Barbara, CA 93106, USA
e-mail: farnoodmb@ece.ucsb.edu

S. Yu (ed.), *Neuro-inspired Computing Using Resistive Synaptic Devices*,
DOI 10.1007/978-3-319-54313-0_6

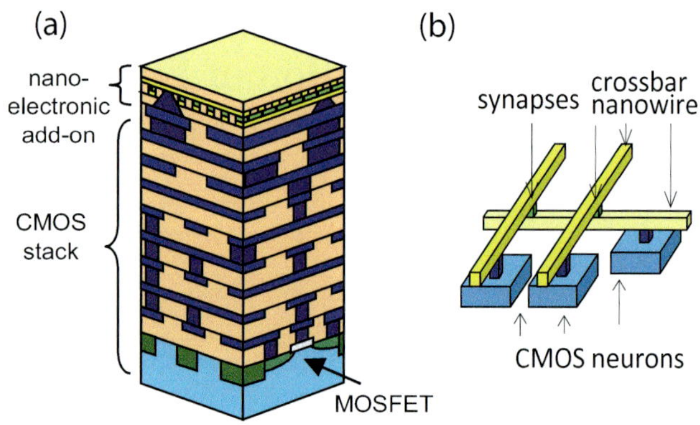

Fig. 6.1 CMOL circuits. (**a**) A cartoon of a hybrid CMOS/memristor integrated circuit. (**b**) The example of three CMOS cells (neurons) interconnected via corresponding crossbar nanowires (dendrites and axons) and cross-point memristive devices (synapses), which are located above CMOS layer

In this chapter, we first show experimentally that a memristor crossbar can facilitate the implementation of vector-by-matrix multipliers used in neural networks. After comprehensive characterization of the utilized memristive devices, we use them to implement single- and multi-layer perceptron and explain how the constructed networks were trained in situ as well as ex situ to classify given patterns [6, 7]. Finally, we conclude the chapter by reviewing some of the challenges in front of integrating memristor crossbars with existing CMOS circuitries in CMOL-based architecture.

6.2 Memristive Devices and Crossbar Circuits

6.2.1 Device Fabrication and Forming

The term "memristive" was introduced by Leon Chua in the 1970s [8] within his work on nonlinear circuits. Here we focus on a subset of memristive systems called resistive RAMs (RRAMs) and in particular on valence change memory (VCM) devices [9, 10]. The inset of Fig. 6.2a shows the material stack of the devices which were used for the implementation of the perceptron networks [11]. In this stack, platinum and titanium are used for the bottom and top electrodes, respectively. The 30-nm TiO_{2-x} oxide is the switching layer, while the thin 4-nm alumina layer is a barrier, which was optimized for the maximum nonlinearity of the device I-V [6].

Normally, a "forming" process (dielectric breakdown) is necessary in order to form one or more conductive filaments, which enable the switching operation. These filaments consist of regions with defects, which promote the drift of oxygen ions. The same forming process creates oxygen vacancies, resulting in the change in

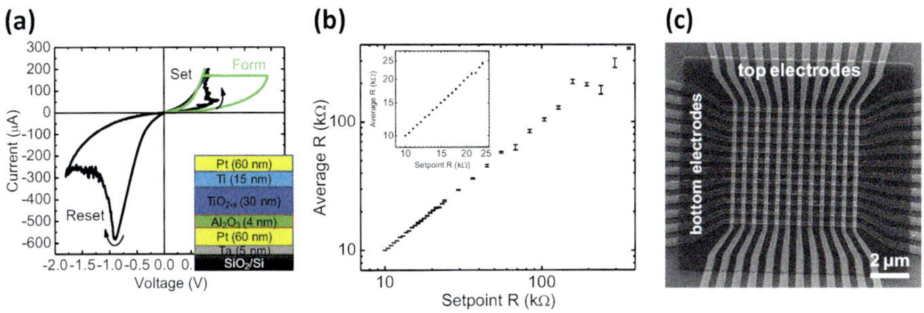

Fig. 6.2 $Al_2O_3/TiO_{2\text{-}x}$ resistive switching devices. (**a**) Typical quasi-dc I-V curves of memristor forming, set and reset, with the inset showing the material stack. (**b**) Analog properties of crossbar-integrated devices: tuning of the resistance measured at a non-disturbing voltage of 0.2 V to various values within the dynamic range. (**c**) Micrograph of a 12 × 12 cross-point crossbar built with DUV lithography

cation valence, effectively doping the material and changing its conductivity. Figure 6.2a shows the typical I-V curve of the forming process (green line), with the forming occurring at around 2 V and 180 μA (note that as explained later, all of this current is not passing through the device, and some part of it is due to the leakage from other already formed devices). In our devices, the stoichiometry of the $TiO_{2\text{-}x}$ is optimized to reduce the forming voltage which would simplify the design of underlying neuromorphic system. Once the forming has occurred, the device jumps to the low-resistance state (LRS), and, from now on, its conductance can be adjusted repeatedly by applying appropriate set and/or reset pulses or voltage sweeps as shown in Fig. 6.2a. The figure shows a "reset" cycle on the negative side of the voltage axis, where the device is gradually turned to its high-resistance state (HRS). The device has a reset threshold near −1 V, meaning that any stimulus with absolute amplitude lower than that does not change the state of the device. Applying "reset" voltages below −1 V but higher than −2 V will turn the device into an intermediate resistive state, while any voltage below −2.0 V turns the device completely off. Similarly, the device can be turned to its low-resistance state again by applying a "set" pulse, i.e., any voltage above the set threshold of the device (which here is around 0.8 V) as shown in Fig. 6.2a, on the positive side of the voltage axis. Due to the existing positive feedback in the setting procedure, this process is much more abrupt than resetting. By the consecutive application of these set and reset voltage pulses, resistance of a device can be tuned to any intermediate/analog state [12] as shown in Fig. 6.2b. This graph and the inset (zoomed lower-resistance region) show the different resistance states to which the device was effectively tuned (y-axis) vs. the desired target resistances (x-axis). The ability to program the device to more than 32 distinct states corresponds to the possibility of storing more than 5 bits of information in each memristor either when it is used as a multilevel memory cell or when it is emulating the functionality of biological synapses.

The memristive devices we used are integrated in a passive 12 rows (bottom electrodes) by 12 columns (top electrodes) crossbar array structure as shown in Fig. 6.2c. Each device in the array has a footprint area of $200 \times 200\ nm^2$, and the gap between the metal lines is 400 nm. Unlike other reported crossbars that use dedicated active devices, such as transistor, as selectors to enable peripheral circuitry to individually access memristors inside the array, in the passive crossbar array, the functionality of the selector is imitated by strengthening the nonlinearity of devices through the engineering of the device material stack. The advantages of such passive structure are the lower fabrication complexity, higher density, and simpler peripheral circuitry. The absence of selector devices has some drawbacks too. In particular, in a passive crossbar array, the integrated memristors must have very high uniformity and low forming voltages to ensure successful forming and switching of all devices in the crossbar arrays.

We measured these two key characteristics, i.e., uniformity and forming voltage, in our crossbar, and the results are presented in Fig. 6.3. In particular, panel 6.3 (a) and 6.3 (b) indicate that, after annealing and before forming, the array has a very high uniformity in terms of nonlinearity and initial resistive states, as indicated by the state histogram and the I-Vs of four different devices across the array. This high uniformity not only plays an important role in facilitating the forming process but also, as we would see later, has a huge impact on the trainability of the network. Thanks to this high uniformity and low forming voltage, we were able to form all devices inside the array (see Fig. 6.3c, d). Note that in order to be able to form all devices in a fully passive crossbar array, memristive devices should satisfy several other conditions as well. First, the resistance state of devices before forming should be close to their state when they are fully turned off through complete resetting. Second, the forming voltage should not be much larger than the twice of the set threshold voltage (see Fig. 6.3c). Thirdly, device-to-device variations, especially when considering the switching voltage thresholds, should be as low as possible.

Regarding the performance of a memristive network, there are two other important issues intrinsic to the passive structure: the so-called sneak-path and half-select disturbance [13]. The former problem is due to the current leaking through unselected devices during the reading or operation phase, which mainly happens because of the resistance of the nonideal metal lines of the crossbar. The later problem is related to the unwanted change in the states of the unselected devices when another device in the array is being programmed. To minimize the effects of sneak-path and half-select disturbance, three different biasing schemes (as shown in Fig. 6.4) were employed to perform following experiments. We formed the devices by applying a forming voltage across the selected device while leaving the rest of unused rows and columns floating (see Fig. 6.4a). Figure 6.4b shows how the effect of sneak paths is minimized during operation (reading the state of a device) by grounding the unselected rows and columns of the crossbar and reading the current from the virtually grounded row or column. Finally, to avoid the half-select disturbance problem during programming or weight updating, the well-known V/2 scheme is utilized (Fig. 6.4c).

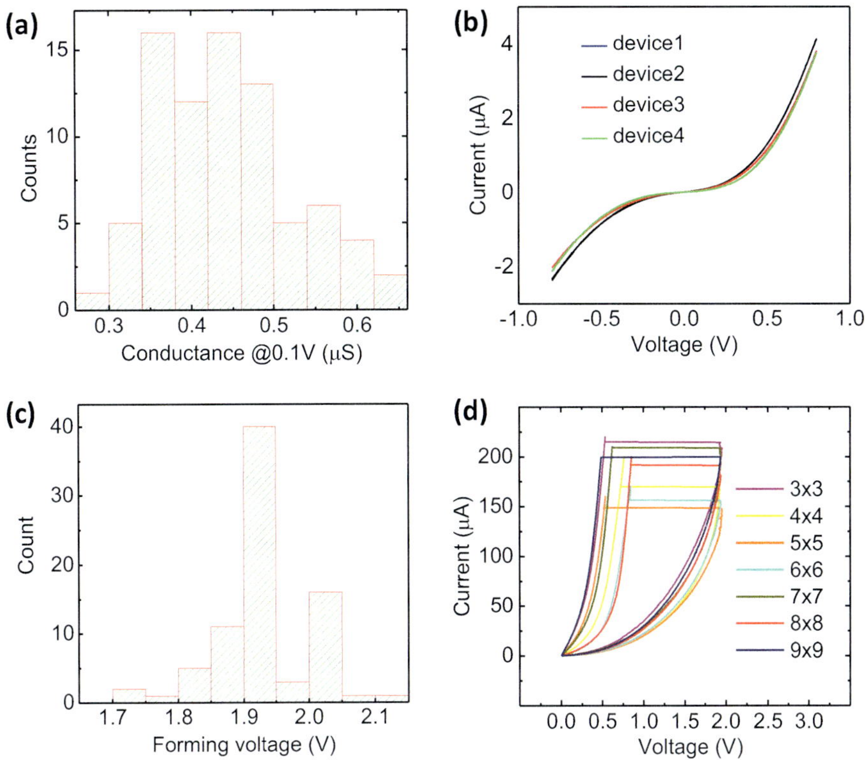

Fig. 6.3 Electrical characteristics of fabricated devices before and after forming. (**a**) Histogram of conductances of all devices in the array before forming. (**b**) I-V curves of four devices before being formed showing the high uniformity inside the fabricated crossbar. (**c**) Histogram of the forming voltages. (**d**) Forming curves of the all diagonal devices. These forming curves show that thanks to the high nonlinearity in device behavior, the forming current stays constant during the forming process – a critical property required for forming large crossbars

6.2.2 *Dynamic Characteristics*

Neural networks are usually trained in software by modifying synaptic weights to minimize the error between the actual and target outputs [14]. However, at each iteration during training, the maximum amount of change applied to every synaptic weight (determined by a parameter called learning rate) is usually kept small in order to avoid the network from diverging or falling into local minima. This is also true for a training that happens in the hardware.

In a memristive network, for example, training at physical level is done by changing the conductance of memristors emulating synaptic weights through the consecutive application of set and/or reset pulses. However, changing the conductance of these devices during training should be well controlled and gradual in order for the underlying network to converge. Any unwanted large change in the

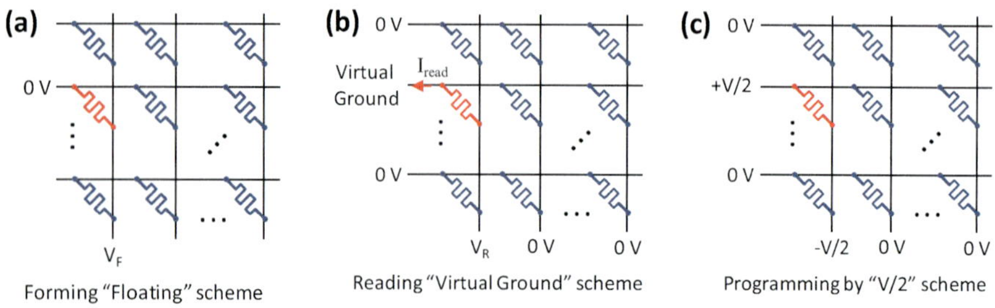

Fig. 6.4 Biasing schemes used during various modes of operation. (**a**) Forming: here a simple "floating" scheme is adopted. Assuming that after each forming event the formed device is reset to HRS, which is slightly lower than preforming resistance, the maximum voltage drop during forming of the device on its nearest neighbors will be between V/3 and V/2. (**b**) Reading: in this case, virtual ground is used on the selected row (a source measure unit is used to generate 0 V and measure current), the reading voltage (V_R) is applied to selected column, and all the other terminals are grounded. In this way, only the current coming from the selected device will be read. (**c**) Programming: a "V/2" scheme is used to guarantee that no voltage higher than |V/2| drops over any device different from the selected one. This is needed in order to reduce the perturbation of unselected or half-selected devices. In all panels, the device marked by red color indicates the selected device

conductance would correspond to having large learning rate in software which increases the risk of divergence or having oscillatory behavior at the outputs of the network. In order to keep the conductance updates small, one possible solution is to decrease the amplitude or width of the weight tuning pulses. Still, it should be noted that this comes with a cost of tightening the working conductance range. This is because, as can be seen in Fig. 6.2a, larger voltages are required when we want to push the device to increasingly lower or higher conductances.

In order to choose a proper voltage pulse (with appropriate width and amplitude) to be used for the training of the memristive neuromorphic hardware, dynamic behavior of memristive devices needs to be characterized. To this end, we mapped the switching voltage thresholds from which we can infer the suitable pulse amplitude. To obtain the map, we kept fixed the pulse duration at 500 μs (due to tools limitations), and then consecutive set or reset pulses with increasing amplitudes were applied to every individual device (originally initialized to a state of 30 μS). The results are summarized in Fig. 6.5a, b. The important information that can be extracted from these switching curves is the histograms of set and reset voltage thresholds as shown in Fig. 6.5c. Despite the existing variations, majority of devices have set and reset voltage thresholds below 1.1 V and higher than −1.3 V, respectively. Note that while narrower histograms are preferred for neuromorphic applications, having devices with different switching voltages makes the selection of the programming pulse for the parallel weight update, during the network in situ training, a challenging problem.

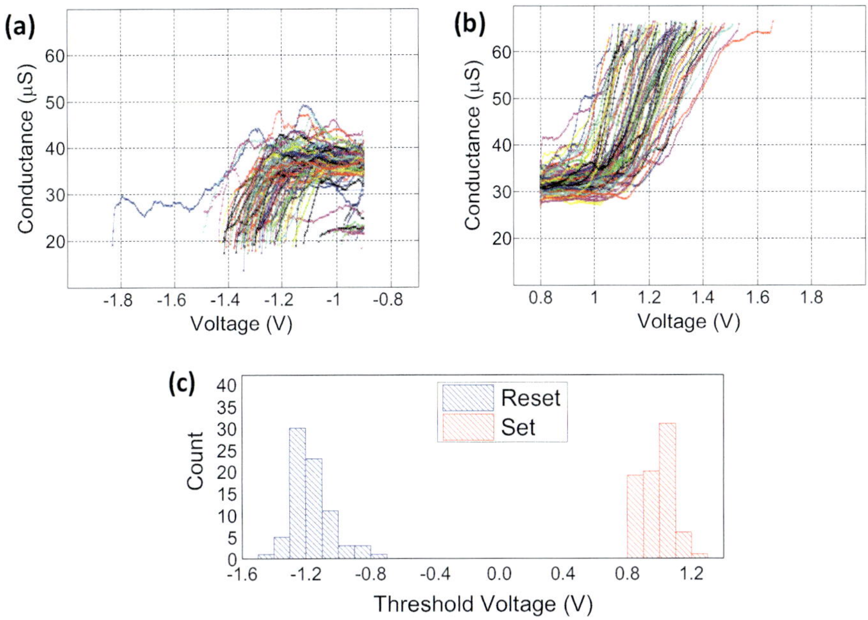

Fig. 6.5 Characterization of dynamic behavior of the fabricated memristor crossbar by applying consecutive 500 μs pulses with increasing amplitudes. For this experiment, at the beginning, all devices were initialized to about 30 μS. (**a**) Resetting. (**b**) Setting. (**c**) Switching voltage thresholds extracted from (**a**, **b**)

As mentioned before, having a successful in situ training in memristive neural network completely depends on how gradually and how smoothly we can update the state of memristors. Due to the existing device-to-device variations, the straightforward solution to tackle this problem might seem to be the fine adjustment of pulse amplitudes for every individual memristor based on its switching dynamics. Unfortunately, this solution has a practicality issue since it requires full mapping and storage of switching voltage thresholds and dynamic behaviors of all devices. The simpler solution, which is also employed here, is to choose set and reset pulses with fixed amplitudes slightly higher than the switching thresholds of the majority of devices in order to reach a trade-off between the number of devices which would get updated and the amplitude of their conductance change. In this case, larger changes can be achieved during in situ training as a result of the consecutive application of these fixed pulses over time. Therefore, in the next step, we have studied how our metal-oxide memristors respond to the sequence of set and reset pulses with fixed amplitudes (i.e., training with a fixed pulse over time). As can clearly be seen in Fig. 6.6, we can achieve very good analog state modulation in wide range of conductances for different pulse amplitudes.

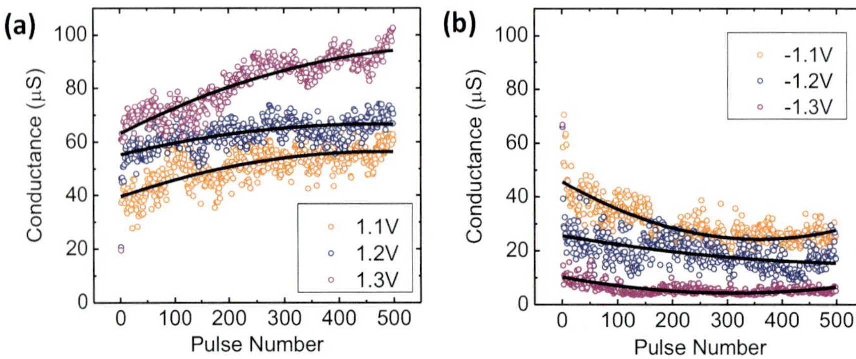

Fig. 6.6 Gradual state modulation of the fabricated memristor through the application of the train of pulses with fixed amplitudes. For set and reset experiments, at the beginning the underlying device was initialized to ~20 μS and ~70 μS, respectively. (**a**) Setting. (**b**) Resetting

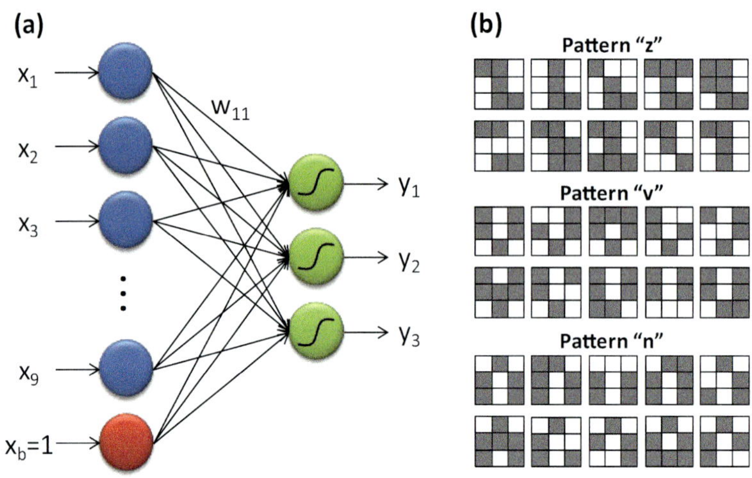

Fig. 6.7 (**a**) Graph diagram of the feed-forward network implemented using a memristive crossbar array. (**b**) Linearly separable training data set used to benchmark the fabricated neural network

6.3 Single-Layer Perceptron

6.3.1 Inference

To experimentally demonstrate potentials of memristive devices in neuromorphic computing, a single-layer neural network (see Fig. 6.7a) was implemented in hardware based on a passive integrated memristive crossbar [6, 7]. The network has ten inputs (including the bias) and three outputs which can be used to classify linearly separable data such as the patterns shown in Fig. 6.7b provided that the network is trained properly. In this type of networks, during the inference phase, neuron outputs can be expressed as a linear combination of inputs:

$$y_j = f\left(\sum_{i=1}^{10} x_i w_{ij}\right), \quad \text{for} \quad 1 \leq j \leq 3 \tag{6.1}$$

or in matrix form as:

$$\boldsymbol{y} = f(\boldsymbol{x}\boldsymbol{W}), \quad \boldsymbol{y} = [y_1, y_2, y_3], \quad \boldsymbol{x} = [x_1, \cdots, x_{10}], \quad \boldsymbol{W} = \{w_{ij}\} \tag{6.2}$$

where x_i is the ith input, y_j is the jth output, $f(\cdot)$ is neuron's activation function, and w_{ij} is the synaptic weight connecting the ith input to the jth output. As shown by these equations too, the main computation in the network is a simple vector-by-matrix multiplication between the input pattern and the weights stored in the synapses. Therefore, the performance of any neuromorphic system would mainly depend on how efficiently it implements this basic but computationally extensive operation. Fortunately, a memristor crossbar structure is shown to be an ideal candidate for the implementation of this task [15].

Indeed, Fig. 6.8 shows how the network in Fig. 6.7a is actually implemented using a single 10 × 6 memristor crossbar array emulating the functionality of all synapses connecting the input layer to output neurons. In this circuit, inputs corresponding to the binary pixel intensities are applied to the rows of the crossbar in the form of ±0.2 V voltage biases. Because all columns of the crossbar are

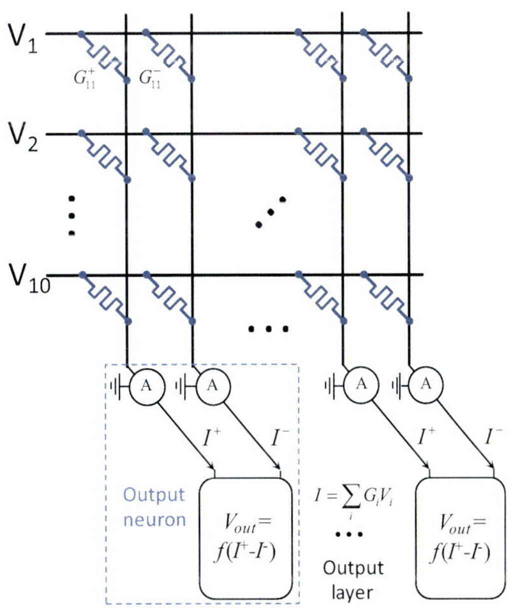

Fig. 6.8 Physical implementation of the network of Fig. 7a using memristor crossbar array. Since the conductance of a memristor is a positive quantity, each synaptic weight is implemented in differential mode using two memristors. In this experiment, the functionality of neurons is emulated in the software

virtually grounded, by ignoring the effects of line resistances, the current flowing through each memristor would be proportional to its conductance and the applied voltage (input). Following Kirchhoff's current law, these currents are summed up automatically on vertical wires of the crossbar as follows:

$$I_j = \sum_{i=1}^{10} V_i G_{ij} \tag{6.3}$$

where G_{ij} is the conductance of a memristor connecting the ith input (with V_i applied) to the jth output. Taking into account that conductances (G_{ij}) can only be positive, in order to represent negative weights, we have to use a pair of devices. In other terms, we need to subtract the currents from two neighbor columns. The total input for the neuron would then be:

$$I_j^{tot} = \sum_{i=1}^{10} V_i \left(G_{ij}^{+} - G_{ij}^{-} \right) \quad for \quad 1 \le j \le 3 \tag{6.4}$$

Again, $\left(G_{ij}^{+} - G_{ij}^{-} \right)$ represents the effective synaptic weight connecting the ith input to the jth output neuron implemented in differential mode to allow producing both positive and negative weights. At physical level, this is done by subtracting the current of the column containing negative weights (i.e., I_j^{-}) from the current coming from the column hosting positive weights (i.e., I_j^{+}). This current is then sensed by a virtual ground implemented by an operational amplifier with R_f as feedback resistor, so the total V_{out} per neuron will be:

$$V\text{out}_j = R_f \left(I_j^{+} - I_j^{-} \right) = R_f \sum_{i=1}^{10} V_i \left(G_{ij}^{+} - G_{ij}^{-} \right), \quad \text{for} \quad 1 \le j \le 3 \tag{6.5}$$

To this V_{out} we will apply, by software, the neuron activation function f, which we selected to be *tanh*. Now that we have the means to implement the network of Fig. 6.7a, it needs to be trained to be able to work as a classifier. By training, we mean that the conductance of memristors should be adjusted so that each output neuron produces highest output voltages (compared to other neurons) only for those input patterns that it represents [11].

6.3.2 Network Training

Supervised training was used to train the implemented memristive neural network to perform the perfect classification of 30 3 × 3-pixel black and white images into three classes, one per each output neuron. In particular, the utilized patterns were the stylized versions of "z", "v," and "n" characters on the 3 × 3 grid. Both the ideal patterns and the noisy ones, created by flipping one pixel in the image, were used for the training due to the small size of the training data set. This resulted in ten patterns per each class for a total of 30 patterns as shown in Fig. 6.7b.

In situ training of the hardware starts by calculating the current classification error of the network in order to determine how to update synaptic weights and consequently the memristors. To do so, all 30 patterns were applied to the network, and outputs were measured as shown in Fig. 6.8 and explained in Eq. (6.5). Based on the measured outputs and known targets, the amount of update for each weight at the end of each epoch can be calculated using a regular delta rule [16] as:

$$\Delta w_{ij} = \alpha \sum_{\substack{\text{training} \\ \text{data}}} \left(t_j - y_j\right) f'\left(\sum_{i=1}^{10} x_i w_{ij}\right) x_i \qquad (6.6)$$

where α is the learning rate, t_i is the desired target output, and $f'(\cdot)$ is the derivative of the utilized activation function. This equation shows that in order to shrink the gap between the actual and target outputs, at the end of each epoch in which direction and for how much each synaptic weight should be fine adjusted. Although the conductance of a memristor could be changed in each iteration by different amounts, as explained before, due to the existing device-to-device variations, this process is not well controlled and deterministic. To relax this problem, we proposed to update all the memristors using the same pulse, independent from the magnitude of Δw, while the direction of the update is determined based on the sign of the calculated gradient [17]. This method, as we would show later, has the advantage of simplifying the crossbar update process by allowing part of the devices to be updated simultaneously, thanks to the use of fixed pulses. However, it should be noted that using only the sign information of the gradient to update all weights requires smaller effective learning rate to avoid the network from diverging during training. To satisfy this requirement, we used the {1.3 V, 500 μs} and {−1.3 V, 500 μs} pulses, respectively, to set and reset memristors during in situ training based on the characterization results presented in Figs. 6.5 and 6.6.

At the end of each epoch, memristors are updated in parallel by the application of a set of four consecutive pulses to every pair of columns of the crossbar (columns representing positive and negative portions of weights). As an example, Fig. 6.9 shows how, in our experimental single-layer perceptron (SLP), the first two columns of the crossbar are updated based on the sign of the hypothetical gradients given in the first column of the presented gradient matrix. For positive(negative) terms in the gradient matrix, the corresponding weight in the network should be decreased(increased). At the physical level, decreasing(increasing) a weight is done by resetting(setting) G^+ and setting(resetting) G^- through the application of fixed amplitude pulses using V/2 scheme. Since the amplitudes of the set or reset pulses are the same for all devices, devices located on the same column with the same gradient signs can be updated all at once [18]. Therefore, each column of the

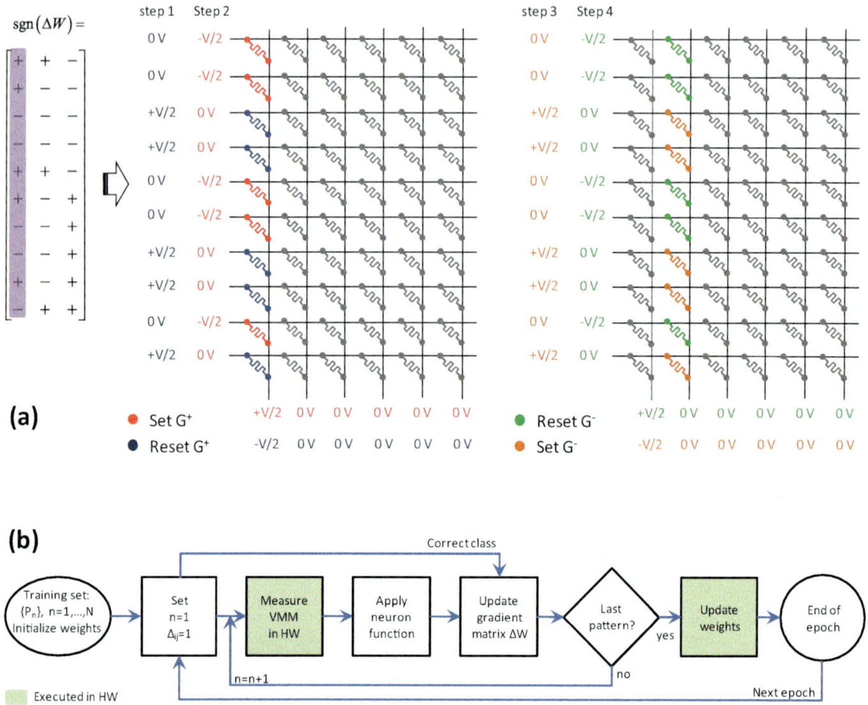

Fig. 6.9 (**a**) The first four steps of crossbar update. The sign of the gradient matrix (on the *left*), which is obtained after one epoch of training, specifies the direction of the state update for each device in crossbar circuit, i.e., whether to incrementally set or reset the device. The update is performed using the V/2 scheme with appropriate chosen voltages (on the *right*). The voltage shown in *red*(*green*) and *blue*(*orange*) are for the first and second steps, respectively. (**b**) Flow chart of the training algorithm. *Gray boxes* show the steps implemented in hardware, while all remaining steps were emulated in software

crossbar can be updated by using only one set and one reset pulses (or with four pulses per each column of weights). This process is shown in Fig. 6.9 using two different colors per each column. Note that to avoid the disturbance of unselected devices, each column of the crossbar needs to be updated separately.

To summarize, in each epoch, and after feeding all training patterns and calculating the sum of gradients, all memristors are updated in parallel (column by column) with fixed set and reset pulses. The training process would then repeat these steps until the stop condition is met (see Fig. 6.9b). Following this procedure, the memristive neuromorphic network of Fig. 6.8 was trained with the training patterns shown in Fig. 6.7b. Prior to training, all memristors were initialized to around 35 μS. The experiment was repeated for six times, and, on average, the network reached the perfect classification of the training set after ~20 epochs as shown in Fig. 6.10a. Histograms of initial and final conductances of all memristors involved in one of the runs are presented in Fig. 6.10b.

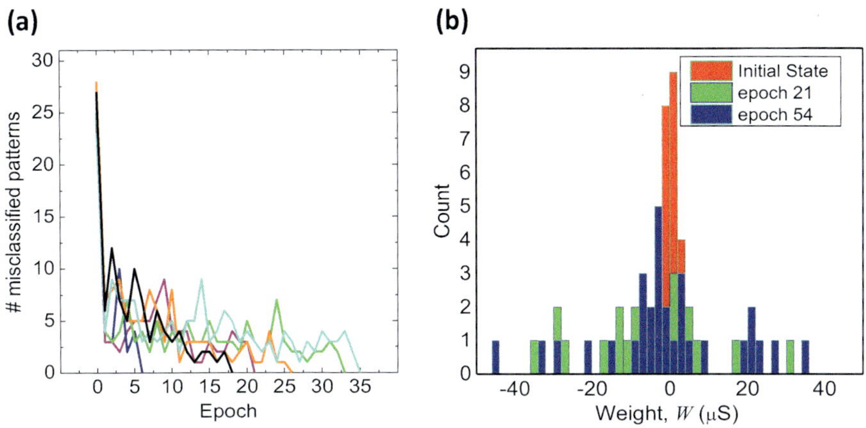

Fig. 6.10 (**a**) Experimental results for six different runs of the training for the single-layer perceptron. The y-axis represents the number of misclassified patterns (out of 30 in total) at each epoch. In all cases perfect classification was reached. (**b**) Histograms of states of all memristive devices at different stages of training (for one of the runs)

6.4 Multilayer Perceptron

While the demonstrated single-layer perceptron (SLP) is valuable because it shows the potentials of this new technology, the network itself is not practical in most of real-world applications, because as we saw before, it can only be used to solve simple problems like classifying linearly separable patterns. To improve the processing power and therefore broadening the possible applications of these artificial networks, we need to move from SLP toward multilayer networks. Each multilayer network can be built by a serial concatenation of two or more single-layer networks and can solve much more advanced and complex problems like classification of linearly inseparable patterns.

Before reaching the point that we can efficiently integrate memristive crossbars with CMOS-based neuron circuitry in one chip, a simpler way to implement multilayer memristive neural network would be to integrate memristive crossbars with neurons implemented using discrete integrated circuits (ICs) on printed circuit boards (PCB) like what is shown in Fig. 6.11a. Indeed, on these two PCBs, we have implemented a MultiLayer Perceptron (MLP) with 17 input, ten hidden, and three output neurons alongside all required peripheral circuitries. Two 20×20 crossbars were used to implement the functionality of synaptic weights connecting input to hidden as well as hidden to output neurons (see Fig. 6.11b for the high-level schematic of the implemented network).

Two different training methods, i.e., ex situ and in situ, were tested experimentally on the constructed network for the application of pattern classification. In ex situ training, the network was modeled and trained in software using the patterns shown in Fig. 6.12a as training data set. Final synaptic weights obtained from

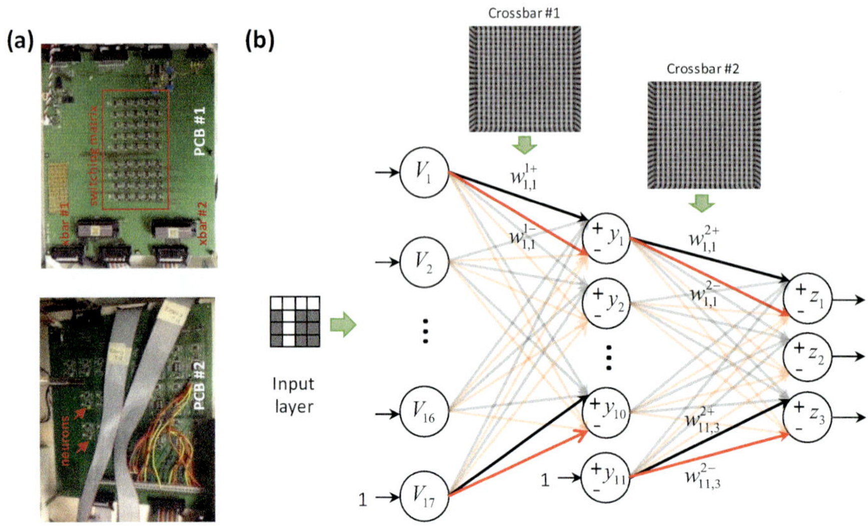

Fig. 6.11 (**a**) The memristive MLP fabricated on two printed circuit boards, one including two 20 × 20 crossbars with peripheral circuitries while the other one implements neurons. (**b**) High-level diagram of the implemented MLP. Each set of weights is implemented with one crossbar

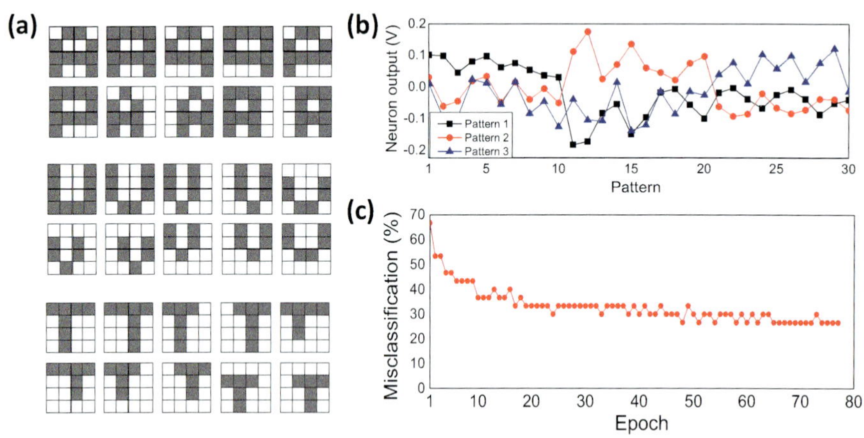

Fig. 6.12 (**a**) Linearly inseparable training data set used to benchmark the fabricated memristive MLP. (**b**) Classification performance of the network when it is trained using the ex situ training algorithm. As can be seen in this figure, all 30 training patterns have been classified correctly. (**c**) Training the memristive MLP inside the hardware. Due to the existing device-to-device variations, the network couldn't classify all the patterns (mainly those patterns which have so many pixels in common)

software simulation were then imported into the hardware by programming corresponding memristive devices using write-and-verify algorithm [12]. As Fig. 6.12b shows, the network trained by ex situ algorithm could successfully classify all training patterns. In the next experiment, we trained the network inside the hardware (also called in situ training) using the same patterns similar to what was done in SLP experiment and explained in the previous section. Although the trained network could classify most of the patterns (see Fig. 6.12c), the performance is not as good as the network trained by ex situ training. This is mainly because of the fact that the complexity of the training data set and the small size of the implemented network require the possibility of fine adjustment of synaptic weights during training. Unfortunately, this is hard to achieve in the utilized memristor crossbars because of the existing device-to-device variations. For further information about the implemented memristive MLP please refer to [19].

Note that the demonstrated SLP and MLP are particularly significant because they represent the basic element of the much more complex and powerful deep learning convolutional neural network [20, 21] which are now widely used in many practical applications [22].

6.5 3D Memristor-CMOS Hybrid Circuits

Although memristive crossbars significantly facilitate the hardware implementation of synaptic weights in neuromorphic systems, neurons still need to be implemented using conventional well-developed CMOS technology. However, due to the large number of input and output terminals in each crossbar, placing crossbars next to CMOS circuitries can extremely complicate the design and waste remarkable area (just for the sake of signal routings). The more reasonable option resides in the monolithic integration of memristive crossbars on top of the CMOS layers implementing neurons (see Fig. 6.13). The critical factors for successful monolithic integration of these two different technologies are (i) to create functional electrical interconnections between the memristive crossbar and the underlying CMOS circuits and (ii) to have compatible fabrication processes for crossbars and CMOS technologies. Moreover, beside these elements, obviously the memristive components need to be electrically compatible with the CMOS circuitry in terms of operational voltage and current requirements.

While memristive components have been previously integrated between two metal layers of a CMOS process [23, 24], vertical monolithic integration requires flexibility of the process so that memristive crossbars can be integrated on any prefabricated CMOS chip. This is much more challenging than simply fabricating a device between CMOS metal layers as the two fabrication processes now would be completely independent of each other. Moreover, requirement to integrate 3D memristive crossbars to achieve higher device density can increase design complexities and fabrication challenges even more. Very few previous reports of vertical monolithic integration on CMOS substrates exist in the literature

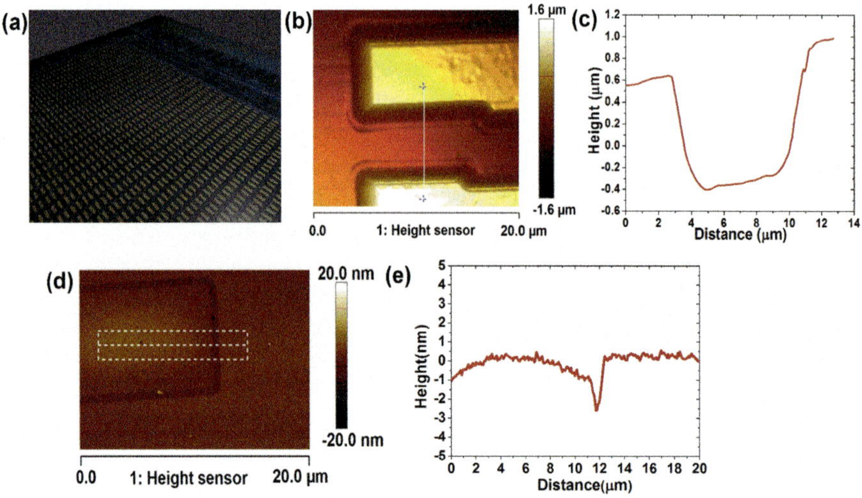

Fig. 6.13 (**a**) A 3D-rendered image of the CMOS chip showing the array of CMOS pads. (**b**) AFM image of a pair of CMOS pads covered under scratch-protect oxynitride on the as-received chip. (**c**) Topography on the chip across the region shown in (**b**). (**d**) AFM image of a CMOS pad after the planarization process. (**e**) Topography of the region shown in (**d**)

[25, 26]. The work done by Xia et al. [25] only demonstrated integration of planar devices, and hybrid 3D CMOL circuits with 3D memristive crossbars monolithically integrated on prefabricated CMOS substrates have not been demonstrated before. We have recently demonstrated successful monolithic integration of 3D memristive crossbars on prefabricated CMOS substrate [27]. The CMOS chip was processed in a foundry with 0.5 μm technology and houses all the necessary circuitry required for read, write, and decode operations [27–29]. The chip also contains an area-distributed interface to provide connectivity for the memristive crossbars to be integrated (Fig. 6.13a). The area-distributed interface contains an array of CMOS cells with each cell having a pair of contact pads [27–29]. The two pads in each cross-point are connected to driver circuitry through two separate transmission gates and can be biased with read/write voltages independently of each other. One of the significant challenges involved with monolithic integration of heterogeneous components lies in overcoming the severe topography of the CMOS substrate. The topography on the as-received CMOS chip in our case was ~1 μm (Fig. 6.13b, c). Understandably this topography is not favorable for monolithic integration of memristive crossbars on top and creation of successful electrical contacts between the memristors and the underlying CMOS. Chemical mechanical planarization (CMP) was employed to remove the surface topography. To perform CMP on the CMOS chip first, the scratch-protect oxynitride layer (1.4 μm) is completely removed, and a SiO_2 layer (2 μm) is deposited by inductively coupled plasma-enhanced chemical vapor deposition (ICP-PECVD). The final topography on the top surface after CMP is less than 10 nm (Fig. 6.13d, e)

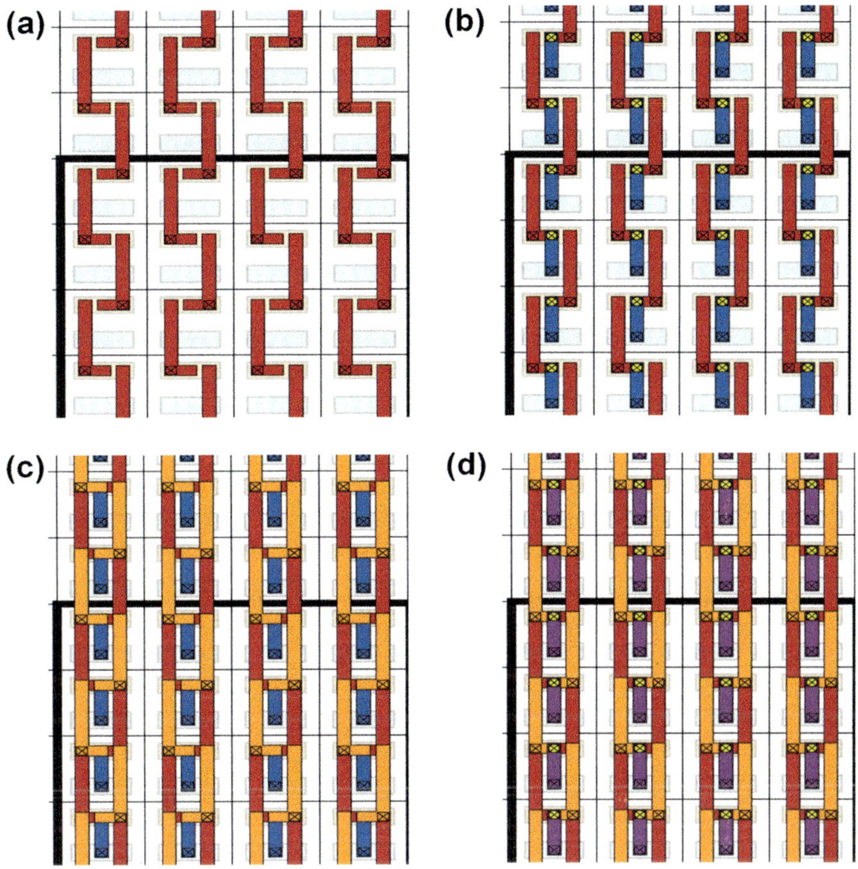

Fig. 6.14 (**a**) Schematic showing bottom electrodes (BEs) of the first layer of crossbars on top of the CMOS substrate. (**b**) Structure of the first layer of crossbars with memristive devices at the cross-points (*yellow circles*) between the bottom (*Red*) and top (*Blue*) electrodes. (**c**) Structure of the 3D crossbars after fabricating the second layer BEs. (**d**) Final structure of the 3D crossbars

[27]. Note that the SiO_2 thickness after planarization is ~1 μm, the oxide is thinned down to ~200 nm after the CMP by dry etching.

Once the chip is planarized, the integration of memristive crossbars is carried out using standard fabrication steps. Figure 6.14a shows the structure of the BEs (red lines) of the layer 1 crossbars fabricated on top of the CMOS substrate. The BEs (Ta/Pt) are connected to the underlying "red" pins in the CMOS layer through the via holes (square boxes in Fig. 6.14a) created through the 200 nm SiO_2 layer. After the deposition of the switching oxide (Al_2O_3/TiO_{2-x}), the top electrode (TE) contact via holes is created on top of the "blue" pads, and then TEs (Ti/Pt) are deposited (blue lines in Fig. 6.14b). An isolation oxide (SiO_2) is deposited after the fabrication of the first layer of crossbars, and then the second layer of crossbars is fabricated using the same steps used for layer 1 crossbar fabrication (i.e., open via holes for

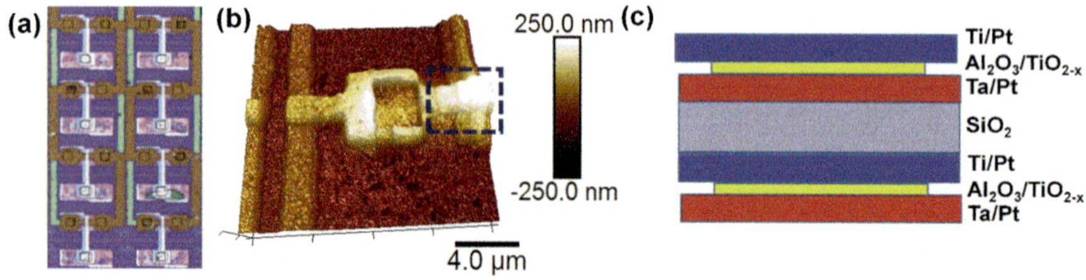

Fig. 6.15 (**a**) Optical image of a section of the 3D crossbars. (**b**) AFM image of a section of the 3D crossbar with the region inside the *blue dotted box* showing two stacked crossbar layers. (**c**) Cartoon of the cross-section of a 3D crossbar showing the different layers

layer 2 BE contact, layer 2 BE deposition, switching oxide deposition, creation of layer 2 TE contact via holes, and layer 2 TE deposition). The final structure of the fabricated 3D crossbars is shown in Fig. 6.14d.

Figure 6.15 (**a**) shows an optical image of a section of the fabricated 3D crossbars integrated on the CMOS substrate. An AFM image showing two stacked layers of crossbars is shown in Fig. 6.15 (**b**). Figure 6.15 (**c**) shows a cross-section of the layers in a stacked crossbar. Note that all the fabrication steps were carried out using standard photolithography with an i-line (365 nm), e-beam evaporation, sputtering, and PECVD. Moreover, all the process steps used are performed at low temperature (< 150 °C) except a final rapid thermal anneal at 300 °C in forming gas ($N_2 + H_2$) for 15 min.

One of the crucial challenges of monolithic hybrid circuits is to establish good electrical contacts between the two heterogeneous layers (in this case memristive crossbars on CMOS substrate). Figure 6.16a shows a histogram of the current measured at all the "red" pins in the CMOS active region. The current was measured at read voltage of 0.2 V. As seen from Fig. 6.16a, majority of the contacts show a good electrical connection between the two layers. The DC switching characteristics of the integrated devices can be seen in Fig. 6.16b. The devices display a forming-free behavior with low (~ 1 V)-voltage operation. To accurately control the states of the devices, a write-and-verify tuning algorithm was used [27] which is similar to the tuning algorithm used in our previous works [12, 30]. We employed 1 μs write pulses (limited by our setup) for the tuning operation. The devices can be tuned to at least eight levels with 10 % accuracy. An example of the tuning procedure is shown in Fig. 6.16c. The tuning algorithm was optimized for the speed of operation (measured as the number of pulses required to tune to each level). Figure 6.16d shows cumulative distributions of the tuning operations for all the levels where each of the eight levels were tuned 1000 times. In most cases the levels can be tuned within 100 tuning pulses. The tuning speed is comparable to the state of the art for programming resistive memory devices [30], whereas we achieve better accuracy due to the deployment of a write-and-verify algorithm. Therefore, the 3D integrated memristive crossbars have the potential for high-density memory applications. We have also demonstrated multiply-add operations using the developed 3D hybrid circuits [27] that can be used for neuromorphic computation applications.

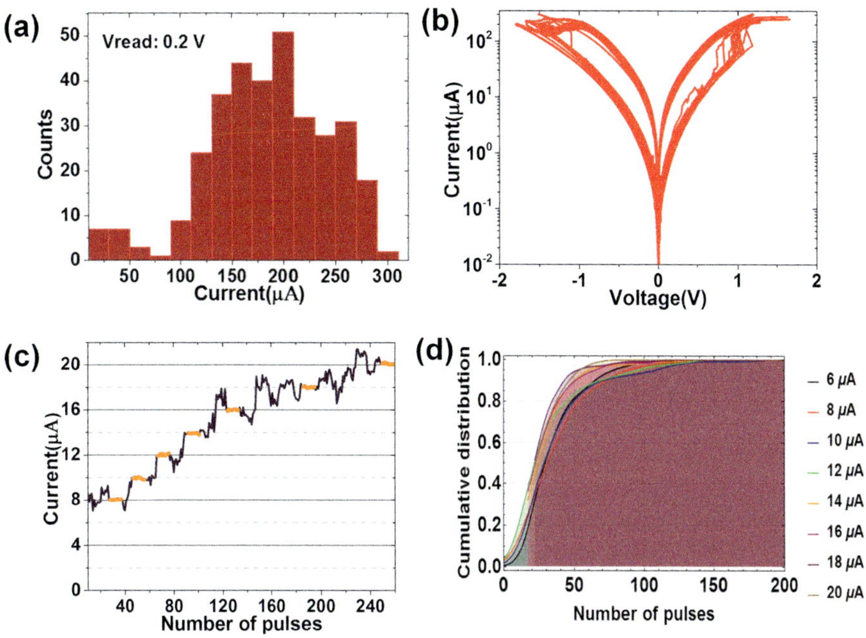

Fig. 6.16 (**a**) Histogram of the contacts currents measured at read voltage of 0.2 V showing the establishment of good electrical contacts between the two heterogeneous layers. (**b**) The DC switching characteristics of the integrated devices after forming. (**c**) An example of the tuning procedure for the integrated devices using the write-and-verify tuning algorithm. (**d**) Cumulative distributions of the tuning operations for all the levels where each of the eight levels were tuned 1000 times

6.6 Challenges and Future Work

Although system-level applications have been demonstrated in this chapter using the memristive crossbar technology that we have developed, many challenges still remain to be addressed, both at the device and system levels. Among them, the optimization of the device performance in terms of device scaling and operating power is the key challenging problem that needed to be taken care of. This is particularly crucial for CMOS integration as the power requirements of the memristive devices can significantly challenge design considerations for CMOS substrate. The devices used in the current generation of crossbars presented in this chapter operate at several hundreds of μA. As a result, a large network of these devices may take up to several mA's of current to operate. The optimization of the Al_2O_3 barrier layer can be a possible way to reduce operating current since having aluminum oxide in material stack is known to result in very low operating current in resistive memory devices [31].

Besides fabrication-related requirements at physical level, many other more system-level problems and issues remain to be solved, even for the realization of simple networks. Starting from a device level, it is now a well-known fact that the switching mechanism on which RRAMs are based has a high level of an intrinsic stochasticity. As a consequence, a wide scattering in the switching dynamics (thresholds, nonlinearity, and speed), even across a single crossbar, is expected. This can give rise to issues during network training or weight importing phases specially when dealing with passive crossbars, in the form of having very different updates (learning rate) among devices or the magnification of unwanted half-select disturbances. In the worst case scenario, this can even cause some devices to be practically treated as being stuck in on or off states [17].

Another difficulty in implementing efficient neuromorphic networks directly comes from memristor crossbar itself. Ultra dense fabrication of memristive devices requires the utilization of very thin nanowires with higher resistivity. Using the nonideal resistive nanowires magnifies the effect of sneak-path currents which can notably degrade the accuracy of analog computation and also can make the forming process much harder specially when dealing with large passive crossbars. Since adding active selector like CMOS-based transistor to every memristor would demolish the whole area benefit of crossbar architectures, one possible solution can be to improve the nonlinearity in devices through the careful engineering of material stack [32]. In the presented work, we introduced such nonlinearity by adding the alumina barrier between the bottom electrode and the switching layer. While this modification was sufficient for our experiment, higher nonlinearity would be desirable for more practical and larger crossbar arrays. An alternative solution is the insertion of passive selector layer into the material stack which is still an open field of research [32–35]. Despite all these challenging problems, the good news is that artificial neural networks (ANNs) are shown to be inherently resilient to malfunctioning synapses and neurons, so these aforementioned problems are expected to be mostly mitigated by the existing redundancies in the well-designed networks.

Lastly, due to the lack of extensive experimental work, there is still much to be learnt about training and operation of ANNs based on memristive devices. The intrinsic nonlinearity and stochasticity of such devices complicate the already complex behavior of ANNs as dynamical systems. Simulation-based studies give some insights to these issues, but without experimental demonstrations, they would lack trustable validation.

Acknowledgments The authors would like to acknowledge useful discussion with G. Adam, K. T. Cheng, B. Hoskins, I. Kataeva, K. K. Likharev, and D. B. Strukov. The work has been supported in part by the Air Force Office of Scientific Research (AFOSR) under the MURI grant FA9550-12-1-0038, DARPA under Contract No. HR0011-13-C-0051UPSIDE via BAE Systems, and National Science Foundation grant CCF 1528205.

References

1. J. Hasler, H.B. Marr, Finding a roadmap to achieve large neuromorphic hardware systems. Front. Neurosci. **7**, 118 (2013)
2. C. Diorio, P. Hasler, A. Minch, C.A. Mead, A single-transistor silicon synapse. IEEE Trans Electron Devices **43**(11), 1972–1980 (1996)
3. S. George et al., A programmable and configurable mixed-mode FPAA SoC. IEEE Trans Very Large Scale Integr (VLSI) Syst **24**(6), 2253–2261 (2016)
4. K.K. Likharev, CrossNets: neuromorphic hybrid CMOS/nanoelectronic networks. Sci. Adv. Mater. **3**(3), 322–331 (2011)
5. K.K. Likharev, Hybrid CMOS/nanoelectronic circuits: Opportunities and challenges. J. Nanoelectron. Optoelectron. **3**(3), 203–230 (2008)
6. M. Prezioso, F. Merrikh-Bayat, B.D. Hoskins, G.C. Adam, K.K. Likharev, D.B. Strukov, Training and operation of an integrated neuromorphic network based on metal-oxide memristors. Nature **521**(7550), 61–64 (2015)
7. M. Prezioso, et al., Modeling and implementation of firing-rate neuromorphic-network classifiers with bilayer Pt/Al2O3/TiO2-x/Pt Memristors, in *IEEE International Electron Devices Meeting (IEDM)* (2015)
8. L.O. Chua, S.M. Kang, Memristive devices and systems. Proc. IEEE **64**(2), 209–223 (1976)
9. R. Waser, M. Aono, Nanoionics-based resistive switching memories. Nat. Mater. **6**(11), 833–840 (2007)
10. M.D. Pickett et al., Switching dynamics in titanium dioxide memristive devices. J. Appl. Phys. **106**(7), 74508 (2009)
11. J. Hertz, A. Krogh, and R. G. Palmer, Introduction to the theory of neural computation (Westview Press, Boulder, Revised edition 1991)
12. F. Alibart, L. Gao, B.D. Hoskins, D.B. Strukov, High precision tuning of state for memristive devices by adaptable variation-tolerant algorithm. Nanotechnology **23**(7), 75201 (2012)
13. M.A. Zidan, H.A.H. Fahmy, M.M. Hussain, K.N. Salama, Memristor-based memory: The sneak paths problem and solutions. Microelectron. J. **44**(2), 176–183 (2013)
14. R. Hecht-Nielsen, Theory of the backpropagation neural network, in *International Joint Conference on Neural Networks (IJCNN)* (1989), pp. 593–605
15. J.J. Yang, D.B. Strukov, D.R. Stewart, Memristive devices for computing. Nat. Nanotechnol. **8** (1), 13–24 (2013)
16. W. Schiffmann, M. Joost, R. Werner, Optimization of the backpropagation algorithm for training multilayer perceptrons, Univ. Koblenz, Inst. Physics, Rheinau (1992), pp. 3–4
17. I. Kataeva, F. Merrikh-Bayat, E. Zamanidoost, D. Strukov, Efficient training algorithms for neural networks based on memristive crossbar circuits, in *International Joint Conference on Neural Networks (IJCNN)* (2015), pp. 1–8
18. F. Alibart, E. Zamanidoost, D.B. Strukov, Pattern classification by memristive crossbar circuits using ex situ and in situ training. Nat. Commun. **4**, 2 (2013)
19. F. M. Bayat, M. Prezioso, B. Chakrabarti, I. Kataeva, and D. B. Strukov, "Advancing memristive analog neuromorphic networks: increasing complexity, and coping with imperfect hardware components," *arXiv preprint arXiv:1611.04465*, 2016
20. E.B. Baum, On the capabilities of multilayer perceptrons. J. Complex. **4**(3), 193–215 (1988)
21. Y. LeCun, Y. Bengio, G. Hinton, Deep learning. Nature **521**(7553), 436–444 (2015)
22. A. Krizhevsky, I. Sutskever, G. E. Hinton, Imagenet classification with deep convolutional neural networks," in Advances in *Neural Information Processing Systems (NIPS)* (The MIT Press, Cambridge, 2012), pp. 1097–1105
23. S.-S. Sheu et al., A 4 Mb embedded SLC resistive-RAM macro with 7.2 ns read-write random-access time and 160 ns MLC-access capability, in *2011 IEEE International Solid-State Circuits Conference (ISSCC)*, (2011)
24. T. Liu et al., A 130.7-2-layer 32-Gb ReRAM memory device in 24-nm technology. IEEE J. Solid State Circuits **49**(1), 140–153 (2014)

25. Q. Xia et al., Memristor- CMOS hybrid integrated circuits for reconfigurable logic. Nano Lett. **9**(10), 3640–3645 (2009)
26. P. Lin, S. Pi, Q. Xia, 3D integration of planar crossbar memristive devices with CMOS substrate. Nanotechnology **25**(40), 405202 (2014)
27. B. Chakrabarti, et al., A multiply-add engine with monolithically integrated 3D memristor crossbar/CMOS hybrid circuits, *submitted*
28. M. A. Lastras-Montano, A. Ghofrani, K.-T. Cheng, Architecting energy efficient crossbar-based memristive random-access memories, in *IEEE/ACM International Symposium on Nanoscale Architectures (NANOARCH)* (2015)
29. M. Payvand et al., A configurable CMOS memory platform for 3D-integrated memristors, in *IEEE International Symposium on Circuits and Systems (ISCAS)* (2015), pp. 1378–1381
30. L. Zhao et al., Multi-level control of conductive nano-filament evolution in HfO 2 ReRAM by pulse-train operations. Nanoscale **6**(11), 5698–5702 (2014)
31. W. Kim et al., Forming-free nitrogen-doped AlOx RRAM with sub-μA programming current, in *Symposium on VLSI Technology* (2011)
32. J.J. Yang et al., Engineering nonlinearity into memristors for passive crossbar applications. Appl. Phys. Lett. **100**(11), 113501 (2012)
33. Y. Huang et al., A new dynamic selector based on the bipolar RRAM for the crossbar array application. IEEE Trans Electron Devices **59**(8), 2277–2280 (2012)
34. J. Woo, D. Lee, E. Cha, S. Lee, S. Park, H. Hwang, Vertically stacked ReRAM composed of a bidirectional selector and CB-RAM for cross-point array applications. IEEE Electron Device Lett **34**(12), 1512–1514 (2013)
35. S. H. Jo, T. Kumar, M. Asnaashari, W. D. Lu, H. Nazarian, 3D ReRAM with field assisted super-linear threshold (FAST) selector technology for super-dense, low power, low latency data storage systems, in *IEEE/ACM Asia and South Pacific Design Automation Conference (ASP-DAC)* (2015), pp. 575–575

Chapter 7
Weight Tuning of Resistive Synaptic Devices and Convolution Kernel Operation on 12 × 12 Cross-Point Array

Ligang Gao and Shimeng Yu

7.1 Introduction

To learn and recognize thousands of objects from millions of images, the neuro-inspired machine learning algorithms (e.g., deep learning with hierarchical neural network architecture [1]) have achieved great success in various intelligent tasks, including image/object classification, speech recognition, etc. One of the most popular and successful deep learning algorithms today is the convolutional neural network (CNN) [2], which is based on multiple convolution layers and pooling layers, followed by a fully connected classification layers. The fully connected layers essentially perform the vector-matrix multiplication. However, deep learning typically requires huge computational recourses for training and inference. This remains a grand challenge for the conventional hardware based on silicon CMOS technology because of its low efficiency as compared to the biological neural systems in dealing with the complex, real-world environment problems [3]. Alternative approaches such as neuro-inspired architecture, with distributed computing and localized storage in neural networks, have attracted great attention due to the tolerance to fault/error and massively parallel computation [4]. The primary goal of neuro-inspired computing is to develop application-specific hardware for tasks such as visual/sensory data classification and autonomous robotics. In the recent years, progresses in building large-scale neuromorphic hardware have been made with silicon CMOS technology, such as SpiNNaker [5], TrueNorth [6], etc. However, such conventional hardware systems are inadequate for fast online training with huge dataset with limitations in the power consumption and scalability. CMOS design with SRAM arrays as the weight matrix has limitations such as binary bit

L. Gao • S. Yu (✉)
School of Electrical, Computer, and Energy Engineering, Arizona State University, 781 E Terrace Rd, ISTB4 room 591, Tempe, AZ 85287, USA
e-mail: shimengy@asu.edu

S. Yu (ed.), *Neuro-inspired Computing Using Resistive Synaptic Devices*, DOI 10.1007/978-3-319-54313-0_7

storage and sequential write/read process. To achieve further speedup and higher density, one promising approach is to realize a fully parallel write/read using a cross-point array architecture [7, 8], where each cross-point is implemented with a resistive synaptic device.

Resistive synaptic device [9, 10] is a type of resistive random access memory (RRAM) [11, 12] or memristor [13] that exhibits analog memory functionality with multilevel tunable weights and emulates the synaptic weight in a neural network. Recently, neuromorphic or machine learning algorithms using resistive synaptic devices have been extensively investigated [14–19], and there are a few experimental demonstrations of simple algorithms on cross-point arrays, e.g., 12 × 12 TiO_x/Al_2O_3 crossbar (simple perceptron algorithm with software neurons) [20] and 32 × 32 PCMO crossbar (off-chip neuron circuits on a board) [21]. Most of these works implemented the weighted sum and/or weight update in a fully connected layer. None of these works discussed how to implement the convolution kernel. In this chapter, we review our recent works that aim to tune the weights of resistive synaptic devices [22] for offline training and implement the key operation of the CNN—the convolution kernel on cross-point arrays [23, 24]. First, we optimized the iterative programming protocol of HfO_x RRAM by adjusting the pulse amplitude incremental steps, the pulse width incremental steps, and the start voltages. Then, we experimentally demonstrated the Prewitt horizontal and vertical kernel operation for the edge detection on the fabricated 12 × 12 cross-point arrays. Both 20 × 20 pixels of black-and-white MNIST handwritten digits and 50 × 50 pixels of grayscale "dog" image were used as the input image for the edge detection.

7.2 Pt/HfO_x/TiN Device Fabrication

Pt/HfO_x/TiN devices were fabricated in the cross-point structure with the active device area (i.e., overlap region between top and bottom electrodes) of about 1 × 1 μm^2, as shown in Fig. 7.1a, b. First, a reactive sputtered TiN (25 nm) bottom electrode was patterned by standard optical lithography techniques on a SiO_2/Si substrate (200 nm and 500 μm, respectively). A blanket HfO_x thin film with the thickness of 6 nm was deposited by using atomic layer deposition (ALD) at 180 °C using tetrakis ethylmethylamino hafnium and water as a precursor and a reactant, respectively. The Pt (25 nm) electrode was grown on top of the HfO_x layer by using e-beam evaporation with a subsequent optical lithography and lift-off process to define the perpendicular cross-point structure. The bottom electrode pads were exposed by optical lithography and buffered HF (BOE) wet etching of the HfO_x layer. Figure 7.1b shows the scanning electron microscopy (SEM) image of the cross-point array area, indicating that the 12 × 12 cross-point array with ~1 μm half-pitch was fabricated successfully. Figure 7.1c shows typical bipolar resistive switching I–V characteristics of the Pt/HfO_x/TiN device with 10 voltage sweeps from 0 to 2 V for set process and 0 to −2.4 V for reset process, respectively.

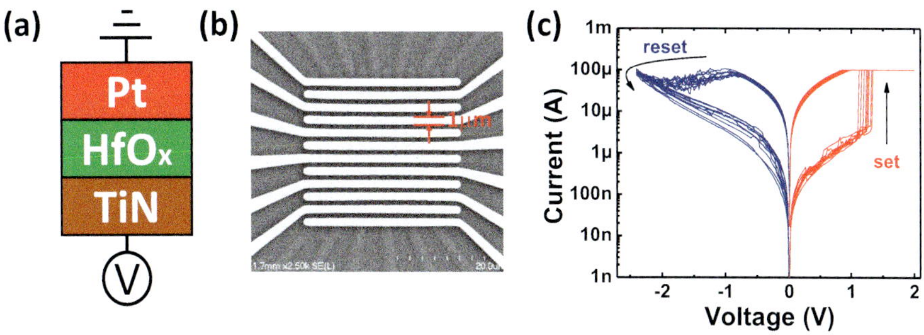

Fig. 7.1 (**a**) The schematic of the Pt/HfO_x/TiN device structure. (**b**) The SEM image of the active cross-point array area showing 1 μm half-pitch. (**c**) Typical I–V curves of the Pt/HfO_x/TiN devices. Superimposed ten full switching cycles

A 100 μA current compliance was applied to prevent the permanent breakdown during the set process.

7.3 Iterative Programming Protocol for Weight Tuning

The conductance of many RRAMs can be tuned continuously by applying voltage/current pulses consecutively [25–27]. Such analog conductance modulation capability is not only interesting for multilevel memory operations but also has attracted extensive attentions to function as synaptic devices for neuro-inspired computing [9]. An adaptable variation-tolerant algorithm to tune TiO_x-based RRAM devices to an intermediate state was proposed by varying pulse widths and amplitudes [28]. However, this approach is time-consuming to reach the target conductance as it is not optimized. Here we develop optimization strategies of the programming protocols by adjusting the pulse amplitude incremental steps, the pulse width incremental steps, and the start voltages on the HfO_x-based RRAM devices. The key limiting factor for the weight tuning convergence is the over-reset in the reset process due to the stochastic nature of the switching dynamics. Therefore, optimizing the incremental steps of voltage, the incremental steps of pulse width, and the start voltages carefully to avoid the over-reset is critical for fast tuning of the device conductance.

In order to characterize the analog switching dynamics, both the set and reset transitions can be gradually tuned by applying short programming pulses. Compared to the reset process, the set process is less controllable with abrupt jumps (see Fig. 7.1c), probably due to a positive feedback between the filament growth and the electric field and/or local temperature [29]. Therefore, the gradual reset process was used for the analog weight tuning. Figure 7.2 shows the optimization flow of the programing protocol. The device conductance state (G) is checked with a small read pulse (0.1 V) after each programming pulse. If the conductance reaches the

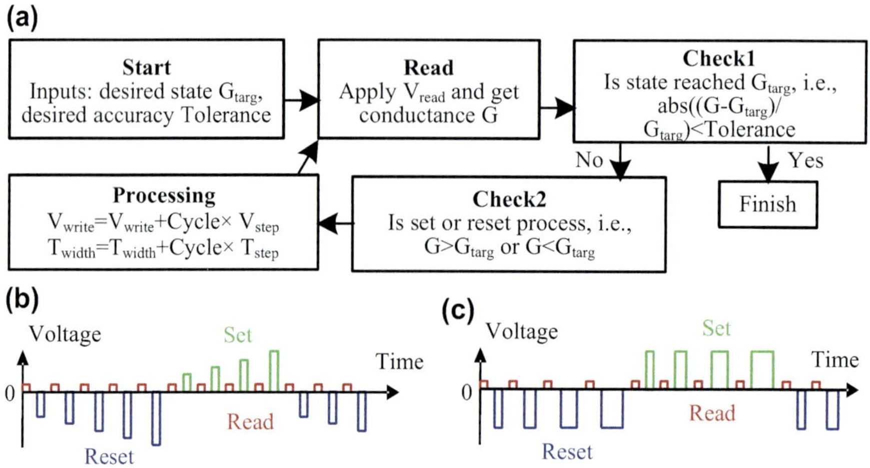

Fig. 7.2 (**a**) Programming protocol flow for tuning device conductance. (**b, c**) show schematically a gradually increased voltage pulse sequence and a gradually increased pulse width sequence for device tuning to a desired conductance state, respectively

target (G_{targ}) with defined tolerance margin, the program stops. If the conductance does not reach G_{targ}, an additional pulse with an increasing incremental amplitude step (V_{step}) or incremental width (T_{step}) is applied. If the conductance over-reset the tolerance of G_{targ}, then, a new voltage ramp of opposite voltage polarity starts. It is expected that larger amplitude pulses may reach a target state faster but with less precision, while smaller amplitude pulses will approach the state more precisely but require an exponentially longer time [28, 29].

Figure 7.3a shows the evolution of the conductance tuning with different incremental amplitude steps. Before the tuning, the devices are set to a low-resistance state with a quasi-DC voltage sweep in order to start the reset transition with the same initial conductance state. In particular, the reset tuning process starts with a −0.6 V voltage pulse, and the amplitude of every programming pulse in a sequence is increased by different V_{step} (−10 mV, −20 mV, −40 mV, −60 mV, and −80 mV) with fixed pulse width of 100 ns. If the conductance over-reset the tolerance of G_{targ}, a new voltage ramp of opposite polarity (set process) is applied with a voltage pulse starting at 0.6 V by different V_{step} (10 mV, 20 mV, 40 mV, 60 mV, and 80 mV) with fixed pulse width of 100 ns. For each V_{step}, the experiments were repeated five times. It is worth noting that a single gradual reset sequence is sufficient at small V_{step}, such as 10 mV, 20 mV, and 40 mV, while application of several reset-set sequences might be needed to converge on a desired state at larger V_{step}, such as 60 mV and 80 mV, as shown in Fig. 7.3a because of the over-reset problem.

Figure 7.4a shows the representative tuning process to illustrate the over-reset problem. Although using pulses with larger V_{step} (e.g., 80 mV) takes shorter time to reach G_{targ}, it runs a risk of over-reset due to the stochastic nature of the atomic

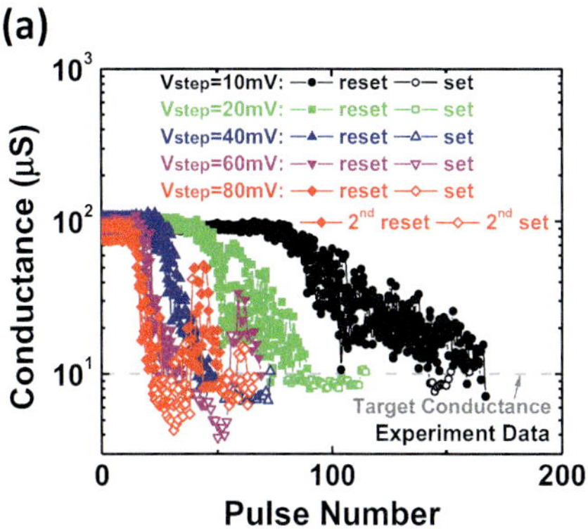

(b)

V_{step} (mV)	Average pulse number	Average number of over-reset	Average cumulative energy (nJ)
10	165	0	1.4
20	129	0	0.9
40	50	0	0.5
60	46	1	0.3
80	55	2	0.3

Fig. 7.3 (**a**) Conductance tuning process from 100 to 10 μS with increasing reset pulse amplitudes at different V_{step} = 10 mV, 20 mV, 40 mV, 60 mV, and 80 mV. Each V_{step} experiment repeats five times. (**b**) Average pulse number needed to reach target conductance, average number of over-reset out of five trials for different V_{step}, and the corresponding average cumulative energy

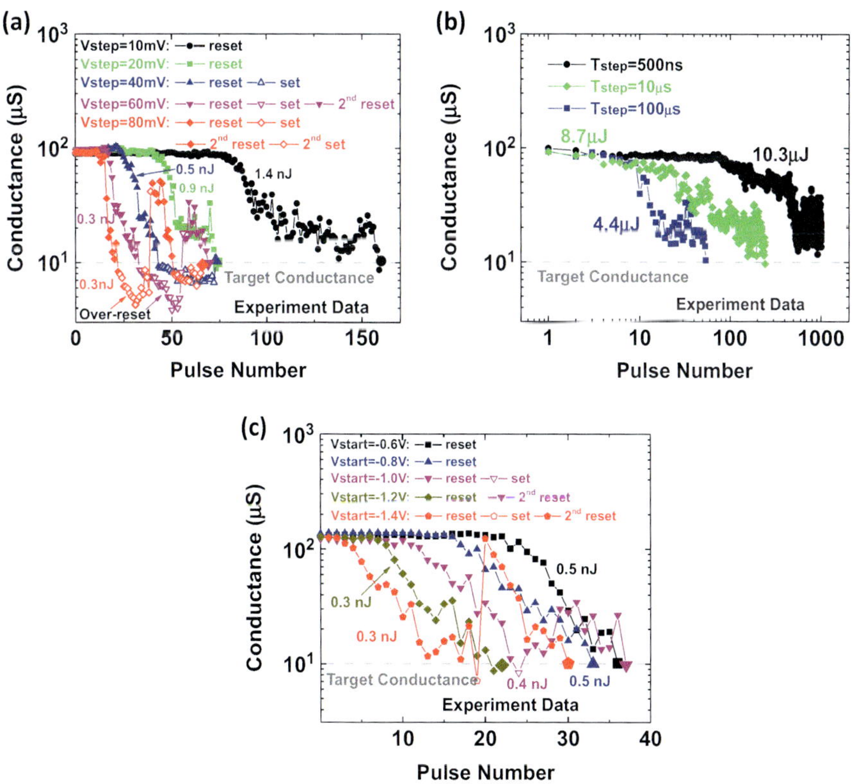

Fig. 7.4 (**a**) Conductance tuning process with increasing reset pulse amplitudes at different V_{step} = 10 mV, 20 mV, 40 mV, 60 mV, and 80 mV. (**b**) Conductance tuning process as a function of variable pulse width, i.e., T_{step} = 500 ns, 10 μs, and 100 μs. (**c**) Conductance tuning process with different V_{start} = −0.6 V, −0.8 V, −1.0 V, −1.2 V, and −1.4 V

oxygen ions and vacancies migration [29]. Once over-reset occurs, then we need to set the device and restart the reset process. Figure 7.3b counts the number of over-reset for different V_{step}, the average pulses needed to reach G_{targ}, and the corresponding average cumulative energy. It is observed that there is an optimal point for achieving minimum average pulses (the time to converge) and also minimum energy consumption with $V_{step} = 60$ mV.

Instead of varying the pulse amplitude, varying pulse width with constant amplitude and varying start voltages are also employed to tune the device conductance. Figure 7.4b shows the tuning process with fixed pulse amplitude (1.2 V) but different pulse widths ($T_{step} = 500$ ns, 10 μs, and 100 μs). However, the tuning time is much longer than that of increasing pulse amplitude, which indicates that tuning V_{step} is more effective. Figure 7.4c shows the tuning process with different start voltages ($V_{start} = -0.6$ V, -0.8 V, -1.0 V, -1.2 V, and -1.4 V) at fixed V_{step} (40 mV) and pulse width (100 ns). The larger the V_{start}, the faster to start changing the conductance, but the riskier of suddenly abrupt jump at the following set process. The cumulative energy consumption of different T_{step} and V_{start} was calculated and marked in Fig. 7.4b, c, respectively, indicating that tuning V_{step} with proper V_{start} is most energy efficient.

To further investigate the over-reset behavior and gain more statistics in the conductance tuning process, we take account the stochastic nature of filament formation and rupture into the resistive memories compact model [30] to capture the conductance variation. In the model, device conductance is determined by the gap distance (g) through electron tunneling conduction mechanism, where g is defined as the average distance between the filament tip and the electrode. We then add the variation into the equation of filament growth rate:

$$\frac{dg}{dt} = F(V,T) + \frac{\alpha \tilde{X} \eta(g)}{\sqrt{dt}} \tag{7.1}$$

$$\eta(g) = \tanh(\beta_1(g - g_{min}))\tanh(\beta_2(g_{max} - g)) \tag{7.2}$$

The first term on the right-hand side of (7.1) is the nominal filament growth rate as a function of applied voltage V and local temperature T. The details of this function can be found in the RRAM model [30]. The second term is the variation of filament growth rate in the conductance tuning process. $\tilde{X}$ is a random number function in standard Gaussian distribution and α is the amplitude. $\eta(g)$ is a simplified window function that suppresses the amount of variation at the minimum and maximum g (g_{min} and g_{max}), which represents the highest and lowest conductance state, respectively. In (7.2), β_1 and β_2 are fitting parameters that can control the shape of $\eta(g)$. The adoption of $\eta(g)$ is physically reasonable as the filament in these two states is completely formed or ruptured; thus, the effect of perturbation on the filament growth rate would not be prominent, while the fluctuation of the conductance is significant in the intermediate states during the switching [29]. To ensure the variability of the tuning process does not change with different simulation time step dt, we add $\sqrt{dt}$ into the variation term of (7.1); otherwise, the variation in dg

will tend to be averaged out if smaller dt is used. In this work, we set $\alpha = 250$ nm/s$^{1.5}$, $\beta_1 = \beta_2 = 1000$ nm^{-1}, $g_{min} = 0.12$ nm, and $g_{max} = 1.7$ nm and apply the same programming protocol in the experiments for the simulation, but we repeated the simulation for 1000 times for statistics. Figure 7.5a shows the representative conductance tuning process with different V_{step} in the simulation. The trend of the tuning process and the conductance variation are similar to the experimental results (see Fig. 7.4a). Figure 7.5b shows the average number of over-reset with different V_{step} in the simulation. The number of over-reset dramatically increases when V_{step} is larger than 100 mV. Figure 7.5c shows the time to converge for the conductance tuning with different V_{step} in simulation. $V_{step} = 60$ mV has the shortest average tuning time, which agrees with the experimental results. If V_{step} is too large, the worst-case average tuning time will start to increase, which is predicted to be as long as 300 μs for $V_{step} = 200$ mV.

Finally, using optimized programming parameters ($V_{start} = -1.4$ V, $V_{step} = 60$ mV, $T_{step} = 0$), we are able to tune the device conductance with 5% tolerance with respect to the predefined target conductance states (i.e., 50 μS, 10 μS, 5 μS, 1 μS) within its dynamic range, as shown in Fig. 7.6a. In particular, we applied increasing pulse amplitude, i.e., $V_{step} = 60$ mV, with fixed 100 ns pulse width starting from −1.4 V and 1.4 V for reset and set sequences, respectively. The device conductance is checked with a read pulse (0.1 V) after each programming pulse, and the voltage ramp is applied until it reaches or over-reset the G_{targ}. If over-reset occurs, a new voltage ramp of opposite polarity begins. Figure 7.6b indicates that all these intermediate states can maintain the conductance over time under stress at 0.1 V, which is promising for functionality as multilevel analog weights.

This proposed protocol can be used in a row-by-row programming scheme for the 1-transistor-1-resistor (1T1R) array as each cell in the same row can be operated independently. However, it can be used only in a bit-by-bit programming scheme

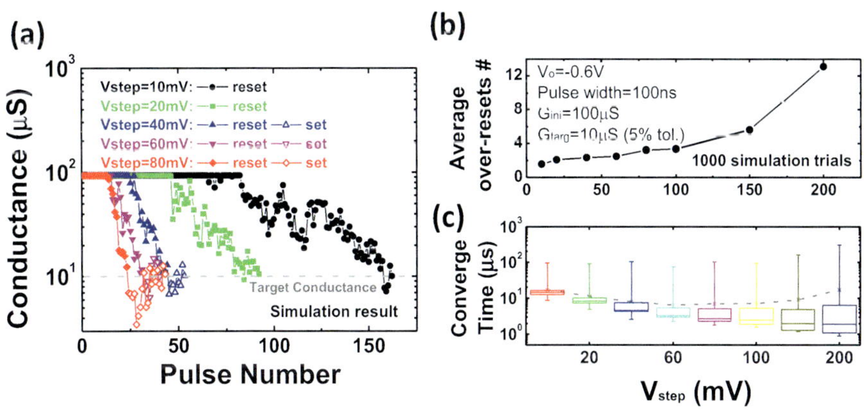

Fig. 7.5 (**a**) Examples of conductance tuning process with different V_{step} in the simulation. (**b**) Average number of over-resets and (**c**) time to converge for conductance tuning process with different V_{step} in the simulation

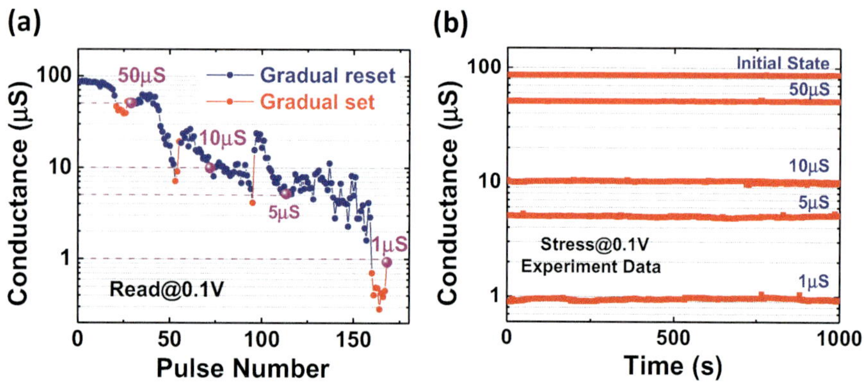

Fig. 7.6 (**a**) Demonstration of using optimized programming parameters ($V_{start} = -1.4$ V, $V_{step} = 60$ mV, $T_{step} = 0$) to tune the conductance state of the device to 50 μS, 10 μS, 5 μS, and 1 μS with 5% tolerance. (**b**) The retention of the tuned conductance states. These intermediate states can be maintained over time under stress at 0.1 V, indicating nonvolatile property in the dynamic range

for the cross-point array as the set and reset cannot be performed simultaneously at the same row in a cross-point array [31]. The overhead of latency and energy consumption associated with the iterative programming is analyzed as follows. Since the proposed protocol can program the state to be an arbitrary target with a tolerance margin in the device's dynamic range, the number of levels is limited by the larger one of the tolerance margin and data retention or readout fluctuation at each state. In our case, the on/off ratio of the device is around 100, and the relatively retention fluctuation for the device states is about 30% in the worst case (i.e., lowest conductance in Fig. 7.6b), which is larger than the tuning protocol tolerance margin (5%). This translates to a 4-bit precision (16 levels) that can be reliably reached in the device. Obviously, trade-offs exist between the high density and the latency/energy consumption. Our results indicate ~40 pulses are needed to achieve the 4-bit/cell, while typically 1 pulse is needed to program the 1-bit/cell. The number of required pulses actually depends on how many bits of the operation needs. With 3-bit/cell or 2-bit/cell, the tolerance margin in the protocol can be relaxed; thus, fewer pulses are needed. For implementing neural network, even more bits per cell is needed, but the neural network may tolerate some degree of the variation or noise of the weights. The detailed discussions on the trade-offs between the learning accuracy and the device non-ideal effects can be referred to [32]. Considering the relatively long latency associated with the iterative programming, the proposed protocol is only good for offline training, which means the synaptic weights (conductance values) should be predefined by means of computer simulations, and the cells are programmed before the actual run time. In addition, the external circuitry for implementing the programming protocol can also be off-chip since it will be one-time programming.

7.4 Convolution Kernel Operation on Cross-point Array

Convolution kernels in the first convolution layer of the CNN are typically edge detection filters. Image edges are important features at the early stages of image recognition, as edges are useful for segmentation and identification of objects in a scene. Figure 7.7a shows the input stage for the CNN architecture. A receptive field in the input image is convoluted with the kernels toward several feature maps. The convolution kernel (also named filter) typically is a 2D matrix, e.g., 3 × 3 or 5 × 5 matrix, as shown in Fig. 7.7b. The convolution operation is performed individually on each pixel of the input image, and it involves three sequential steps: Firstly, the operation starts when the convolution kernel is overlapped on the input image in such a manner that the center pixel of the kernel aligns with the single pixel to be convoluted in the input image. Secondly, each pixel value in the receptive field of the input image is multiplied by the corresponding value in the kernel. Thirdly, the sum of products from the second step is computed, and this weighted sum becomes the pixel value in the feature map. To perform a convolution on an entire image, the receptive field must be shifted and repeated to scan out the input image pixel by pixel.

To implement the convolution kernel on resistive cross-point array, we propose a dimensional reduction of 2D kernel matrix into 1D column vector (e.g., a column of the array), and the multiple kernels for different feature maps form multiple columns of the array. With one receptive field unrolled into the row input, the voltage representing the input image is applied to the rows in parallel; thus, the readout current from multiple columns in one array calculates the convolution results in several feature maps in parallel, as shown in Fig. 7.7c, thereby greatly speeding up the convolution operation in the CNN architecture.

Considering the offline training of the CNN, the values of convolution kernels will be pre-trained by the software and loaded into the hardware by one-time programming. In this work, a well-defined simple edge detector will be used, i.e., the Prewitt kernel, to demonstrate the convolution kernel on resistive cross-point array as a proof-of-concept. The Prewitt kernel is a 3 × 3 edge detector and it has two templates, one detects the horizontal edge and the other detects the vertical

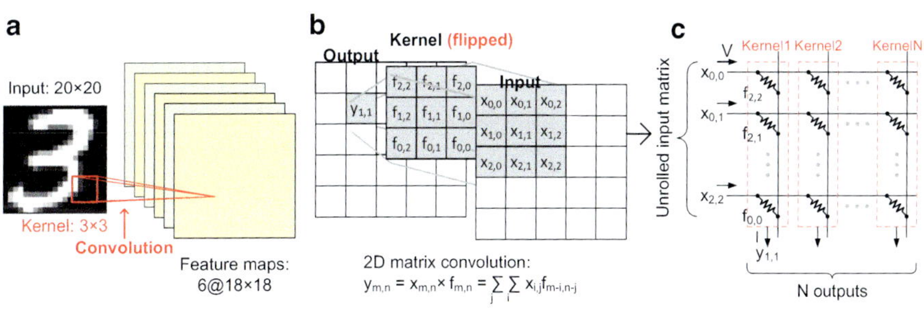

Fig. 7.7 (**a**) Input stage for CNN architecture. (**b**) 2D matrix convolution. (**c**) Implementation of convolution for multiple feature maps into a cross-point array architecture by reduction of 2D kernel matrix into 1D column vector

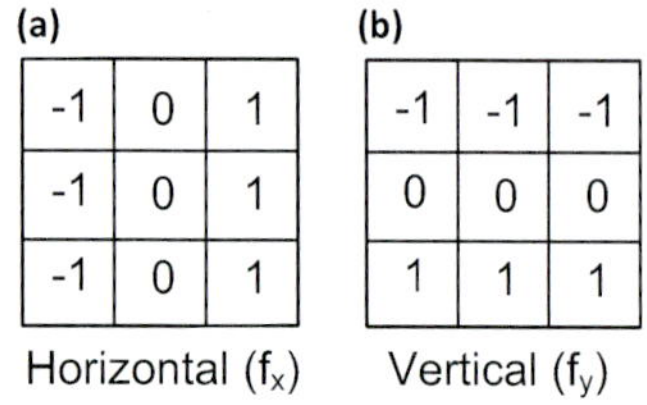

Fig. 7.8 Prewitt edge detector. (**a**) Horizontal kernel (f_x). (**b**) Vertical kernel (f_y)

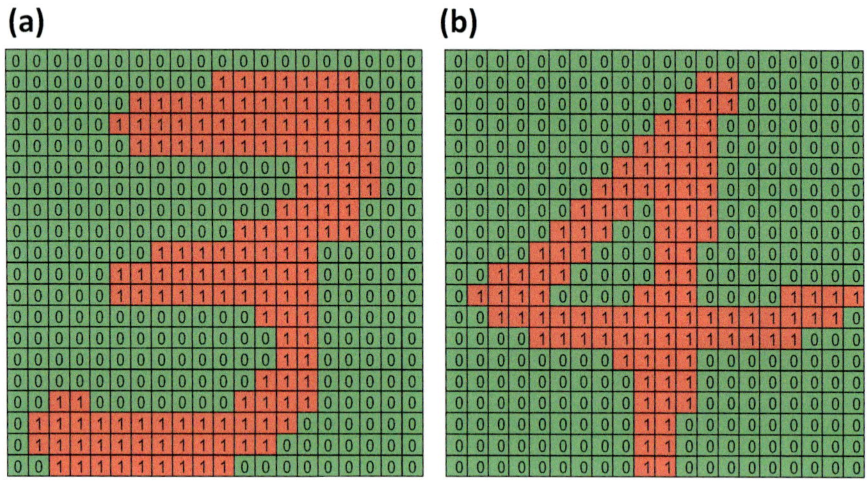

Fig. 7.9 Original 20 × 20 pixels of black-and-white MNIST handwritten digits: (**a**) "3" and (**b**) "4"

edge. The values of the Prewitt horizontal kernel (f_x) and vertical kernel (f_y) are shown in Fig. 7.8a, b, respectively.

To implement the horizontal kernel (f_x) and vertical kernel (f_y), the 20 × 20 pixels of black-and-white MNIST handwritten digits "3" and "4" are used as the examples, as shown in the Fig. 7.9a, b, respectively. The software simulation results of the horizontal kernel (f_x) and vertical kernel (f_y) operation, for the MNIST handwritten digit "3" and digit "4," are shown in Fig. 7.10a–d, respectively. Next, the Prewitt kernel operation is experimentally implemented using Pt/HfO_x/TiN-based resistive cross-point array. The fabrication procedure is described earlier, but with a major difference of shrinking the HfO_x thickness down to 3 nm, thereby successfully reducing the electroforming voltage to around 3 V, which is a critical step for the following switching operation in cross-point array. Figure 7.11a shows the microscopic top-view image of our fabricated 12 × 12 cross-point array: six contact pads are located at the four edges, and resistive Pt/HfO_x/TiN stack (from top to bottom) are formed at the cross-point junctions of the rows and columns.

Figure 7.11b shows how to map the Prewitt horizontal kernel (f_x) and vertical kernel (f_y) on the resistive cross-point array. To implement the negative weight value in the kernels, each weight is represented as the difference of a pair of cells

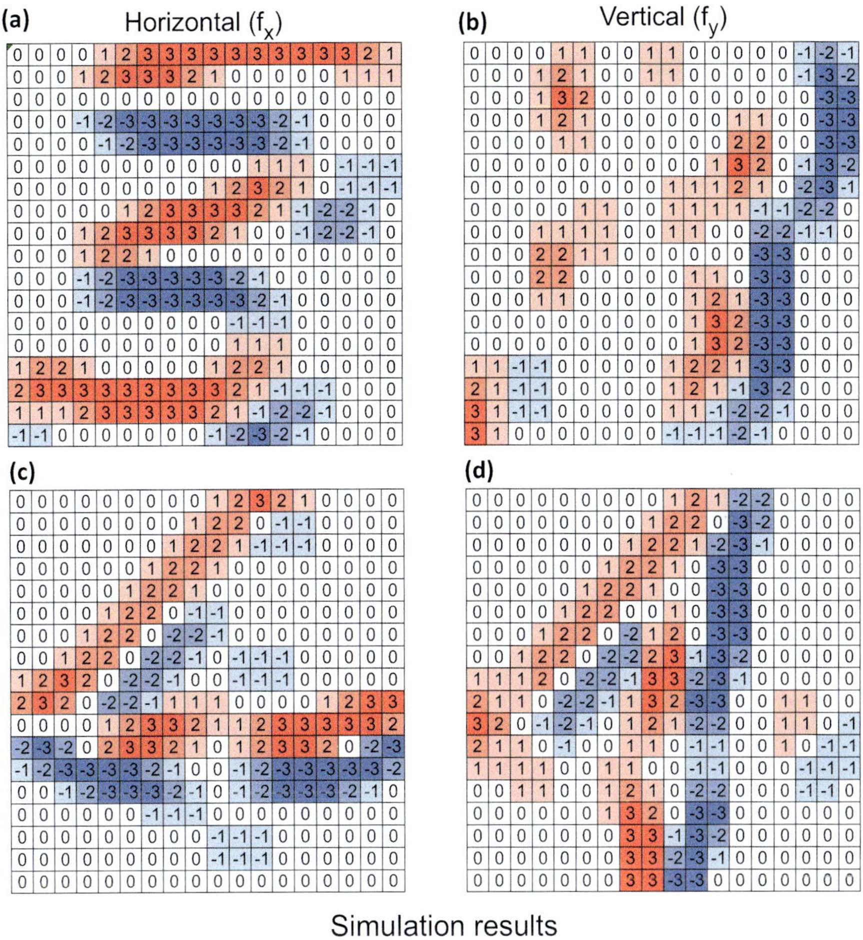

Fig. 7.10 The software simulation results of (**a**) Prewitt horizontal kernel operation and (**b**) vertical kernel operation for digit "3." (**c**) Prewitt horizontal kernel operation and (**d**) vertical kernel operation for digit "4"

with current I_i^+ and I_i^- in two adjacent columns ($f_i = I_i^+ - I_i^-$). Therefore, using the principle of a dimensional reduction of 2D kernel matrix into 1D column vector discussed earlier, the green cells are programmed to on-states, while the blue cells are kept at off-states. In this way, the horizontal kernel (f_x) is programmed into columns 1 (C_1) and 2 (C_2), while the vertical kernel (f_y) is programmed into columns 3 (C_3) and 4 (C_4). Finally, a differential readout of the two-column current represents the convolution results.

We prepare such on-state data pattern in Fig. 7.11b before the kernel operation. We used the optimized iterative programming protocol to achieve the fast converge of the weight to an arbitrary target level in TiN/HfO_x/Pt RRAM devices. The current of the green cells at on-states in Fig. 7.11b is programmed to 5 μA at read

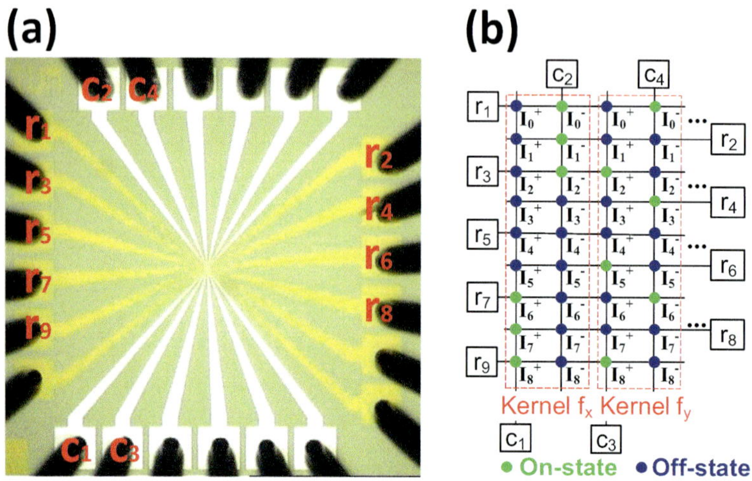

Fig. 7.11 (**a**) The microscopic top-view image of the fabricated 12 × 12 cross-point array. The probe card tips have touched on the pads. (**b**) The implementation of the Prewitt horizontal kernel (f_x) by programming the cells of C_1 and C_2 columns into on-state and off-state and the Prewitt vertical kernel (f_y) by programming the cells of C_3 and C_4 columns into on-state and off-state

voltage of 0.2 V with 20% tolerance using the half biasing technique, where the selected cell's row and column are biased at the write voltage V_w and ground, respectively. All the unselected rows and columns are biased at $V_w/2$. Therefore, only the selected cell sees a full V_w, while the half-selected cells along the selected row or column see a half V_w, and all the other unselected cells in the array see zero voltage. Note that no current compliance is used in the pulse programming. As we sequentially program each cell in the array, the previous programmed cells in the array are not disturbed when we use this half biasing technique to tune the next cell to the target weight. The current of blue cells in Fig. 7.11b remains in the initial off-states, which is typically less than 10 nA at read voltage of 0.2 V. During the iterative programming, a read scheme is used to read out the cell's current and compare with the target level. In the read scheme, the selected row is biased at the read voltage of 0.2 V; the unselected rows and all the columns are connected to the ground. Therefore, only the cells of the selected row see a read voltage, and all the other unselected cells see zero voltage. After the programming, the weights are stable with negligible fluctuations over time.

The 20 × 20 pixels of black-and-white MNIST handwritten digits "3" and "4" are used to experimentally demonstrate the Prewitt horizontal (f_x) and vertical (f_y) kernel operations in parallel. Each pixel x_i of the 3 × 3 receptive field in the input image is applied to the corresponding row r_i of the cross-point array. Therefore, the output column current is calculated as $I_c = \sum_{i=0}^{8} x_i \cdot r_i$. If the pixel x_i is 1, a read voltage of 0.2 V is applied to that row; if the pixel x_i is 0, the row is grounded. All other unused rows and columns are also grounded to avoid the sneak path when we read the column current out. For example, considering one 3 × 3 receptive field, the output pixel value I_{out} for horizontal kernel f_x and vertical kernel f_y is calculated as

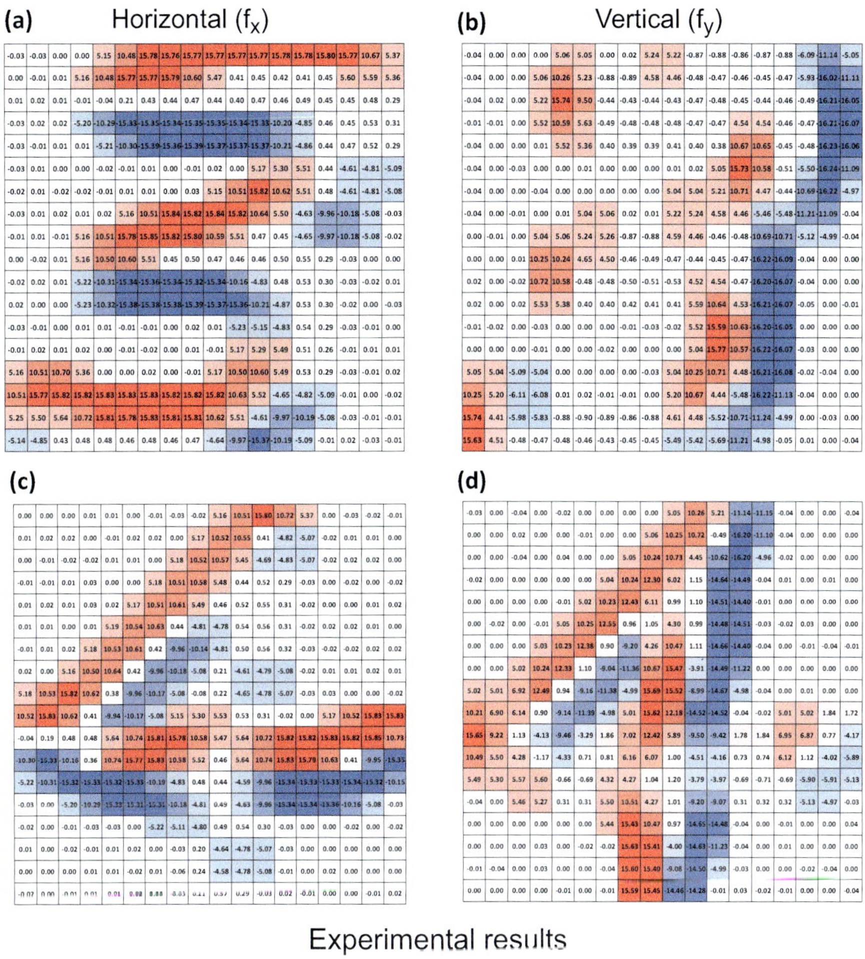

Fig. 7.12 The experimental results of (**a**) Prewitt horizontal kernel operation and (**b**) vertical kernel operation for digit “3.” (**c**) Prewitt horizontal kernel operation and (**d**) vertical kernel operation for digit “4.” The value of each pixel is the current in the unit of μA while read at 0.2 V

$I_{\text{out},fx} = I_{c1} - I_{c2}$ and $I_{\text{out},fy} = I_{c3} - I_{c4}$, respectively. Figure 7.12a shows the experimental results of the horizontal kernel (f_x) operation for the input MNIST handwritten digit “3.” The value of each pixel is the current with the μA unit read at 0.2 V. The gradient grayscale pattern is exactly the same as the simulation pattern (see Fig. 7.10a) if we digitalize the readout current levels (15 μA = 3, 10 μA = 2, 5 μA = 1). The horizontal kernel (f_y) operation result for digit 3 is shown in Fig. 7.12b. Similarly, the kernel operations of f_x and f_y for digit “4” are presented in Fig. 7.12c, d, respectively. The experimental results of these kernel operations perfectly match the simulation results (if we compare Figs. 7.10 and 7.12).

By using the fabricated 12 × 12 cross-point arrays, the Prewitt horizontal and vertical kernel operations for the edge detection in the 20 × 20 pixels of black-and-white MNIST handwritten digits "3" and "4" have been successfully implemented. The proposed methodology can be applied to implementing larger kernels in deeper CNN architecture on chip. Next, we will demonstrate a more general and more challenging task: demonstration of kernel operation for the edge detection in the 50 × 50 pixels of 8-level grayscale "dog" image (see Fig. 7.13a, b) on the 12 × 12 cross-point array. Figure 7.13c, d shows the software simulation results of the horizontal kernel (f_x) and vertical kernel (f_y) operations for 8-level grayscale "dog" image. Similarly, to demonstrate the convolution kernel on resistive cross-point array, the offline training was used to prepare the well predefined Prewitt kernel. We prepared such on-state data pattern in Fig. 7.11b before the kernel operation using the optimized programming protocol to achieve the fast converge of the weight to an arbitrary target level. The current of the green cells at on-states in Fig. 7.11b is programmed to 2 μA at read voltage of 0.1 V with 20% tolerance using the half biasing technique in the array. The current of blue cells in Fig. 7.11b remains in the initial off-states, which is typically less than 10 nA at read voltage of

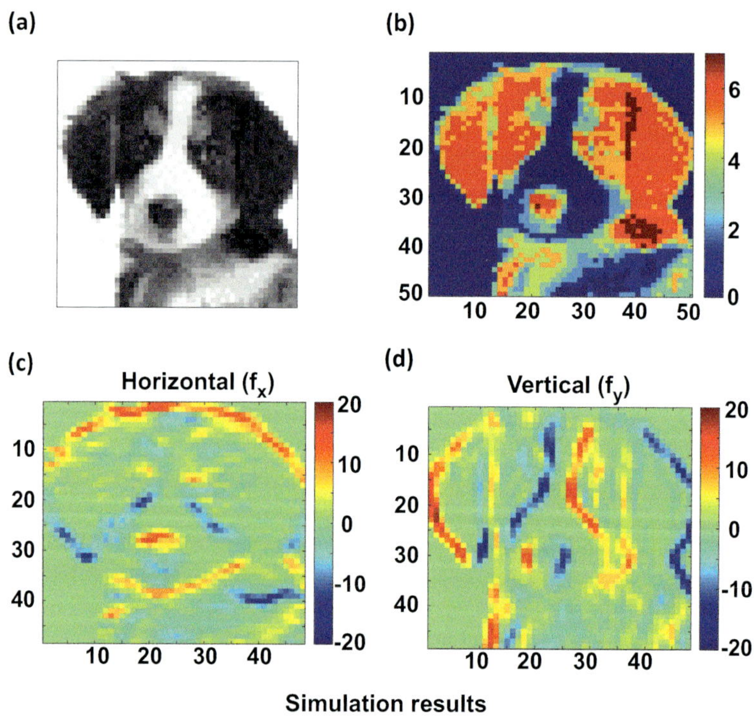

Fig. 7.13 (**a**) Original 50 × 50 pixels of 256-level grayscale "dog" image. (**b**) 50 × 50 pixels of 8-level grayscale "dog" image. Color is used to show the intensity though the image is grayscale. The software simulation results of (**c**) Prewitt horizontal kernel operation and (**d**) vertical kernel operation for grayscale "dog" image

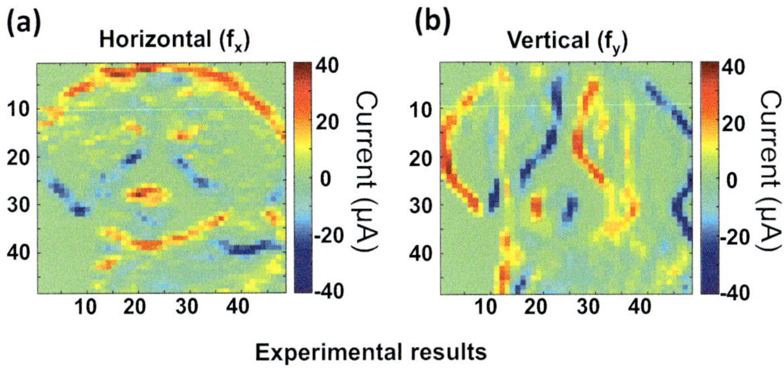

Fig. 7.14 The experimental results of (**a**) Prewitt horizontal kernel operation and (**b**) vertical kernel operation for 8-level grayscale "dog" image

0.1 V. During the iterative programming, a read scheme is used to read out the cell's current and compare with the target conductance G_{targ}. The 50 × 50 pixel of 8-level grayscale "dog" image is used to experimentally demonstrate the Prewitt horizontal (f_x) and vertical (f_y) kernel operations in parallel. Each pixel of the 3 × 3 receptive field in the input image is applied to the corresponding row r_i of the cross-point array. To reduce the number of input pulse generation, the 8-level grayscale image was encoded to 3-bit binary data "1" or "0." If the encoded binary data x_i is 1, a read voltage 0.1 is applied to that row; if the encoded binary data x_i is 0, that row is grounded. All other unused rows and columns are also grounded to avoid the sneak path when we read the column current out. Therefore, the output column current is calculated as $I_c = \sum_{i=0}^{8} x_i \cdot r_i . I_c = \sum_{i=0}^{8} x_i \cdot r_i$. Then, the output pixel value I_{out} for horizontal kernel f_x and vertical kernel f_y is calculated as $I_{\text{out},fx} = I_{c1} - I_{c2}$ and $I_{\text{out},fy} = I_{c3} - I_{c4}$, respectively. Figure 7.14a, b shows the experimental results of the horizontal kernel (f_x) and vertical kernel (f_y)'s I_{out} pattern for the input 8-level grayscale "dog" image. The value of each pixel is the current with the μA unit read at 0.1 V. Comparing to the simulation results (see Fig. 7.13c, d), the experimental results of these kernel operations match the simulation results well.

7.5 Conclusions

In this chapter, we firstly described the optimization strategies of iterative programming protocols by adjusting the pulse amplitude incremental steps, the pulse width incremental steps, and the start voltages, to tune the weights of RRAM devices. Then, we proposed a dimensional reduction of 2D kernel matrix into 1D column vector for implementing the convolution kernel on resistive cross-point array. By using the optimized programming protocol, we experimentally demonstrated the Prewitt horizontal and vertical kernel operation for the edge detection in both the 20 × 20 pixels of black-and-white MNIST handwritten digits and the 50 × 50 pixels of 8-level grayscale "dog" image on the 12 × 12 cross-point array. The experimental results of the Prewitt kernel operation perfectly match the simulation results, indicating that the proposed implementation methodology could be applied to implementing larger kernels in a deeper CNN architecture on chip.

References

1. Y. LeCun, Y. Bengio, G. Hinton, Deep learning. Nature **521**, 436–444 (2015)
2. A. Krizhevsky, I. Sutskever, G. E. Hinton, ImageNet classification with deep convolutional neural networks, *Advances in Neural Information Processing Systems (NIPS)* (2012)
3. C.S. Poon, K. Zhou, Neuromorphic silicon neurons and large-scale neural networks: challenges and opportunities. Front. Neurosci. **5**, 108 (2011)
4. J. Hasler, B. Marr, Finding a roadmap to achieve large neuromorphic hardware systems. Front. Neurosci. **7**, 118 (2013)
5. S.B. Furber, F. Galluppi, S. Temple, L.A. Plana, The SpiNNaker project. Proc. IEEE **102**, 652–665 (2014)
6. P. Merolla, J. Arthur, R. Alvarez-Icaza, A. Cassidy, J. Sawada, F. Akopyan, B. Jackson, N. Imam, C. Guo, Y. Nakamura, B. Brezzo, I. Vo, S. Esser, R. Appuswamy, B. Taba, A. Amir, M. Flickner, W. Risk, R. Manohar, D. Modha, A million spiking-neuron integrated circuit with a scalable communication network and interface. Science **345**(6197), 668–673 (2014)
7. M. Hu, H. Li, Q. Wu, G. S. Rose Y. Chen, Memristor crossbar based hardware realization of BSB recall function, in *International Joint Conference on Neural Networks (IJCNN)* (2012)
8. P. -Y. Chen, D. Kadetotad, Z. Xu, A. Mohanty, B. Lin, J. Ye, S. Vrudhula, J. Seo, Y. Cao S. Yu, Technology-design co-optimization of resistive cross-point array for accelerating learning algorithms on chip, in *IEEE/ACM Design, Automation & Test in Europe Conference (DATE)* (2015)
9. D. Kuzum, S. Yu, H.-S.P. Wong, Synaptic electronics: materials, devices and applications. Nanotechnology **24**, 382001 (2013)
10. D.S. Jeong, I. Kim, M. Ziegler, H. Kohlstedt, Towards artificial neurons and synapses: a materials point of view. RSC Adv. **3**, 3169–3183 (2013)
11. R. Waser, M. Aono, Nanoionics-based resistive switching memories. Nat. Mater. **6**, 833–840 (2007)
12. H.-S.P. Wong, H.-Y. Lee, S. Yu, Y.-S. Chen, Y. Wu, P.-S. Chen, B. Lee, F.T. Chen, M.-J. Tsai, Metal–oxide RRAM. Proc. IEEE **100**(6), 1951–1970 (2012)
13. J.J. Yang, D.B. Strukov, D.R. Stewart, Memristive devices for computing. Nat. Nanotechnol. **8**, 13–24 (2013)
14. D. Garbin, E. Vianello, O. Bichler, Q. Rafhay, C. Gamrat, G. Ghibaudo, B. DeSalvo, L. Perniola, HfO2-based OxRAM devices as synapses for convolutional neural networks. IEEE Trans Electron Devices **62**(8), 2494–2501 (2015)
15. I-T. Wang, Y.-C. Lin, Y.-F. Wang, C.-W. Hsu, T.-H. Hou, 3D synaptic architecture with ultralow sub-10 f. energy per spike for neuromorphic computation, in *IEEE International Electron Devices Meeting (IEDM)* (2014)
16. B. Gao, Y. Bi, H.Y. Chen, R. Liu, P. Huang, B. Chen, L. Liu, X. Liu, S. Yu, H.S.P. Wong, J. Kang, Ultra-low-energy three-dimensional oxide-based electronic synapses for implementation of robust high-accuracy neuromorphic computation systems. ACS Nano **8**(7), 6998–7004 (2014)
17. M. Hu, H. Li, Y. Chen, Q. Wu, G. Rose, R. Linderman, Memristor crossbar based neuromorphic computing system: a case study. IEEE Trans Neural Netw Learn Syst **25**(10), 1864–1878 (2014)

18. M. Suri, V. Parmar, G. Sassine F. Alibart, OXRAM based ELM architecture for multi-class classification applications, in *International Joint Conference on Neural Networks (IJCNN)* (2015)
19. D. Querlioz, O. Bichler, A.F. Vincent, C. Gamrat, Bioinspired programming of memory devices for implementing an inference engine. Proc. IEEE **103**(8), 1398–1416 (2015)
20. M. Prezioso, F. Merrikh-Bayat, B.D. Hoskins, G.C. Adam, K.K. Likharev, D.B. Strukov, Training and operation of an integrated neuromorphic network based on metal-oxide memristors. Nature **521**, 61–64 (2015)
21. S. Park, M. Chu, J. Kim, J. Noh, M. Jeon, B.H. Lee, H. Hwang, B. Lee, B.-G. Lee, Electronic system with memristive synapses for pattern recognition. Sci. Rep. **5**, 10123 (2015)
22. L. Gao, P.-Y. Chen, S. Yu, Programming protocol optimization for analog weight tuning in resistive memories. IEEE Electron Device Lett **36**(11), 1157–1159 (2015)
23. L. Gao, P.-Y. Chen, S. Yu, Demonstration of convolution kernel operation on resistive cross-point array. IEEE Electron Device Lett **37**(7), 870–873 (2016)
24. L. Gao, P.-Y. Chen S. Yu, Weight tuning of resistive memories and convolution kernel operation on cross-point array for neuro-inspired computing, in *IEEE International Conference on Solid-State and Integrated Circuit Technology (ICSICT)* (2016)
25. S.H. Jo, T. Chang, I. Ebong, B.B. Bhadviya, P. Mazumder, W. Lu, Nanoscale memristor device as synapse in neuromorphic systems. Nano Lett. **10**(4), 1297–1301 (2010)
26. D. Garbin, O. Bichler, E. Vianello, Q. Rafhay, C. Gamrat, L. Perniola, G. Ghibaudo B. DeSalvo, Variability-tolerant convolutional neural network for pattern recognition applications based on OxRAM synapses, in *IEEE International Electron Devices Meeting (IEDM)* (2014)
27. S. Park, H. Kim, M. Choo, J. Noh, A. Sheri, S. Jung, K. Seo, J. Park, S. Kim, W. Lee, J. Shin, D. Lee, G. Choi, J. Woo, E. Cha, J. Jang, C. Park, M. Jeon, B. Lee, B. H. Lee H. Hwang, RRAM-based synapse for neuromorphic system with pattern recognition function, in *IEEE International Electron Devices Meeting (IEDM)* (2012)
28. F. Alibart, L. Gao, B. Hoskins, D.B. Strukov, High precision tuning of state for memristive devices by adaptable variation-tolerant algorithm. Nanotechnology **23**, 075201 (2012)
29. S. Yu, Y. Wu, H.-S.P. Wong, Investigating the switching dynamics and multilevel capability of bipolar metal oxide resistive switching memory. Appl. Phys. Lett. **98**, 103514 (2011)
30. X. Guan, S. Yu, H.-S.P. Wong, A SPICE compact model of metal oxide resistive switching memory with variations. IEEE Electron Device Lett **33**(10), 1405–1407 (2012)
31. C. Xu, X. Dong, N. P. Jouppi Y. Xie, Design implications of memristor-based RRAM cross-point structures, in *IEEE/ACM Design, Automation & Test in Europe Conference (DATE)* (2011)
32. P.-Y. Chen, B. Lin, I.-T. Wang, T.-H. Hou, J. Ye, S. Vrudhula, J.-S. Seo, Y. Cao, S. Yu, Mitigating effects of non-ideal synaptic device characteristics for on-chip learning, in *IEEE/ACM International Conference on Computer-Aided Design (ICCAD)* (2015)

Chapter 8
Spiking Neural Network with 256 × 256 PCM Array

SangBum Kim

8.1 Spiking Neural Network

As it has been briefly discussed in Chap. 1, neural networks can be roughly categorized into two types. Almost all of commercial machine learning applications based on artificial neural networks use digital representation and the back-propagation learning rule. These artificial neural networks (ANNs) operate in frames (i.e., discrete-time step), and all the neurons are forced to update their states (i.e., representation) exactly once in each frame. On the other hand, spiking neural networks (SNNs) based on spike representation operate out of frame, and each neuron might or might not update its state (i.e., by generating spikes) multiple times for any given period of time. These key differences between ANN and SNN are summarized in Table 8.1.

In terms of the maturity of technology development, ANNs are far more mature than SNNs. In fact, machine learning algorithms based on ANNs already successfully tackled many problems unsolvable before by other means [1, 2] and are currently opening doors to many commercial applications at scale. However, development of machine learning algorithms based on SNNs is at the very early stage because it is not clear even how to encode information using spikes [3], and scalable and efficient learning algorithms have not been found for SNNs. Nevertheless, SNNs are sometimes referred as the third-generation neural network, while ANNs are referred as the second-generation neural network [4, 5]. This classification which insinuates that SNNs are potentially more advanced than ANNs can be supported by the fact that (1) SNNs are more biologically realistic than ANNs, (2) SNNs can perform certain computations more efficiently than ANNs [4],

S. Kim (✉)
IBM T. J. Watson Research Center, 1101 Kitchawan Road, Yorktown Heights, NY 10598, USA
e-mail: SangBum.Kim@us.ibm.com

S. Yu (ed.), *Neuro-inspired Computing Using Resistive Synaptic Devices*,
DOI 10.1007/978-3-319-54313-0_8

Table 8.1 Comparison of key characteristics of artificial neural network and spiking neural network

	Artificial neural network	Spiking neural network
I/O	Real-valued numbers	Spikes
Temporal steps	Discrete step	Continuous time
Neuron operation	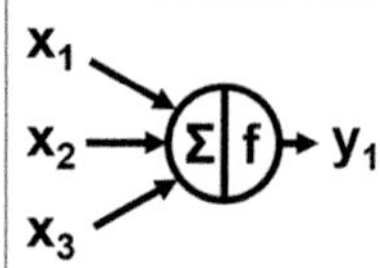	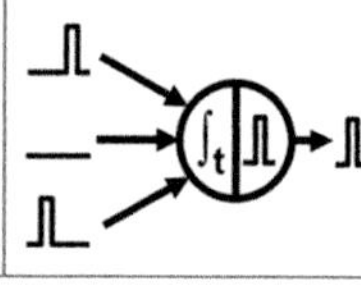

(3) many algorithms based on ANNs can be mapped onto SNNs [6] but not in vice versa, (4) SNNs can use information in time for better computation capability [3, 5], and (5) SNNs can be more power efficient because it can enable event-driven computation [7].

In addition to developing scalable and powerful algorithms for SNNs, another important topic to consider regarding SNNs is what kind of hardware is required to run algorithms based on SNNs. Majority of operations needed to run ANNs are vector-to-vector and matrix-to-vector multiplications, which can be accelerated by massively parallelizing these operations using not only neuromorphic chips based on emerging resistive synaptic devices but also more conventional technologies such as GPUs and CMOS ASICs. However, operations in SNNs can hardly be simplified as vector-to-vector or matrix-to-vector multiplications, and building a tailored neuromorphic hardware specifically for SNNs becomes even more important to fully realize potential benefits of SNNs. In this chapter, we will discuss one such neuromorphic hardware demonstration [8] which implemented 256 neurons along with 65,536 (= 64 k) synaptic phase change memory (PCM) cells based on leaky integrate-and-fire (LIF) and spike-timing-dependent plasticity (STDP) spiking neuron models.

8.2 Spiking Neuron Model

There are various spiking neuron models that can compose a SNN [9]. The PCM neuromorphic core implemented a point neuron model based on leaky integrate-and-fire (LIF) and spike-timing-dependent plasticity (STDP). This LIF and STDP point neuron model is one of the simplest neuron models, but it is sophisticated enough to successfully implement the restricted Boltzmann machine [10].

The LIF neuron integrates its input currents over time. The input current is generated when a presynaptic neuron spikes. Input currents are integrated on the capacitor, the voltage of which represents the membrane potential. The integration capacitor loses stored charges due to the leakage applied to it until it reaches the resting potential. The amount of the input current is linearly proportional only to the synaptic weight (current-based synapse) or to both the synaptic weight and voltage

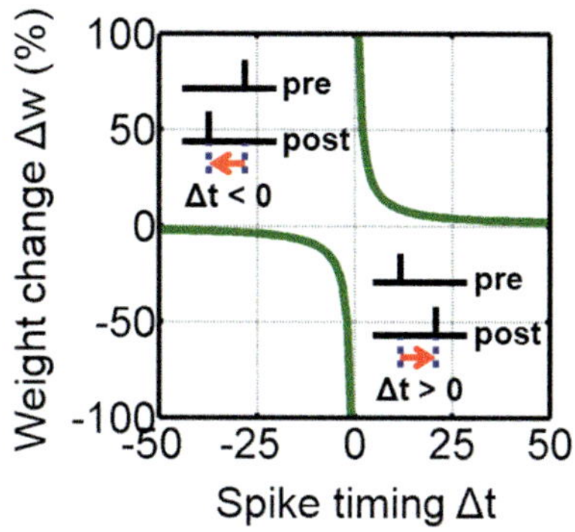

Fig. 8.1 Synaptic weight change as a function of relative timing of pre- and postsynaptic spikes ($\Delta t = t_{\mathrm{post}} - t_{\mathrm{pre}}$)

across the synapse (conductance-based synapse) [11]. When membrane potential increases over the threshold voltage, the neuron generates spikes (i.e., fires), its membrane potential goes back to the resting potential, and the neuron goes into the refractory period, during which the membrane potential is held at the resting potential not allowing the neuron to fire. After the refractory period, the neuron resumes integrating its input currents.

Synaptic weight plays an important role in determining how the LIF neuron behaves. Then, how these synaptic weights are updated? In fact, effectiveness of the synaptic weight update rule is a key factor which will determine the performance and usefulness of neural networks. Various types of weight update rules are being explored for SNNs, and spike-timing-dependent plasticity (STDP) is one of the most frequently used update rules. STDP originates its name because the amount of synaptic weight update depends on timing of relevant spikes. In an asymmetric form of STDP shown in Fig. 8.1, the synaptic weight increases if postsynaptic neuron fires after presynaptic neuron fires and decreases if presynaptic neuron fires after postsynaptic neuron fires. STDP demonstrates how SNNs might use timing information for computation. Since its discovery in biological systems [12], various efforts have been made to understand STDP as a learning algorithm for practical systems because synaptic plasticity, the change of synaptic weight, is believed to play a key role in how learning and memory are achieved in the brain [13]. Efficient neuromorphic hardware can facilitate such study.

Without using nonvolatile memory (NVM) as a synaptic device storing synaptic weights, various CMOS-based implementations of SNN have been demonstrated. Depending on the level of details in the implemented synapse model, the size of synapse circuit varies significantly. For example, a synaptic weight itself is stored in various analog and digital devices such as charge-storing capacitors [14, 15], 4-bit SRAM [16], 8-transistor transposable SRAM [17], 1-bit SRAM [18], and off-die DRAM [19]. To implement STDP and LIF with these types of synaptic devices, special circuits are often implemented in each synapse circuit instead of implementing them in the neuron circuit. This can result in larger overall array size because typical neural networks have significantly more number of synapse circuits than neuron circuits in an array.

Resistive-type NVM has shown a potential in reducing the synapse circuit size compared to CMOS-based implementations. The key idea is to implement STDP by designing programming pulses tailored for each NVM device and move STDP learning circuits from the synapse circuit to the neuron circuit. In the next section, we will introduce how this has been achieved with two-transistor one-phase change memory resistor (2T-1R) unit cell [8].

8.3 2T-1R Unit Cell with Phase Change Memory (PCM) Synaptic Cell

We utilized a phase change memory (PCM) cell as a synaptic device storing synaptic weights. PCM is one of the most mature emerging memory devices [20–22] as evidenced by high-density prototype chips [23, 24] as well as commercial products. PCM satisfies many requirements of synaptic devices including nonvolatility, large on/off ratio, partial programmability, and scalability. Programming current of PCM cells can be relatively larger than other types of resistive memory devices such as resistive random-access memory (RRAM) and spin-torque transfer magnetic RAM (STT-MRAM). However, as a synaptic device, a reasonable trade-off can be made to reduce the programming current while increasing resistance of low-resistance state (LRS). This trade-off is not desirable for conventional memory applications such as storage-class memory [25] because increased LRS resistance slows down the read speed. In synaptic applications, massive parallel operation allows slowdown in operating speed of each device [7] while achieving the number of synaptic operations per second per power similar to the conventional von Neumann architectures.

We designed a two-transistor one-PCM resistor (2T-1R) unit cell which can efficiently implement leaky integrate-and-fire (LIF) and various local synaptic weight update rule such as spike-timing-dependent plasticity (STDP). Transistors are added to each unit cell to enable each PCM cell to operate asynchronously and in parallel from others in the same array. Figure 8.2 shows the circuit diagram for a 2T-1R unit cell and the transmission electron microscopy (TEM) image. The layout can be efficiently done because PCM cell is placed on top of the shared diffusion region between two transistors and two wordlines and two bitlines run in parallel.

Figure 8.3a, b describes how the unit cell performs LIF and STDP along with peripheral circuits representing postsynaptic and presynaptic neurons. Only one memory cell is shown in Fig. 8.3, but multiple memory cells in the array form (Fig. 8.4) can operate in the same manner without interfering with each other. When a spike input comes in, the corresponding presynaptic axon driver goes into the LIF mode and activates the LIF transistor in all its downstream synapses, discharging the capacitors in the associated (postsynaptic) neuron circuits to the STDP bitline held at ground. The amount of discharging current per a PCM cell is not large enough to program the PCM cell because neither the LIF wordline pulse amplitude

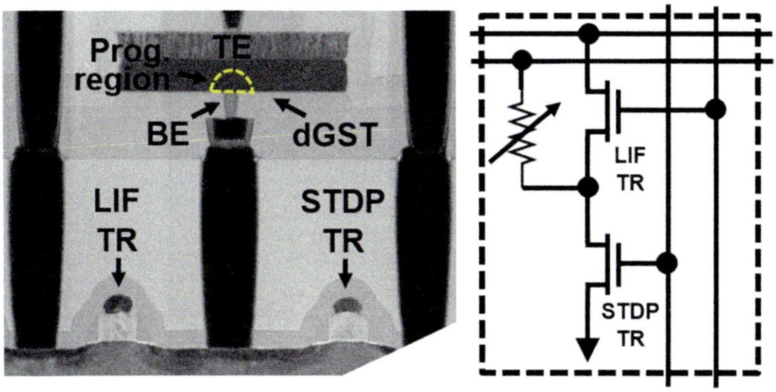

Fig. 8.2 TEM image of synaptic phase change memory cell in 2T-1R configuration. Doped GeSbTe (dGST) is used as phase change material

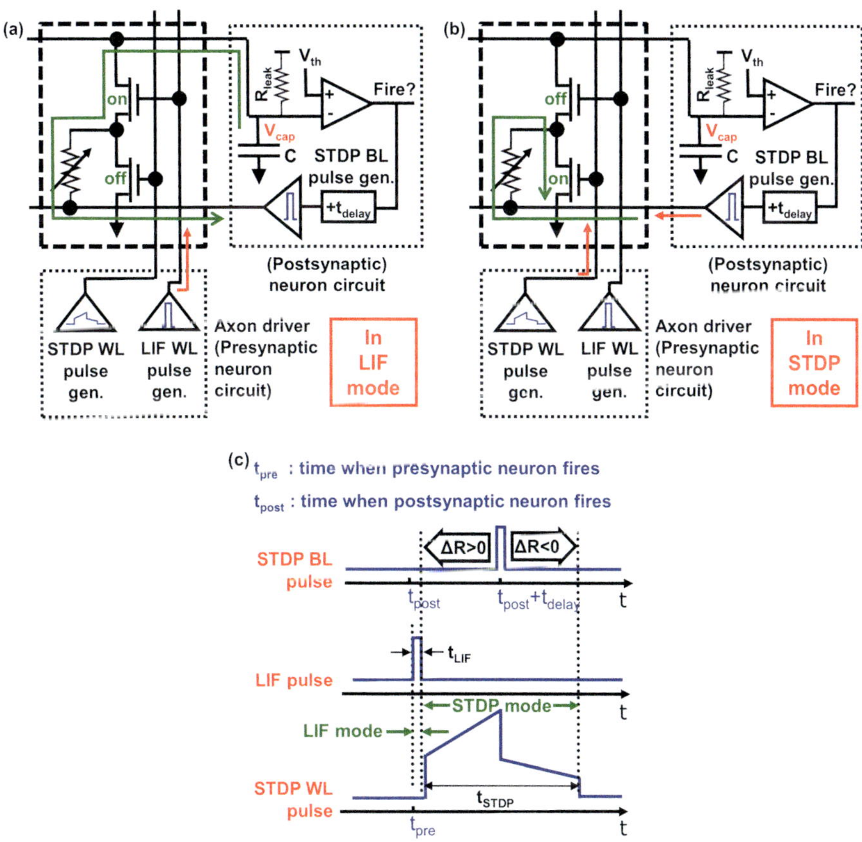

Fig. 8.3 LIF and STDP operations of 2T-1R PCM synaptic unit cell

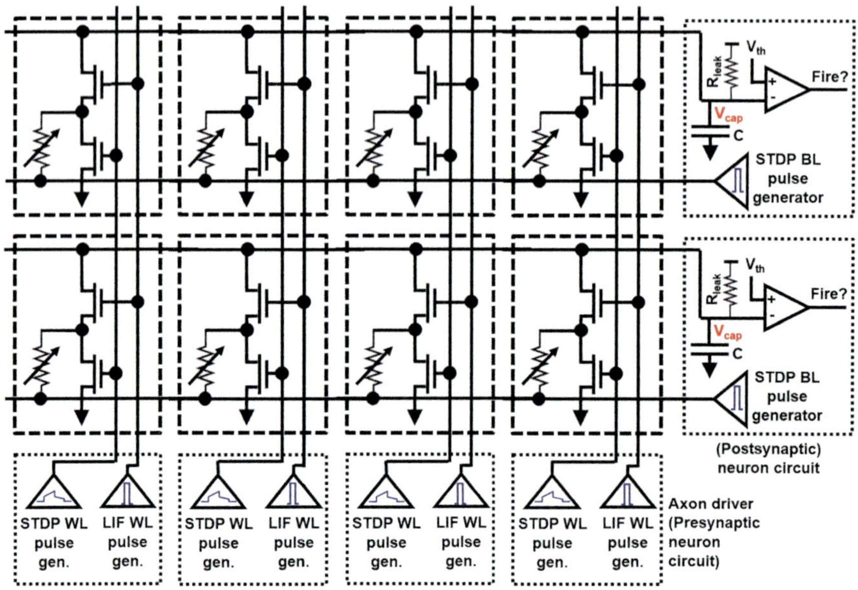

Fig. 8.4 2T-1R PCM synaptic unit cell array with axon drivers and neuron circuits on periphery

nor the membrane potential is high enough. If any membrane potential decreases below the threshold due to the capacitor discharges through LIF transistors, the corresponding neuron circuit generates an immediate spike output and subsequently generates STDP bitline pulse after predefined delay. This delay plays an important role in the STDP operation. During refractory period which starts at the moment when neuron circuit fires, the membrane potential is held at the resting potential so that any incoming LIF discharging pulses can be ignored. Refractory period plays an important role by setting maximum firing rate. In addition, all the coincidental collisions between LIF wordline pulse and STDP bitline pulse are invalidated by the refractory period because STDP bitline pulse delay is shorter than the refractory period which ensures all the collisions to occur during the refractory period.

After the axon driver switches off the LIF transistor, it continues to activate the STDP transistor with a slowly varying waveform. If the neuron circuit generates STDP bitline pulse, large amount of programming current will flow through the PCM cell and change its resistance. The polarity and magnitude of the resistance change will be controlled by the voltage on the gate of the STDP transistor which is in the saturation mode. Therefore, the relative timing of STDP bitline pulse and STDP wordline pulse will determine the polarity and magnitude of the resistance change as shown in Fig. 8.3c.

Slowly varying waveform on STDP wordline can be tailored to produce the target STDP behavior. Assuming that the STDP bitline pulse is generated at t_{post}, STDP wordline waveform can be formulated as $f^{-1}(\mathrm{h}(t - t_{\mathrm{pre}}))$ where $f(V_{\mathrm{WL}})$ is a function which describes the amount of conductance change of the PCM cell for the given STDP BL pulse width and amplitude as a function of STDP WL pulse amplitude, V_{WL}, and $h(\Delta t)$ is a function which describes the target STDP behavior as a function of $\Delta t = t_{post} - t_{pre}$. t_{post} and t_{pre} are time when post- and presynaptic neuron fired, respectively. Finally, both STDP wordline waveform and STDP bitline pulse are delayed to ensure that STDP wordline waveform is generated only after t_{pre}. Without the delay, axon drivers will need to somehow make a forecast for incoming input spikes and start generating STDP wordline pulses in advance.

The 2T-1R synaptic unit cell uses a different path for LIF and STDP. This leads to several design benefits. First of all, pulse width for LIF wordline pulse and STDP wordline pulse can be independently controlled, while those in one transistor and one resistor (1T-1R) have to be the same. By shortening the LIF wordline pulse width, a much smaller capacitor can be used as an integration capacitor for given membrane voltage. At the same time, longer STDP wordline pulse width enables the core on-chip learning function to work with sparse input spikes. Typical pulse width parameters which are reconfigurable on chip are shown in Table 8.2 where typical STDP wordline pulse width is 1000× longer than LIF wordline pulse width. Secondly, additional elements providing computational capability can be added to LIF bitlines without interrupting STDP bitlines. For example, by adding RC delay components on LIF bitlines, one can implement the temporal sequence discrimination.

Figure 8.5 summarizes the PCM cell operation mode which is determined by the operation modes of the corresponding axon driver and neuron circuit. In addition, operation of each PCM cell does not interfere with any other PCM cells in the same

Table 8.2 On-chip reconfigurable timing parameters

LIF WL pulse width	30–240 ns
STDP WL pulse width	27–290 μs
τ_{leak} ($\approx R_{leak} \cdot C$)	30–324 μs
STDP BL pulse width	10–162 ns
STDP BL pulse delay	20–164 μs
Refractory time	10–784 μs

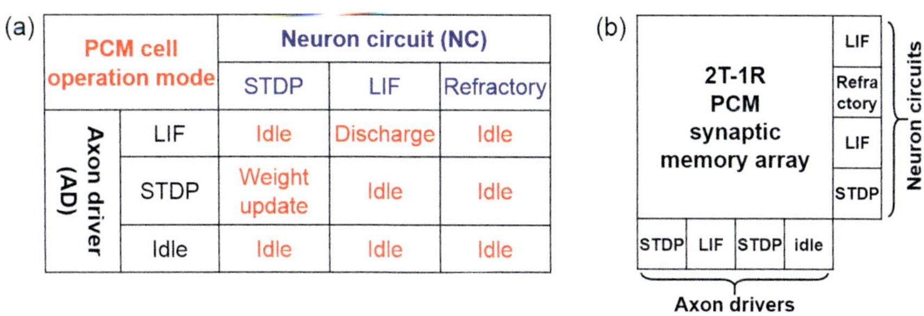

Fig. 8.5 (**a**) PCM cell operation mode matrix; (**b**) axon drivers, neuron circuits, and 2T-1R PCM synaptic cells can operate in asynchronous and parallel fashion

Fig. 8.6 Die micrograph of the core with 64 k-cell PCM array with 256 axon drivers and 256 neuron circuits

array even when they are in different modes of operation or at different stage in the same operation. Therefore, axon drivers and neuron circuits can operate in fully asynchronous and parallel fashion. This capability can enable an efficient design which saves clock power and avoids complicated timing design.

Figure 8.6 shows the die micrograph of the core with 64 k 2T-1R PCM cells with 256 axon drivers and 256 neuron circuits fabricated at IBM 90 nm node.

8.4 Experimental Results

Figure 8.7 shows experimental results where multiple PCM cells as many as 64 bits are successfully programmed in parallel between high-resistance RESET state and low-resistance SET state. Multiple PCM cells were programmed by generating on-chip STDP pulses on multiple wordlines and bitlines. Programming conditions were selected by externally triggering not only axon drivers but also neuron circuits with controlled timing difference, and on-chip axon drivers and neuron circuits generated STDP wordline pulses and STDP bitline pulses accordingly. By triggering axon drivers to fire before triggering neuron circuits to fire, PCM cells were programmed to SET states. On the other hand, by triggering axon drivers after neuron circuits, PCM cells were programmed to RESET state.

To confirm various aspects of LIF operation, experiment shown in Fig. 8.8a was conducted with the fabricated neuromorphic core. One axon driver was periodically

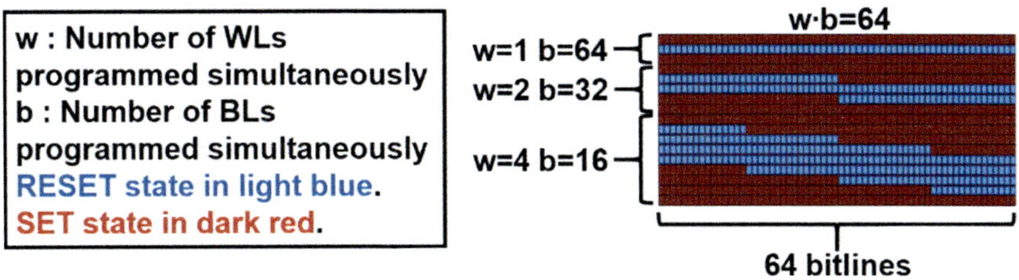

Fig. 8.7 Programming results of multiple bits programmed simultaneously

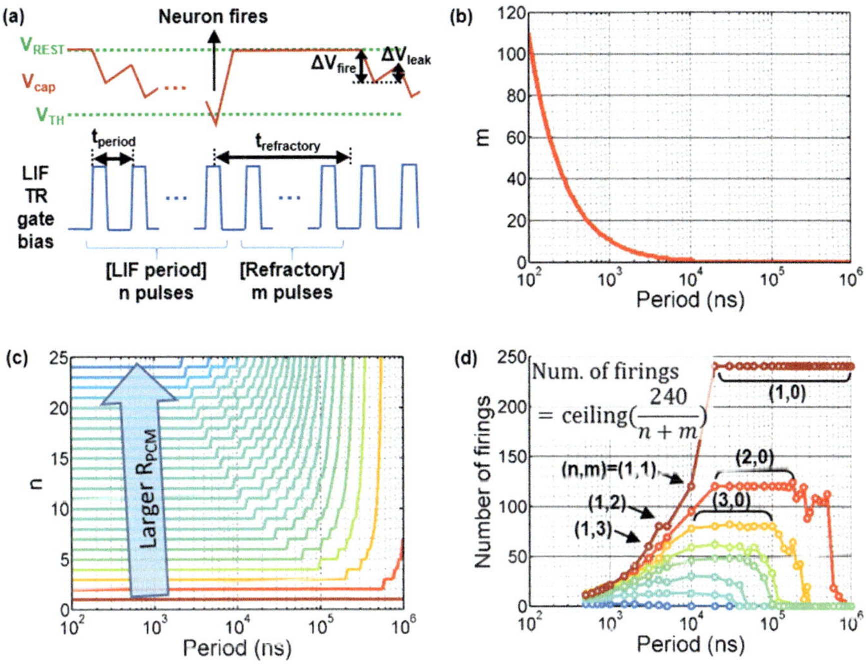

Fig. 8.8 Experimental verification of on-chip LIF operation. (**a**) Description of the experiment. (**b**) Analytical model for m, number of pulses ignored during refractory time. (**c**) Analytical model for n, number of pulses required to trigger a spike. (**d**) Hardware measurement data

triggered for 240 times. The number of pulses required to bring down the membrane potential (V_{cap}) below the threshold voltage (V_{TH}), n, can be approximated by the following equation:

$$n = \text{ceiling}\left(\frac{V_{REST} - V_{TH}}{\Delta V_{fire} - \Delta V_{leak}}\right)$$

$$\Delta V_{leak} = t_{period} \cdot k_{leak}$$

$$\Delta V_{\text{fire}} \propto t_{\text{LIF}}/R_{\text{PCM}}$$

where V_{REST} is the resting potential, ΔV_{fire} is the change in membrane potential from a single LIF pulse, ΔV_{leak} is the change in membrane potential due to the leakage during one period (t_{period}), k_{leak} is the leakage term, t_{LIF} is the LIF wordline pulse width, and R_{PCM} is the PCM cell resistance. Figure 8.8c shows the dependence of n on R_{PCM} and t_{period}. Longer t_{period} strengthens the recovery of membrane potential during one period and leads to larger n value. Larger R_{PCM} decreases ΔV_{fire} and leads to larger n value.

After neuron circuit fires, it goes into the refractory period ($t_{refractory}$). During the refractory period, the membrane potential is held at the resting potential, and incoming LIF pulses have no effect on the membrane potential. The total number of LIF pulses that are ignored during the refractory time, m, is

$$m = floor\left(\frac{t_{refractory}}{t_{period}}\right)$$

Figure 8.8b shows the dependence of m on t_{period} where longer period leads to smaller m. The number of output spikes per 240 input spikes will be given by $240/(n + m)$, and Fig. 8.8d shows the number of output spikes which is experimentally measured with various PCM cell resistance levels and various t_{period}. In terms of t_{period}, m and n have opposite dependence on t_{period}. Experimental results on Fig. 8.8d clearly show such trends where the number of output spikes is smaller at shorter and longer t_{period}. Smaller number of output spikes at shorter t_{period} is due to the larger m, which confirms that the refractory period is successfully implemented on chip. On the other hand, smaller number of output spikes at longer t_{period} is due to the larger n which is attributed to the leakage current and confirms the implementation of leakage on chip. Larger R_{PCM} decreases n, which can be seen by comparing peaks at the intermediate t_{period}, which confirms that the synaptic discharging current is indeed controlled by resistance of the PCM cell.

8.5 Conclusion

We discussed hardware implementation of a neuromorphic core with 64 k-cell PCM array and 256 on-chip neuron circuits for spiking neural network based on leaky integrate-and-fire (LIF) neuron model and spike-timing-dependent plasticity (STDP) weight update rule. Parallel programmability of low-power PCM cells along with 2T-1R design enables each synaptic PCM cell to operate in LIF, STDP, or idle mode asynchronously and independently from other synaptic PCM cells in the same array. Asynchronous operation of the cores guarantees that multiple cores can easily form a large-scale neuromorphic system once neuron circuits are either physically or virtually wired to the corresponding axon drivers. The complexity of wiring multiple cores remains to be addressed and necessitates novel wiring schemes. This successful hardware demonstration suggests that the array size (number of neuron circuits, axon drivers, and synaptic memory devices) could potentially be increased to even match the biological scale, where each neuron may be associated with thousands of synapses. The demonstrated neuromorphic core shows the feasibility of power- and area-efficient hardware implementation for machine learning algorithms based on spiking neural network.

References

1. Y. LeCun, Y. Bengio, G. Hinton, Deep learning. Nature **521**(7553), 436–444 (2015)
2. D. Silver, A. Huang, C.J. Maddison, A. Guez, L. Sifre, G. van den Driessche, J. Schrittwieser, I. Antonoglou, V. Panneershelvam, M. Lanctot, S. Dieleman, D. Grewe, J. Nham, N. Kalchbrenner, I. Sutskever, T. Lillicrap, M. Leach, K. Kavukcuoglu, T. Graepel, D. Hassabis, Mastering the game of Go with deep neural networks and tree search. Nature **529**(7587), 484–489 (2016)
3. A. Grüning, S. M. Bohte, (2014), in *European Symposium on Artificial Neural Networks, Computational Intelligence and Machine Learning*. Spiking neural networks: principles and challenges, ISBN 978-287419095-7, http://www.i6doc.com/fr/book/ GCOI=28001100432440
4. W. Maass, Networks of spiking neurons: the third generation of neural network models. Neural Netw. **10**(9), 1659–1671 (1997)
5. W. Maass, To spike or not to spike: that is the question. Proc. IEEE **103**(12), 2219–2224 (2015)
6. D. Garbin, O. Bichler, E. Vianello, Q. Rafhay, C. Gamrat, L. Perniola, G. Ghibaudo, B. DeSalvo, in *IEEE International Electron Devices Meeting (IEDM)*. Variability-tolerant convolutional neural network for pattern recognition applications based on OxRAM synapses (IEEE, San Francisco, 2014)
7. P.A. Merolla, J.V. Arthur, R. Alvarez-Icaza, A.S. Cassidy, J. Sawada, F. Akopyan, B.L. Jackson, N. Imam, C. Guo, Y. Nakamura, B. Brezzo, I. Vo, S.K. Esser, R. Appuswamy, B. Taba, A. Amir, M.D. Flickner, W.P. Risk, R. Manohar, D.S. Modha, A million spiking-neuron integrated circuit with a scalable communication network and interface. Science **345**(6197), 668–673 (2014)
8. S. Kim, M. Ishii, S. Lewis, T. Perri, M. BrightSky, W. Kim, R. Jordan, G. W. Burr, N. Sosa, A. Ray, J.-P. Han, C. Miller, K. Hosokawa, C. Lam, in *IEEE International Electron Devices Meeting (IEDM)*. NVM neuromorphic core with 64 k-cell (256-by-256) phase change memory synaptic array with on-chip neuron circuits for continuous in-situ learning (IEEE, Washington, DC, 2015)
9. W. Gerstner, W.M. Kistler, *Spiking Neuron Models: Single Neurons, Populations, Plasticity* (Cambridge University Press, Cambridge/New York, 2002)
10. E. Neftci, S. Das, B. Pedroni, K. Kreutz-Delgado, G. Cauwenberghs, Event-driven contrastive divergence for spiking neuromorphic systems. Front. Neurosci. **7**, 1–14 (2014)
11. P.U. Diehl, M. Cook, Unsupervised learning of digit recognition using spike-timing-dependent plasticity. Front. Comput. Neurosci. **9**, 99 (2015)
12. G.Q. Bi, M.M. Poo, Synaptic modifications in cultured hippocampal neurons: dependence on spike timing, synaptic strength, and postsynaptic cell type. J. Neurosci. **18**(24), 10464–10472 (1998)
13. A. Morrison, M. Diesmann, W. Gerstner, Phenomenological models of synaptic plasticity based on spike timing. Biol. Cybern. **98**(6), 459–478 (2008)
14. G. Indiveri, E. Chicca, R. Douglas, A VLSI array of low-power spiking neurons and bistable synapses with spike-timing dependent plasticity. IEEE Trans. Neural Netw. **17**(1), 211–221 (2006)
15. G. Indiveri, F. Corradi, N. Qiao, in *IEEE International Electron Devices Meeting (IEDM)*. Neuromorphic architectures for spiking deep neural networks (IEEE, Washington, DC, 2015)

16. K. Meier, in *IEEE International Electron Devices Meeting (IEDM)*. A mixed-signal universal neuromorphic computing system (IEEE, Washington, DC, 2015)
17. J. Seo, B. Brezzo, Y. Liu, B. D. Parker, S. K. Esser, R. K. Montoye, B. Rajendran, J. A. Tierno, L. Chang, D. S. Modha, D. J. Friedman, in *IEEE Custom Integrated Circuits Conference (CICC)*. A 45 nm CMOS neuromorphic chip with a scalable architecture for learning in networks of spiking neurons (IEEE, San Jose, 2011)
18. F. Akopyan, J. Sawada, A. Cassidy, R. Alvarez-Icaza, J. Arthur, P. Merolla, N. Imam, Y. Nakamura, P. Datta, G. Nam, B. Taba, M. Beakes, B. Brezzo, J.B. Kuang, R. Manohar, W.P. Risk, B. Jackson, D.S. Modha, TrueNorth: design and tool flow of a 65 mW 1 million neuron programmable neurosynaptic chip. IEEE Trans. Comp. Aided. Des. Integr. Circuits Syst. **34**(10), 1537–1557 (2015)
19. G. Orchard, X. Lagorce, C. Posch, S. B. Furber, R. Benosman, F. Galluppi, in *IEEE International Symposium on Circuits and Systems (ISCAS)*. Real-time event-driven spiking neural network object recognition on the SpiNNaker platform (IEEE, Lisbon 2015)
20. H.-S.P. Wong, S. Raoux, S. Kim, J. Liang, J.P. Reifenberg, B. Rajendran, M. Asheghi, K.E. Goodson, Phase change memory. Proc. IEEE **98**(12), 2201–2227 (2010)
21. G.W. Burr, M.J. Breitwisch, M. Franceschini, D. Garetto, K. Gopalakrishnan, B. Jackson, B. Kurdi, C. Lam, L.A. Lastras, A. Padilla, B. Rajendran, S. Raoux, R.S. Shenoy, Phase change memory technology. J. Vac. Sci. Technol. B **28**(2), 223 (2010)
22. G.W. Burr, M.J. BrightSky, A. Sebastian, H.-Y. Cheng, J.-Y. Wu, S. Kim, N.E. Sosa, N. Papandreou, H.-L. Lung, H. Pozidis, E. Eleftheriou, C.H. Lam, Recent progress in phase-change memory technology. IEEE J. Emerging. Sel. Top. Circuits Syst. **6**(2), 146–162 (2016)
23. Y. Choi, I. Song, M.-H. Park, H. Chung, S. Chang, B. Cho, J. Kim, Y. Oh, D. Kwon, J. Sunwoo, J. Shin, Y. Rho, C. Lee, M. G. Kang, J. Lee, Y. Kwon, S. Kim, J. Kim, Y.-J. Lee, Q. Wang, S. Cha, S. Ahn, H. Horii, J. Lee, K. Kim, H. Joo, K. Lee, Y.-T. Lee, J. Yoo, G. Jeong, in *IEEE International Solid-State Circuits Conference (ISSCC)*. A 20 nm 1.8 V 8Gb PRAM with 40 MB/s program bandwidth (IEEE, San Francisco, 2012), pp. 46–48
24. C. Villa, D. Mills, G. Barkley, H. Giduturi, S. Schippers, D. Vimercati, in *IEEE International Solid-State Circuits Conference (ISSCC)*. A 45 nm 1Gb 1.8 V phase-change memory (IEEE, San Francisco, 2010), pp. 270–271
25. G.W. Burr, B.N. Kurdi, J.C. Scott, C.H. Lam, K. Gopalakrishnan, R.S. Shenoy, Overview of candidate device technologies for storage-class memory. IBM J. Res. Dev. **52**(4.5), 449–464 (2008)

Part III
Circuit, Architecture and Algorithm-Level Design of Resistive Synaptic Devices Based Neuromorphic System

Chapter 9
Peripheral Circuit Design Considerations of Neuro-inspired Architectures

Deepak Kadetotad, Pai-Yu Chen, Yu Cao, Shimeng Yu, and Jae-sun Seo

9.1 Neuron Circuit Design Considerations

Emerging resistive nonvolatile memory (NVM) bit cells could provide much higher density and lower leakage compared to the SRAM bit cells in the most scaled CMOS technology [1–5]. This becomes a promising technology in neuro-inspired computing, where memory typically consumes a large portion of area and power of the overall neural network [6]. Since a full memory array consists of not only a two-dimensional array of bit cells but also peripheral circuits that include word line (WL) drivers, sense amplifiers, and additional nonlinear circuitries, it is essential to enhance and optimize the design of peripheral circuits for the overall system density and power efficiency.

In terms of density, array efficiency, defined as (bit cell area) × (number of bits)/(total area), is an important measure that captures the relative area of peripheral circuits. Ideally, the array efficiency should be a high value (e.g., >80%) such that the bit cells, which are responsible for the actual storage, consume most of the memory area. In resistive synaptic devices, the bit cell area reduces substantially compared to the SRAM bit cell area, but the peripheral circuits of the resistive NVM arrays might not scale at a similar ratio, which then makes the NVM array area possibly be more dominated by the peripheral circuits, worsening the array efficiency of resistive NVM arrays.

In terms of power, dynamic and static power are both important. Dynamic power consumption occurs during read and write operations, and typically write operations (resistive synapse weight update) incur higher power due to the large voltage requirement. For sparse workloads, since the resistive synaptic devices will

D. Kadetotad • P.-Y. Chen • Y. Cao • S. Yu • J.-s. Seo (✉)
School of Electrical, Computer, and Energy Engineering, Arizona State University,
781 E Terrace Rd, ISTB4 room 591, Tempe, AZ 85287, USA
e-mail: jaesun.seo@asu.edu

S. Yu (ed.), *Neuro-inspired Computing Using Resistive Synaptic Devices*,
DOI 10.1007/978-3-319-54313-0_9

consume very low leakage power, static power consumption of the peripheral circuits, e.g., current mode circuits, bias currents, etc., will become crucial for the overall system power.

Figure 9.1 shows high-level block diagrams of representative resistive array architectures [7]. Figure 9.1a illustrates the 1-transistor-1-resistor (1T1R) memory array, where resistive bit cells are read or written on a row-by-row basis. Instead of a transistor, a selector device, either an exponential I–V selector [8] or a threshold I–V selector [9], can be used, which can be vertically stacked with resistive devices. In Fig. 9.1b, a pseudo-crossbar array is shown, where the 1T1R array is employed with a rotated bit line (BL) structure (perpendicular to select line (SL)), to enable parallel operation of multiple rows. In a more aggressive design, Fig. 9.1c illustrates a true crossbar without any selector transistors, toward a fully parallel read and write operation. If the read or write operation is not fully parallel, sneak path issues can exist.

In conventional memory arrays, peripheral circuits perform read and write operations. When resistive NVM arrays are applied to neuro-inspired computing, the read operation of peripheral circuits essentially becomes the weighted sum operation, and the write operation is mapped to synapse weight update operation. These are essentially functionalities of neurons in neural networks; thus without losing generality, we will interchangeably use neuron circuits as the peripheral circuits of resistive NVM arrays. Let us take the true crossbar array (Fig. 9.1c) as an example. For the read operation, with BLs driven as virtual ground, a relatively small voltage is applied on the WLs. This read voltage is divided by the resistance of the cross-point resistive device and results in BL current. In other words, the read voltage is multiplied by the conductance of the resistive device, which can be mapped to the multiply-and-accumulate or inner-product operations in neural networks. When multiple cross-point devices in multiple rows are connected to the same BL, the currents from multiple rows are naturally summed. The peripheral circuit will sense this output current, convert to a digital value, compare against a certain threshold, or perform a nonlinear function (e.g., sigmoid, tanh, etc.). On the

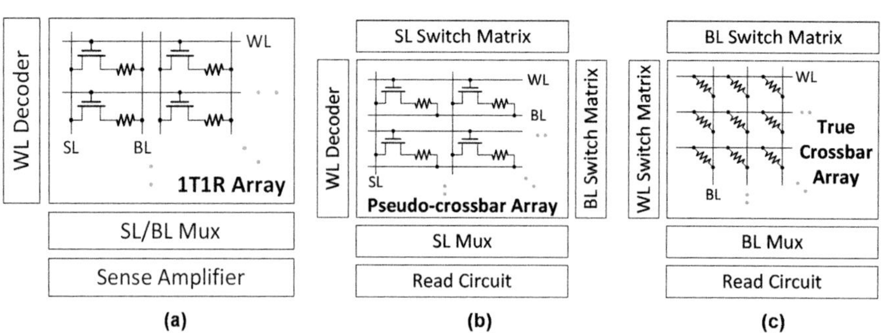

Fig. 9.1 Representative resistive NVM arrays with peripheral circuits are shown. (**a**) 1T1R array with row-by-row operation. (**b**) Pseudo-crossbar array implemented by rotating BL in 1T1R memory array. (**c**) True crossbar array without selector transistors for array-level parallelism

other hand, the write operation, or the synapse weight update, involves driving the WL and/or BL for a certain duration of time with a sufficiently large voltage that can change the value of the resistive devices. The weight update in array structures such as Fig. 9.1a will occur in a row-by-row fashion, while array structures such as Fig. 9.1b or Fig. 9.1c can perform the weight update in a fully parallel manner in the array level.

One important design parameter in neuro-inspired computing using resistive devices, which also affects peripheral circuit design, is the weight representation of resistive synaptic devices. As in conventional memory bit cells, each resistive synaptic device can be programmed into a high (maximum) resistance state (HRS) or a low (minimum) resistance state (LRS), such that each resistive device holds a binary value. In this binary scheme, to represent a n-bit weight, n resistive devices will be employed. On the other hand, since intermediate resistance values between HRS and LRS exist, each resistive device can be programmed accordingly and represent an analog or multi-bit value. While this analog scheme will result in higher density than the binary scheme, nonlinear resistance increase/decrease and temporal/spatial variability of resistive devices make it challenging to reliably program and retain an accurate analog resistance value in a large array of resistive devices, as discussed in Chaps. 11 and 12. It can be seen that these two schemes present a trade-off between area/power and accuracy.

Depending on the level of parallelism of resistive array operation (e.g., row-by-row, multiple rows, etc.) and the level of representation of resistive synaptic devices (e.g., binary, analog, etc.), the peripheral circuits need to be designed accordingly. For example, while a conventional WL decoder output is a one-hot signal, the peripheral circuit design in Fig. 9.1b, c requires a switch matrix for SL/BL/WL to turn on multiple rows/columns in parallel. In addition, if a resistive device only holds a binary value in the row-by-row operation scheme, a simpler sense amplifier with a single reference could be used, but if a resistive device holds an analog value or the current from multiple rows is accumulated, a certain type of analog-to-digital conversion becomes necessary.

To implement these read and write functionalities, there have been approaches to (1) use the CMOS circuits for the neuron design, and interface with the resistive NVM array, or (2) use NVM or other non-CMOS devices for the neuron nodes toward a more compact design. CMOS peripheral circuits have been well established for on-chip SRAM arrays, and such well-known sense amplifier or analog-to-digital converter (ADC) designs could be adopted to interface with resistive memory arrays. During fabrication, the CMOS neuron circuits and the resistive memory arrays could be integrated into the same chip. However, today's CMOS integrate-and-fire neuron typically requires tens of transistors. Such complex CMOS neuron causes the column pitch matching problem: multiple columns have to share one neuron, thereby reducing the parallelism as the time multiplexing is needed to sequentially read out all the weighted sum from the array. To that end, researchers have been also exploring dense non-CMOS devices to implement the neuron functionality.

In this chapter, we will discuss recent neuron circuit designs that have been used for read and write operations of resistive NVM arrays, covering both CMOS and non-CMOS implementations. Design parameters and technical aspects will be discussed, including performance, power, area, reliability, variability, and design challenges.

9.2 CMOS Peripheral Circuits for Read Operation

CMOS peripheral circuits for read operation involve the WL driver circuits, the neuron circuits that convert the analog array current into binary or multi-bit digital values, and a nonlinear activation function (e.g., step function, rectified linear unit (ReLU), sigmoid, tanh, etc.).

WL Driver Assuming that multi-bit digital input signals are given to the input of resistive NVM array at different WLs, several methods exist to encode these input signals for read operation. First, an input value can be encoded into a voltage level, which will directly drive the WL, resulting in different array output current for a given conductance of resistive devices. Alternatively, while keeping the same voltage level, different input values can be encoded by the number of pulses applied at the WLs.

While the voltage-mode scheme will enable read operation for a given multi-bit input value in a single cycle, generating such analog voltage values, either through digital-to-analog converters or DC-DC converters, can be expensive in terms of transistor count, area, and power. Furthermore, the existence of noise and I–V nonlinearity of resistive devices hamper the voltage-mode scheme from obtaining high accuracy in the weighted sum operation. On the other hand, the pulse mode scheme will result in longer latency due to the fact that, depending on the bit resolution of input signals, multiple clocks will be required to process the entire input signals. However, the pulse mode scheme is not affected by the I–V nonlinearity of resistive devices because a pulse with the same voltage is always applied.

Sense Amplifiers Conventional sense amplifiers (SA) have been employed in a number of resistive arrays that have been targeted for memory applications [10–12]. In resistive arrays, typically the BLs are single ended; thus a reference BL that can be shared with all BLs is employed for differential sensing to avoid common-mode noise. In [13], a body-drain-driven sense amplifier was proposed to suppress the voltage headroom at lower voltages (e.g., 0.5 V) and provide resistance-aware dynamic BL bias voltage, as shown in Fig. 9.2a.

The authors of [14] employed a series resistor at the end of the column as the sensing circuit to convert the output current into voltage, as illustrated in Fig. 9.2b. The authors also explored the grounding (GN) or floating (FT) scheme for the unselected WLs and BLs. For a single-row read, the converted output voltage can be compared against a reference voltage. On the other hand, for read

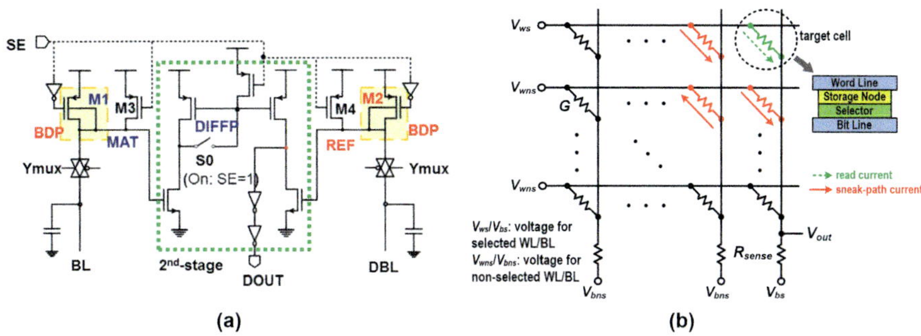

Fig. 9.2 (**a**) Low-voltage body-drain-driven sense amplifier [13]. (**b**) Series resistor-based sensing of column current [14]

operations corresponding to multiple rows, the voltage across the resistor can be conveyed to a voltage-mode ADC, such as the design implemented in [15], to finally output multi-bit digital values.

Analog-Current-to-Digital Converters While different types of voltage-mode ADCs can be employed for the read operation of resistive arrays in neuro-inspired computing, it can be more efficient to directly convert the analog current from the resistive arrays into multi-bit digital values.

Figure 9.3 shows the peripheral neuron circuit proposed in [16] for read operation, which is essentially a current-to-digital converter, where the output current at each resistive NVM array column is sensed and converted into digital pulses (spikes). In principle, this output response is similar to that of a biological neuron model, namely, integrate-and-fire (IF) neuron [17]. Starting from a reset voltage, the output current is integrated on the finite capacitance of each RRAM column; when the voltage charges up above a certain threshold, the output switches and the capacitance is discharged back to the reset voltage. The read property of an RRAM array further poses a constraint that the reset voltage and the threshold voltage should be very close to each other; otherwise the output current does not represent the correct weighted sum. In a 65 nm design, the reset voltage and the threshold voltage are 500 and 530 mV, respectively. To meet this constraint, an asynchronous comparator with high sensitivity to small changes in voltages was required, and a differential Schmitt trigger was employed to generate output spikes proportional to the input current. The differential pair allows for precise calibration of the voltage at which the reset pulse should be activated. The hysteresis loop is created by the cross-linked tail to the differential pair.

To analyze the array-level functionality of the peripheral read circuits, the differential read circuits were simulated with an RRAM array with 100 rows and 100 columns, where the RRAM device model was obtained from experimental data in [18] and calibrated with data from [19]. An array was formed with Verilog-A models

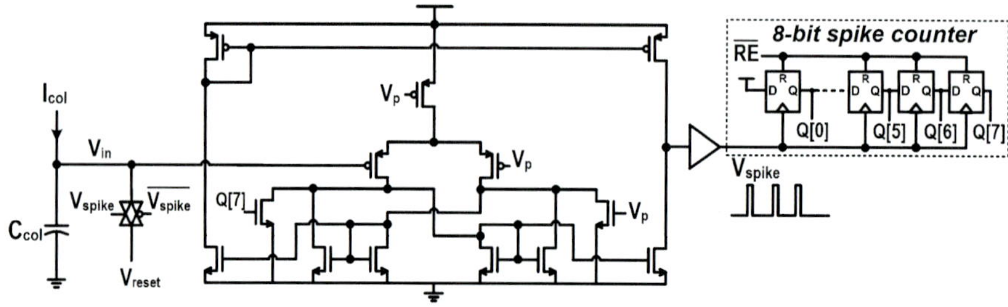

Fig. 9.3 CMOS neuron circuit for read operation, which converts analog weighted sum current into multi-bit digital values [16]

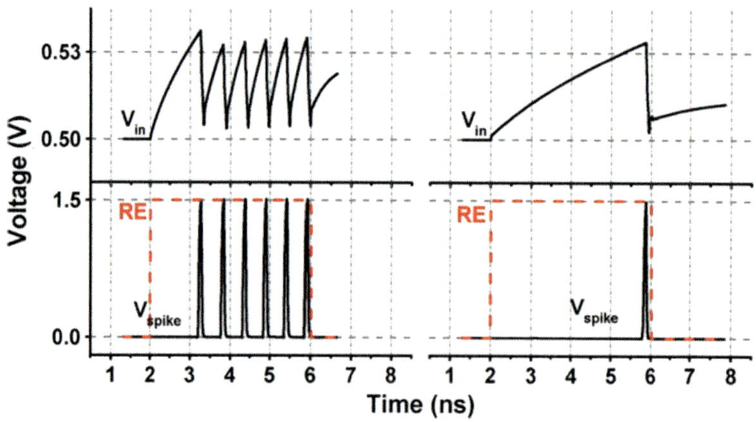

Fig. 9.4 Peripheral read circuit simulation with a 100 × 100 array for two input currents: (*left*) $I_{col} = 5.75$ μA and (*right*) $I_{col} = 1$ μA; the corresponding number of spikes is 6 and 1

of the RRAM devices together with wire parasitics. Figure 9.4 shows the simulation results of array-level read operations, which shows linear increase in the number of spikes for the increase in the input current from the array.

9.3 CMOS Neuron Circuits for Write Operation

In neuro-inspired computing, the synapse weight update operation corresponds to a write operation of the RRAM array. To change the conductance of an RRAM cell, the voltage across the cell should be a high value (e.g., Vdd of >1 V), and the RRAM devices that have Vdd/2 or less across the two terminals should only undergo negligible change on due to its strong nonlinearity in the voltage dependence.

Analog Circuits for Spike Timing-Based Update Toward learning implementation in spiking neural networks based on resistive synaptic devices and spiking neurons, analog peripheral circuits that can mimic the spike timing-dependent plasticity (STDP) behavior have been presented [20]. As shown in Fig. 9.5, the synaptic element consists of a diode connected in series with the RRAM device, and the objective is to enable small conductance changes with each neuron spike for analog resistive synapses toward learning in spiking neural networks. Whenever the presynaptic neuron or postsynaptic neuron spikes, a programmable pulse with exponential decay characteristics is generated, as shown in Fig. 9.5b. The waveform that consists of two exponential signals, mimicking the shape of the biological action potentials, is generated by the circuit in Fig. 9.5a, which consists of predriver and driver stages. The predriver stage incorporates current source I0, switches S1 and S2, and capacitors C1 and C2, which generates the double-exponential waveform. Transmission gates are used for the switches. The two time constants required in the proposed waveform are generated by the current integration on capacitors C1 and C1 + C2. The driver stage incorporating transistor M1 and switches S3–S5 drives the axons and dendrites. Transistor M1 roughly transfers the wave shape of the internal signal V_{int} to output V_{out}. Switches S4 and S5 set the peak amplitudes V_a and V_b of the two decaying shapes in the proposed waveform. During the initialization of the driver output, switch S3 disconnects the predriver stage from the driver output. The comparator and the pulse generation block generate the pulsed control signals for all the switches S1–S5 from the STDP enable signal, the start, and the output signal V_{out}. The pulse generation block is constructed using simple digital logic circuits.

The initial part of the waveform has a peak amplitude V_a of 250 mV and an exponential decay time constant of 1 μs, which is followed by a second part of the waveform that has peak amplitude V_b of −250 mV and a decay time constant of

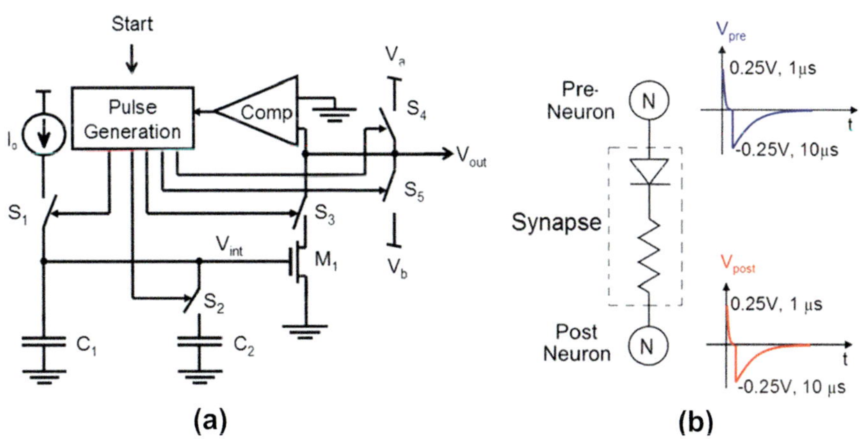

Fig. 9.5 (**a**) Peripheral circuit that generates programming pulses for resistive synapses. (**b**) When either the presynaptic neuron or postsynaptic neuron spikes, an exponentially decaying pulse is generated [20]

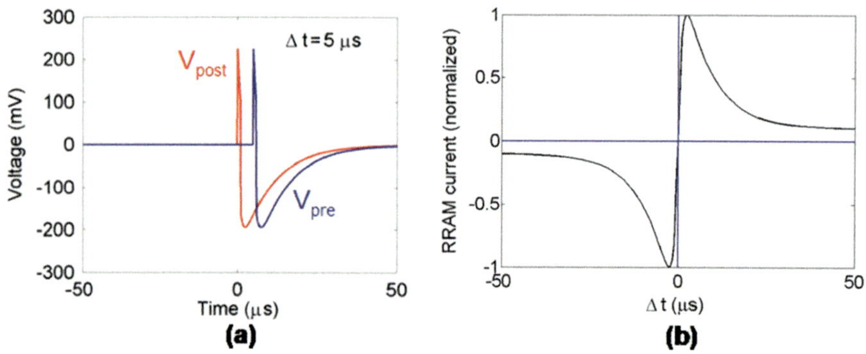

Fig. 9.6 (**a**) Overlapping pulses from presynaptic and postsynaptic neurons. (**b**) Effective current flow through the RRAM device for various overlap times as a function of spike timing difference [20]

10 μs. For positive Δt, a maximum current flows through the RRAM at the instant when the positive peak of V_{pre} coincides with the negative portion of V_{post}, and the magnitude of this current depends on the instantaneous value of V_{post} (Fig. 9.6a). The effective current flowing through the RRAM and, hence, the final state of the synapse depends on the time of the overlap between the waveforms, as shown in Fig. 9.6b. Different plasticity rules could be obtained by simple circuit modifications of the characteristics of these waveforms.

Digital Circuits for Pulse Width/Number Modulation The conductance G of resistive devices can also be digitally programmed by the overlap time between the digital output signals of local column-side neurons and row-side neurons. The authors of [16] implemented the write circuit for the row-side neuron (denoted as Z) that generates a pulse with a duty cycle proportional to the neuron value, while a spike train is generated at the column-side neuron (denoted as r) with the firing rate proportional to the neuron value while fixing the pulse width at 1 ns. Wherever the column-side and row-side neuron pulses are overlapped, it creates a full Vdd voltage across the resistive synaptic devices. Therefore, the total programming time equals to the overlap between the two neuron signals.

In the positive period, Z is 0 for a certain time proportional to Z; then it switches to Vdd/2. The overlap time between $Z = 0$ and $r = \text{Vdd}$ tunes the RRAM conductance. A similar scenario is designed for the negative period, with a symmetric polarity. Figure 9.7 shows the digital design to generate such a pulse pattern. A 16-bit shift register converts the parallel input $Z[15:0]$ into a sequential output. The time when the output is 1 is proportional to the value of Z. The output of the shift register is connected back to the first stage of itself in order to recycle the data Z. With 32 clock cycles allocated for one write period, the Z write circuit generates two identical pulses with the duty cycle proportional to the value of Z. These two identical pulses are connected to multiplexors to generate different programming voltages for the positive period and the negative period.

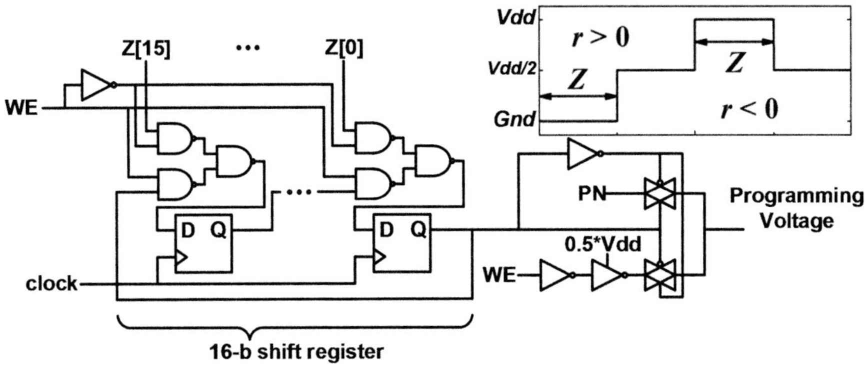

Fig. 9.7 CMOS neuron on the row-side for write operation

The train of pulses generated at r has its pulse number proportional to the value of r, where each pulse has a fixed width for a fixed RRAM programming period. The pulses are evenly distributed across the write period in order to minimize the quantization error. Figure 9.8 presents the design for generating the r signal. It consists of various delay elements forming a configurable ring oscillator (RO) with the polarity control by the sign bit of r. The number of pulses during the write period (i.e., the firing rate) is varied by adding switches into the oscillation loop, which determines the total gate delay in the ring oscillator. The control signal of the switches is generated from the r value, ensuring that only one switch is on for a particular value of r. When $r = 0$, no change in the RRAM conductance is allowed. In total, 15 buffer stages (d1–d15) are implemented with different delay values, such that the number of pulses generated in each write cycle is proportional to the r value. From each rising edge of the RO output, the pulse generator generates a pulse with fixed pulse width. This ensures that the total programming time is proportional to the pulse number for the RRAM technology. The sign bit of r and the write phase (PN) finally select the output signal among Vdd, 0, the pulse generator output or the inversion of it. The proposed r circuit can provide 16 different pulses during one write cycle. To enable higher levels of granularity for the weight update (e.g., 64), the aforementioned write operation can be repeated multiple (e.g., 4) times using fine-grain pulse widths, with corresponding Z and r values.

Figure 9.9 shows the timing diagram of the parallel programming operation, which is simulated with a 100 × 100 resistive device array. When the write enable (WE) signal turns on, both Z and r write circuitries start generating the pulses based on the values of Z and r, thus changing the value of D during the overlap time. Figure 9.9 demonstrates the write operation when r is positive, where the programming occurs in the positive period and the value of D decreases; when r is negative, the programming takes place in the negative period, and the value of D increases. Regarding the array-level simulation of the write operation, one important concern is the IR drop caused by the wire resistance which in turn results in different voltages across devices in same column or row. This effect is most noticeable for

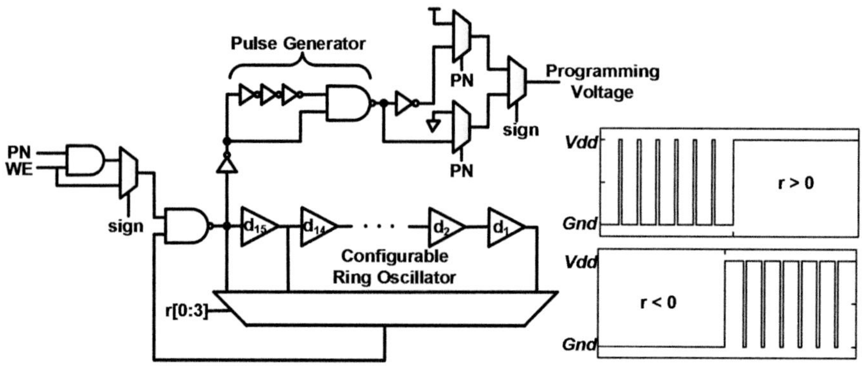

Fig. 9.8 CMOS neuron on the column-side for write operation

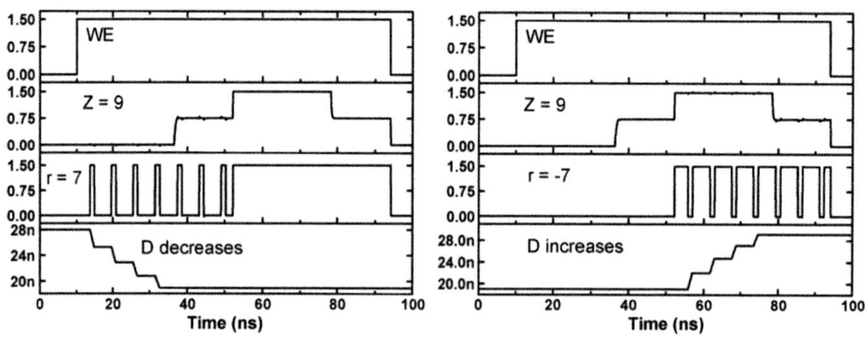

Fig. 9.9 Peripheral write circuit simulation with a 100 × 100 array for decreasing *D* weights (*left*) and increasing the *D* weights (*right*) [16]

the RRAM devices furthest from the periphery where the write circuits for *Z* and *r* are present. To avoid this problem, the write circuits should be equipped with buffers which can drive the capacitive load accordingly presented by the RRAM array.

Despite the array-level parallelism advantage, the fully parallel update [16, 21] has specific challenges, because (1) it is difficult to deliver the large amount of power to update all the RRAM cells simultaneously and (2) the circuits to generate *Z* and *r* consume a large area, so that the array cannot have one write circuit for every row/column. Therefore, time multiplexing is likely needed, and a block of the array (instead of the entire array) can be updated in parallel using a limited number of peripheral write circuits.

Using Pulse Amplitude for Enhanced Programming While the aforementioned neuron circuit designs for write operation or conductance programming have used different number of pulses with a fixed voltage level, modulating the voltage amplitudes of the programming pulses has also been explored. For instance, the authors of [22] investigated the possibility of back-propagation-based training of artificial neural networks, by using nonidentical pulse amplitudes for fine-grain programming of the PCMO-based resistive synaptic devices. As shown in Fig. 9.10, by using progressively increasing voltage amplitudes from −1.5 to −3.5 V (to increase conductance) or from 0.5 to 2.5 V (to decrease conductance), the nonlinear programming characteristics (Fig. 9.10, top) are greatly improved, such that close to linear programming of fine-grain conductance values becomes possible. However, the use of nonidentical training pulses (Fig. 9.10, bottom) requires multiple supply voltages or more complex peripheral circuits such as DC-DC converters, which was not implemented in [22]. Furthermore, the resistive array system has to read the conductance of the current synaptic device before

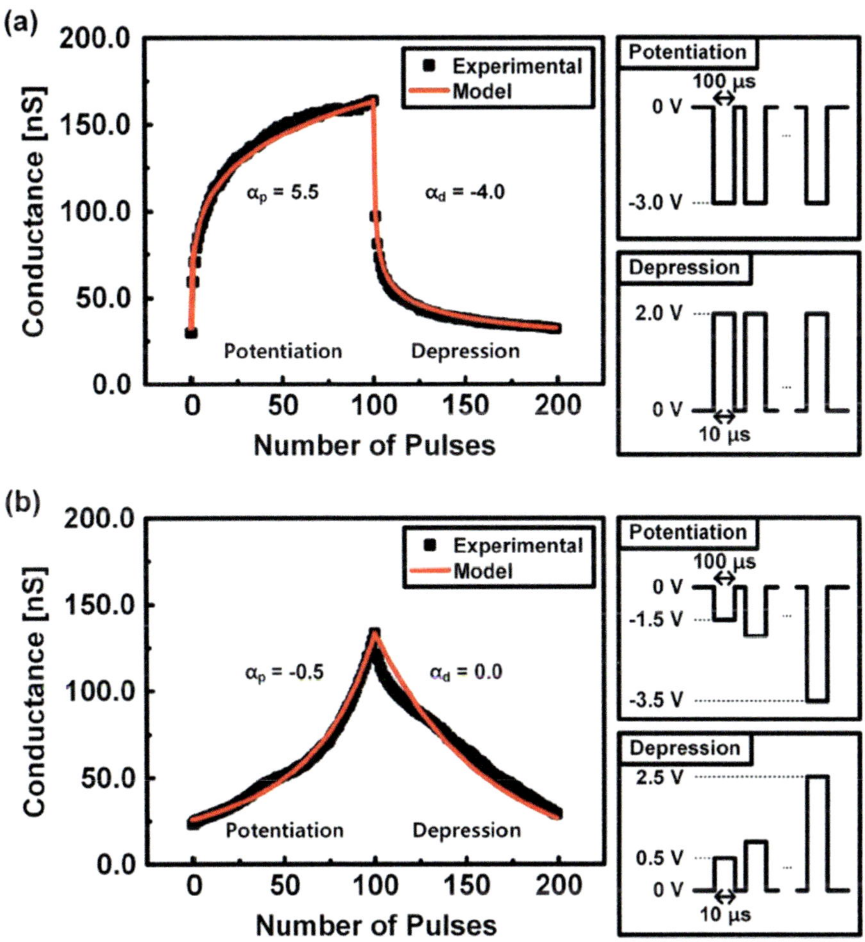

Fig. 9.10 Potentiation and depression characteristics of the experimental TiN/PCMO resistive memory and the resistive memory-based synaptic device model in [22] when (**a**) identical pulses are applied and (**b**) nonidentical (increasing amplitude) pulses are applied

programming it, in order to identify which nonidentical training pulse to apply. Thus, there is a trade-off between the higher accuracy and the resulting lower chip-area efficiency, higher power, and longer training time associated with the need to repeatedly measure individual conductance values.

9.4 Prospects of Non-CMOS Neuron Circuits

Prior RRAM designs [23–25] mostly focused on the synaptic array core instead of the peripheral neuron node. The CMOS peripheral neuron circuit, whether it is a sense amplifier or a current-to-digital converter, typically requires tens of transistors. Such complex CMOS neuron causes the column pitch matching problem and could possibly nullify the form factor benefits of the resistive synaptic devices. To that end, there have been ongoing research on investigating non-CMOS neuron devices and circuits that have small area, exhibit oscillating characteristics, and can directly interface with resistive synaptic arrays.

NbO_2 MIT Oscillator Neuron with PCMO Synapse A recent experimental work demonstrated the oscillation neuron with a 11 kb synaptic array [26]. The authors developed room temperature growth of NbO_2 film and observed metal-insulator-transition (MIT) oscillation characteristics. Coupling Mo/PCMO synapse with NbO_2 oscillator, the feasibility of neuromorphic system with pattern recognition capability was confirmed. For varying electrode size from 30 to 15.0 nm for the MIT device, relatively constant threshold voltage of ~1 V has been measured. Only when the neuron input voltage is above critical threshold voltage (~1.3 V) (Fig. 9.11a) and when the PCMO synapse resistance is within the range of 5–13 kΩ, the MIT oscillation behavior is observed (Fig. 9.11b). The oscillation

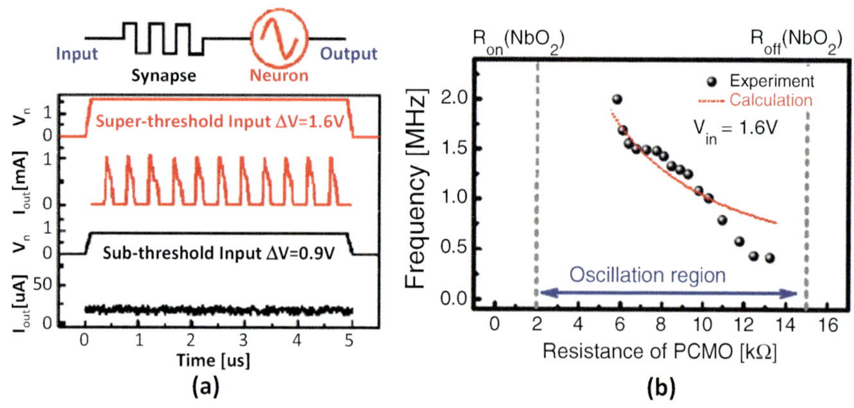

Fig. 9.11 (**a**) Input pulses and the corresponding output current responses. (**b**) Frequency of NbO_2 oscillator as a function of the resistance of Mo/PCMO synaptic device [26]

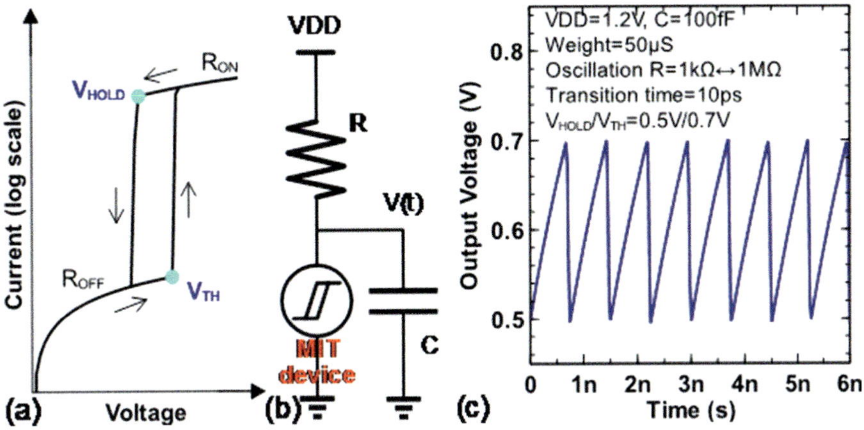

Fig. 9.12 (**a**) Hysteresis I–V characteristics of a MIT device. (**b**) Circuit configuration of an oscillation neuron node with MIT device and RRAM synaptic weight. (**c**) SPICE simulation waveform of the oscillation neuron

frequency or the spiking rate could be used as the neuron value, which may require further circuitry such as a digital counter for binary representation.

Array-Level Analysis of MIT Oscillator Neuron The authors of [27] investigated the feasibility of a compact oscillation neuron using the metal-insulator-transition (MIT) device for replacing the CMOS neuron and for correctly sensing the weighted sum current of large-scale resistive synaptic arrays.

Figure 9.12a shows the hysteresis I–V characteristic of a typical MIT device. A Verilog-A behavior model was built to capture the switching characteristics with parameters such as the resistance in the ON/OFF state (R_{ON}/R_{OFF}), the threshold voltage (V_{TH}), and the hold voltage (V_{HOLD}). The MIT device is initially in the OFF state, and it will switch to the ON state once the applied voltage exceeds V_{TH}. When the applied voltage across the MIT device is smaller than V_{HOLD}, it will switch back to the OFF state. Therefore, the resistive switching in the MIT device is essentially "volatile," unlike the "nonvolatile" resistive switching in the RRAM.

To eliminate the sneak path current that causes interference between columns in the cross-point array and thus can hamper the functionality of the oscillator neuron, a transistor can be added in series with the RRAM device as in conventional 1T1R array architecture for memory applications. As shown in Fig. 9.13, the 1T1R array is extended by adding one more transistor adjacent to the existing transistor, constructing a 2-transistor-1-resistor (2T1R) array architecture. The additional transistor is controlled by the inverting WL signal with its drain floating. In this way, the additional transistor serves as a complementary parasitic capacitance for the SL column. Each cell will contribute one drain and two source parasitic capacitance independent of WL signal as one of transistors will be turned on with the other one turned off.

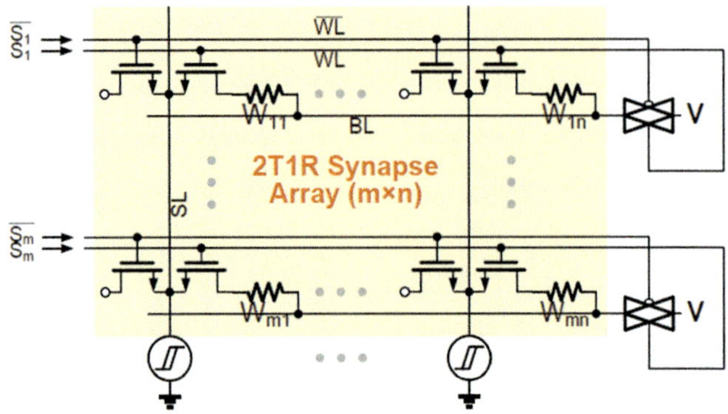

Fig. 9.13 Schematic of 2-transistor-1-resistor (2T1R) array architecture. The transistor in series with RRAM could cut off interference paths between columns. The other transistor with floating drain helps eliminate the capacitance variation when a different number of rows are activated

Table 9.1 Array-level benchmark (1 weighted sum task) of CMOS neuron and oscillation neuron designs

	Array with CMOS neuron [16]	Array with oscillation neuron [27]	Reduction
Total area	36,918 μm^2	35,571 μm^2	~4%
Latency	144 ns	99.2 ns	>30%
Energy consumption	693.2 pJ	139.5 pJ	~5×
Leakage power	1.73 mW	44.12 μW	~40×

The performance of the proposed oscillation neuron is benchmarked with that of the CMOS neuron [16] in the array level at the 65 nm technology node. Table 9.1 shows the array-level benchmark results. The synaptic array size is set to be 128 × 128 and there are four pulse cycles for the input vector. In practical array designs, multiple columns usually share one neuron to improve the area efficiency. From the array's point of view, the oscillation neuron does not gain much benefit in total area (synapse array area + peripheral neuron area) because the total area is still dominated by the array core. However, the oscillation neuron outperforms the CMOS neuron in latency. As the oscillation neuron is more compact, the number of columns shared by one neuron can be reduced from eight to four, thereby increasing the parallelism.

9.5 Summary and Future Prospects

In summary, we discussed the peripheral neuron circuit design for the resistive synaptic arrays for neuro-inspired applications. We reviewed the recent CMOS neuron designs implemented in analog and digital circuits for both read (weighted sum current) and write (synapse weight update) operations. Even if we multiplex the CMOS neurons, due to the large number of transistors, the peripheral circuits can still considerably affect the array efficiency. To that end, non-CMOS neuron designs such as NbO_2 metal-insulator-transition oscillation neurons are being investigated for compact neuron implementation. Further research and development will be needed for compact yet accurate neuron circuit design, which needs to be closely integrated with the resistive synaptic arrays for system-level enhancement in density, power, and performance.

References

1. H.-S.P. Wong, H.-Y. Lee, S. Yu, Y.-S. Chen, Y. Wu, P.-S. Chen, B. Lee, F.T. Chen, M.-J. Tsai, Metal–oxide RRAM. Proc. IEEE **100**(6), 1951–1970 (2012)
2. D. Kuzum, S. Yu, H.-S.P. Wong, Synaptic electronics: materials, devices and applications. Nanotechnology **24**, 382001 (2013)
3. S. Yu, P.-Y. Chen, Emerging memory technologies: recent trends and prospects. IEEE Solid State Circuits Mag. **8**(2), 43–56 (2016)
4. A. Kawahara, R. Azuma, Y. Ikeda, K. Kawai, Y. Katoh, K. Tanabe, T. Nakamura, Y. Sumimoto, N. Yamada, N. Nakai, S. Sakamoto, Y. Hayakawa, K. Tsuji, S. Yoneda, A. Himeno, K. Origasa, K. Shimakawa, T. Takagi, T. Mikawa, K. Aono, An 8 Mb multi-layered cross-point ReRAM macro with 443 MB/s write throughput, in *IEEE International Solid-State Circuits Conference* (2012)
5. T.-Y. Liu, T. H. Yan, R. Scheuerlein, Y. Chen, J. K. Lee, G. Balakrishnan, G. Yee, H. Zhang, A. Yap, J. Ouyang, T. Sasaki, S. Addepalli, A. Al-Shamma, C.-Y. Chen, M. Gupta, G. Hilton, S. Joshi, A. Kathuria, V. Lai, D. Masiwal, M. Matsumoto, et al., A 130.7 mm2 2-layer 32Gb ReRAM memory device in 24 nm technology, in *IEEE International Solid-State Circuits Conference* (2013)
6. P.A. Merolla, J.V. Arthur, R. Alvarez-Icaza, A.S. Cassidy, J. Sawada, F. Akopyan, B.L. Jackson, N. Imam, C. Guo, Y. Nakamura, B. Brezzo, I. Vo, S.K. Esser, R. Appuswamy, B. Taba, A. Amir, M.D. Flickner, W.P. Risk, R. Manohar, D.S. Modha, A million spiking-neuron integrated circuit with a scalable communication network and interface. Science **345**(6197), 668–673 (2014)
7. S. Yu, P.-Y. Chen, Y. Cao, L. Xia, Y. Wang, H. Wu, Scaling-up resistive synaptic arrays for neuro-inspired architecture: challenges and prospect, In *IEEE International Electron Devices Meeting* (2015)
8. S. Lee, S. Lee, K. Moon, J. Park, B. Kim, H. Hwang, Comprehensive methodology for ReRAM and selector design guideline of cross-point array, in *International Memory Workshop (IMW)* (2015)
9. S.H. Jo, T. Kumar, S. Narayanan, H. Nazarian, Cross-point resistive RAM based on field-assisted superlinear threshold selector. IEEE Trans. Electron Devices **62**(11), 3477–3481 (2015)
10. A. Conte, G.L. Guidice, G. Palumbo, A. Signorello, A high-performance very low-voltage current sense amplifier for nonvolatile memories. IEEE J. Solid State Circuits (JSSC) **40**(2), 507–514 (2005)
11. X. Xue, W. Jian, J. Yang, F. Xiao, G. Chen, S. Xu, Y. Xie, Y. Lin, R. Huang, Q. Zou, J. Wu, A 0.13 μm 8 Mb logic-based CuxSiyO ReRAM with self-adaptive operation for yield enhancement and power reduction. IEEE J. Solid State Circuits (JSSC) **48**(5), 1315–1322 (2013)
12. R. Fackenthal, M. Kitagawa, W. Otsuka, K. Prall, D. Mills, K. Tsutsui, J. Javanifard, K. Tedrow, T. Tsushima, Y. Shibahara, G. Hush, A 16Gb ReRAM with 200 MB/s write and

1GB/s read in 27 nm technology, in *IEEE International Solid-State Circuits Conference (ISSCC)* (2014)
13. M.-F. Chang, C.-W. Wu, C.-C. Kuo, S.-J. Shen, S.-M. Yang, K.-F. Lin, W.-C. Shen, Y.-C. King, C.-J. Lin, Y.-D. Chih, A low-voltage bulk-drain-driven read scheme for sub-0.5 V 4 Mb 65 nm logic-process compatible embedded resistive RAM (ReRAM) macro. IEEE J. Solid State Circuits (JSSC) **48**(9), 2250–2259 (2013)
14. J. Zhou, K.-H. Kim, W. Lu, Crossbar RRAM arrays: selector device requirements during read operation. IEEE Trans. Electron Devices **61**(5), 1369–1376 (2014)
15. S.-Y. Kim, J.-M. Baek, D.-J. Seo, J.-K. Park, J.-H. Chun, K.-W. Kwon, Power-efficient fast write and hidden refresh of ReRAM using an ADC-based sense amplifier. IEEE Trans. Circuits Syst. II **60**(11), 776–780 (2013)
16. D. Kadetotad, Z. Xu, A. Mohanty, P.-Y. Chen, B. Lin, J. Ye, S. Vrudhula, S. Yu, Y. Cao, J. Seo, Parallel architecture with resistive crosspoint array for dictionary learning acceleration. IEEE J. Emerg. Selected Top. Circuits Syst. (JETCAS) **5**(2), 194–204 (2015)
17. L.F. Abbott, Lapique's introduction of the integrate-and-fire model neuron (1907). Brain Res. Bull. **50**(5/6), 303–304 (1999)
18. S. Park, A. Sheri, J. Kim, J. Noh, J. Jang, M. Jeon, B. Lee, B.R. Lee, B.H. Lee, H. Hwang, Neuromorphic speech systems using advanced ReRAM-based synapse, in *IEEE International Electron Devices Meeting* (2013)
19. J. Lee, J. Shin, D. Lee, W. Lee, S. Jung, M. Jo, J. Park, K.P. Biju, S. Kim, S. Park, H. Hwang, Diode-less nano-scale ZrOx/HfOx RRAM device with excellent switching uniformity and reliability for high-density cross-point memory applications, in *IEEE International Electron Devices Meeting (IEDM)* (2010)
20. B. Rajendran, Y. Liu, J. Seo, K. Gopalakrishnan, L. Chang, D.J. Friedman, M.B. Ritter, Specifications of nanoscale devices and circuits for neuromorphic computational systems. IEEE Trans. Electron Devices **60**(1), 246–253 (2013)
21. L. Gao, I.T. Wang, P.-Y. Chen, S. Vrudhula, J. Seo, Y. Cao, T.-H. Hou, S. Yu, Fully parallel write/read in resistive synaptic array for accelerating on-chip learning. Nanotechnology **26**, 455204 (2015)
22. J.-W. Jang, S. Park, G.W. Burr, H. Hwang, Y.-H. Jeong, Optimization of conductance change in $Pr_{1-x}Ca_xMnO_3$-based synaptic devices for neuromorphic systems. IEEE Electron Device Lett. **36**(5), 457–459 (2015)
23. M. Hu, H. Li, Y. Chen, Q. Wu, G.S. Rose, R.W. Linderman, Memristor crossbar-based neuromorphic computing system: a case study. IEEE Trans. Neural Netw. Learn. Syst. **25** (10), 1864–1878 (2014)
24. D. Chabi, Z. Wang, W. Zhao, J.-O. Klein, On-chip supervised learning rule for ultra high density neural crossbar using memristor for synapse and neuron, in *International Symposium on Nanoscale Architectures (NANOARCH)* (2014)
25. B. Li, Y. Wang, Y. Wang, Y. Chen, H. Yang, Training itself: mixed-signal training acceleration for memristor-based neural network, in *Asia and South Pacific Design Automation Conference (ASP-DAC)* (2014)
26. K. Moon, E. Cha, J. Park, S. Gi, M. Chu, K. Baek, B. Lee, S. Oh, H. Hwang, High density neuromorphic system with Mo/Pr0.7Ca0.3MnO3 synapse and NbO2 IMT oscillator neuron, in *IEEE International Electron Devices Meeting (IEDM)* (2015)
27. P.-Y. Chen, J. Seo, Y. Cao, S. Yu, Compact oscillation neuron exploiting metal-insulator-transition for neuromorphic computing, in *IEEE/ACM International Conference on Computer-Aided Design (ICCAD)* (2016)

Chapter 10
Processing-In-Memory Architecture Design for Accelerating Neuro-Inspired Algorithms

Ping Chi, Shuangchen Li, and Yuan Xie

10.1 Introduction

Recent work on accelerating neuro-inspired algorithms on the platforms of GPU, FPGA, and ASIC lies in the traditional processor-coprocessor architecture, in which data are accessed from main memory in a conventional way, as shown in Fig. 10.1a. Because many neural network (NN) applications require high-memory bandwidth to transfer large-size synaptic weights and input and output data, the data movement between the processor and the memory becomes both time-consuming and energy-consuming. In a recent ASIC design, i.e., DianNao [1], it is reported that DRAM accesses take 95% of the total energy consumption. To address this challenge, more memory is put on chip to store synaptic weights in recent ASIC studies [2–5]. However, the issue still exists due to the limited on-chip memory size as well as the large-size input and output data transfer.

In this chapter, we introduce a novel processing-in-memory (PIM) architecture based on the resistive crossbar array structure for accelerating neuro-inspired algorithms, called PRIME [6]. As the computing paradigm shifts from the computation-centric to the data-centric, PIM or near data computing (NDC) has become a promising solution to address the "memory wall" challenges by placing computation capabilities in or near memory. PRIME is a PIM architecture based on RRAM main memory design. It moves the computing resources to the memory side by adapting a portion of RRAM crossbar arrays in the main memory for efficient weighted sum and weight update operations in neuro-inspired algorithms. Moreover, it provides a set of circuit and microarchitecture design as well as system-level support to implement various deep neural networks (DNNs). Benefiting from

P. Chi • S. Li • Y. Xie (✉)
Department of Electrical and Computer Engineering, University of California, Santa Barbara, 5159 Harold Frank Hall, Santa Barbara, CA 93106, USA
e-mail: yuanxie@ece.ucsb.edu

S. Yu (ed.), *Neuro-inspired Computing Using Resistive Synaptic Devices*,
DOI 10.1007/978-3-319-54313-0_10

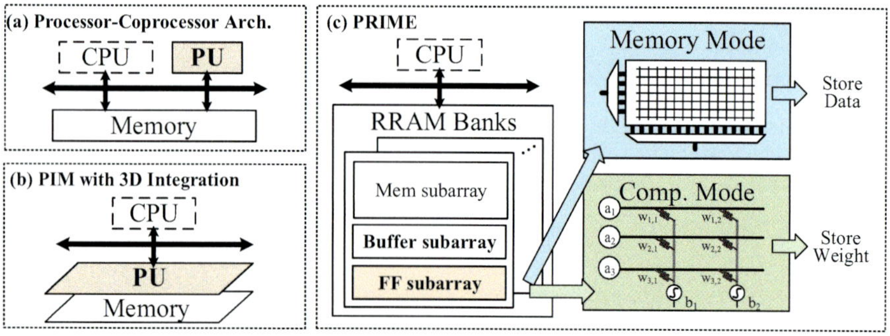

Fig. 10.1 (**a**) Traditional processor-coprocessor architecture with shared memory; (**b**) PIM architecture using 3D integration technologies; (**c**) PRIME design

both the PIM architecture and the efficiency of using RRAM crossbar arrays for neuro-inspired computing, PRIME can achieve significant performance improvement and energy saving compared with prior hardware acceleration work.

10.2 PIM Related Work

In order to reduce the data movement overhead, PIM has been proposed and studied since 1990s [7–10]. Most early PIM work put various logic into the DRAM chip, including simple ALU [11], vectorization [8], SIMD [12], general-purpose processors [13], and FPGA [14]. However, the idea of integrating performance-optimized logic with density-optimized memory did not succeed due to the cost sensitivity of the memory industry [15]. Recently, PIM or NDC is resurgent driven by the data-intensive applications and the 3D-stacking technology. We have seen a lot of effort from industry besides academia, e.g., IBM [16], AMD [17], and Samsung [18]. With the 3D-stacking technology, we can decouple logic and memory designs in different dies, adopting a logic layer that encapsulates processing units to perform computation, as shown in Fig. 10.1b [17, 19–22].

PRIME is a distinct solution from either early or recent PIM work. Instead of adding logic to memory, PRIME utilizes the memory arrays themselves for computing; hence, area overhead is very small. The add-on hardware in PRIME to enable the computation function consists of simple modifications of the existing memory peripheral circuits, which are more manufacturing friendly than integrating complex logic into the memory die. Moreover, PRIME does not rely on 3D-stacking technology, exempt from its high cost and thermal problems. Also, while previous work focused on database and graph-processing applications [20, 21], PRIME aims at accelerating neural network (NN) applications.

Recent work also employs nonvolatile memory technologies including RRAM, PCM, and STT-RAM, to build ternary content addressable memories (TCAMs), which exploits memory cells to perform associative search operations

[23–25]. However, to support such search operations, it requires redesigns of their memory cell structures which makes the cell sizes larger and inevitably increases the memory cost. Compared to these TCAM designs, PRIME obviates memory cell redesign and can support more sophisticated computation than TCAMs.

10.3 PRIME Architecture Overview

We propose processing in RRAM-based main memory, PRIME, which efficiently accelerates NN computation by leveraging RRAM's computation capability and the PIM architecture. Figure 10.1c depicts an overview of our design. While most previous NN acceleration approaches require additional processing units (PU) as shown in Fig. 10.1a, b, PRIME directly leverages RRAM cells to perform computation without the need for extra PUs. To achieve this, as shown in Fig. 10.1c, PRIME partitions a RRAM bank into three regions: memory (Mem) subarrays, full function (FF) subarrays, and Buffer subarrays.

The Mem subarrays only have data storage capability (the same as conventional memory subarrays). Their microarchitecture and circuit designs are similar to a recent design of performance-optimized RRAM main memory [26]. The FF subarrays have both computation and data storage capabilities, and they can operate in two modes. In memory mode, the FF subarrays serve as conventional memory; in computation mode, they can execute NN computation. There is a PRIME controller to control the operation and the reconfiguration of the FF subarrays. The Buffer subarrays serve as data buffers for the FF subarrays, and we use the memory subarrays that are closest to the FF subarrays as Buffer subarrays. They are connected to the FF subarrays through private data ports, so that buffer accesses do not consume the bandwidth of the Mem subarrays. While not being used as data buffers, the Buffer subarrays can also be used as normal memory. From Fig. 10.1c, we can find that for NN computation, the FF subarrays enjoy the high bandwidth of in-memory data movement and can work in parallel with CPU, with the help of the Buffer subarrays.

Next, we describe the details of our microarchitecture and circuit designs of the FF subarrays, the Buffer subarrays, and the PRIME controller. These designs are independent of the technology assumptions for RRAM-based computation. For generality, we assume that the input data have P_{in} bits, the synaptic weights have P_w bits, and the output data have P_o bits. With practical assumptions, the precision of RRAM-based NN computation is a critical challenge. We discuss the precision issue and propose a scheme to overcome it in Sect. 10.7. Finally, more details are given about implementing NN algorithms with our hardware design.

10.4 FF Subarray Design

The design goal for FF subarray is to support both storage and computation with a minimum area overhead. To achieve this goal, we maximize the reuse of peripheral circuits for both storage and computation.

10.4.1 Microarchitecture and Circuit Design

To enable the NN computation function in FF subarrays, we modify decoders and drivers, column multiplexers (MUX), and sense amplifiers (SA) as shown in Fig. 10.2.

Decoder and Driver We add several components in decoders and drivers marked as light green in Fig. 10.2A. First, we attach multilevel voltage sources to the wordlines to provide accurate input voltages. NN computation requires that all input data are simultaneously fed into the corresponding wordline. Therefore, we add a latch to control the input voltage. The control signals determine the combination of voltage sources that provide the demanding input voltage. Second, to drive the analog signals transferring on the wordlines, we employ a separate current amplifier on each wordline. Third, rather than two voltage levels used in the memory mode (for read and write, respectively), NN computation requires $2^{P_{in}}$ levels of input voltages. We employ a multiplexer to switch the voltage driver between memory and computation modes. Finally, we employ two crossbar arrays that store positive and negative weights, respectively, and allow them to share the same input port.

Column Multiplexer In order to support NN computation, we modify the column multiplexers in RRAM by adding the components marked in light green in Fig. 10.2B. The modified column multiplexer incorporates two analog processing units: an analog subtraction unit and a nonlinear threshold (sigmoid) unit [27]. The sigmoid unit can be bypassed in certain scenarios, e.g., when a large NN is mapped to multiple crossbar arrays. In addition, in order to allow FF subarrays to switch bitlines between memory and computation modes, we attach a multiplexer to each bitline to control the switch. Since a pair of crossbar arrays with positive and negative weights require one set of such peripheral circuits, we only need to modify half of the column multiplexers. After analog processing, the output current is sensed by local SAs.

Sense Amplifier Figure 10.2C shows the SA design with the following modifications as marked in light green in the figure. First, NN computation requires SAs to offer much higher precision than memory does. We adopt a P_o-bit ($P_o \leq 8$) precision reconfigurable SA design that has been tested through fabrication [28]. Second, we allow SA's precision to be configured as any value between 1-bit and P_o-bit, controlled by the counter as shown in Fig. 10.2C. The result is stored in the output

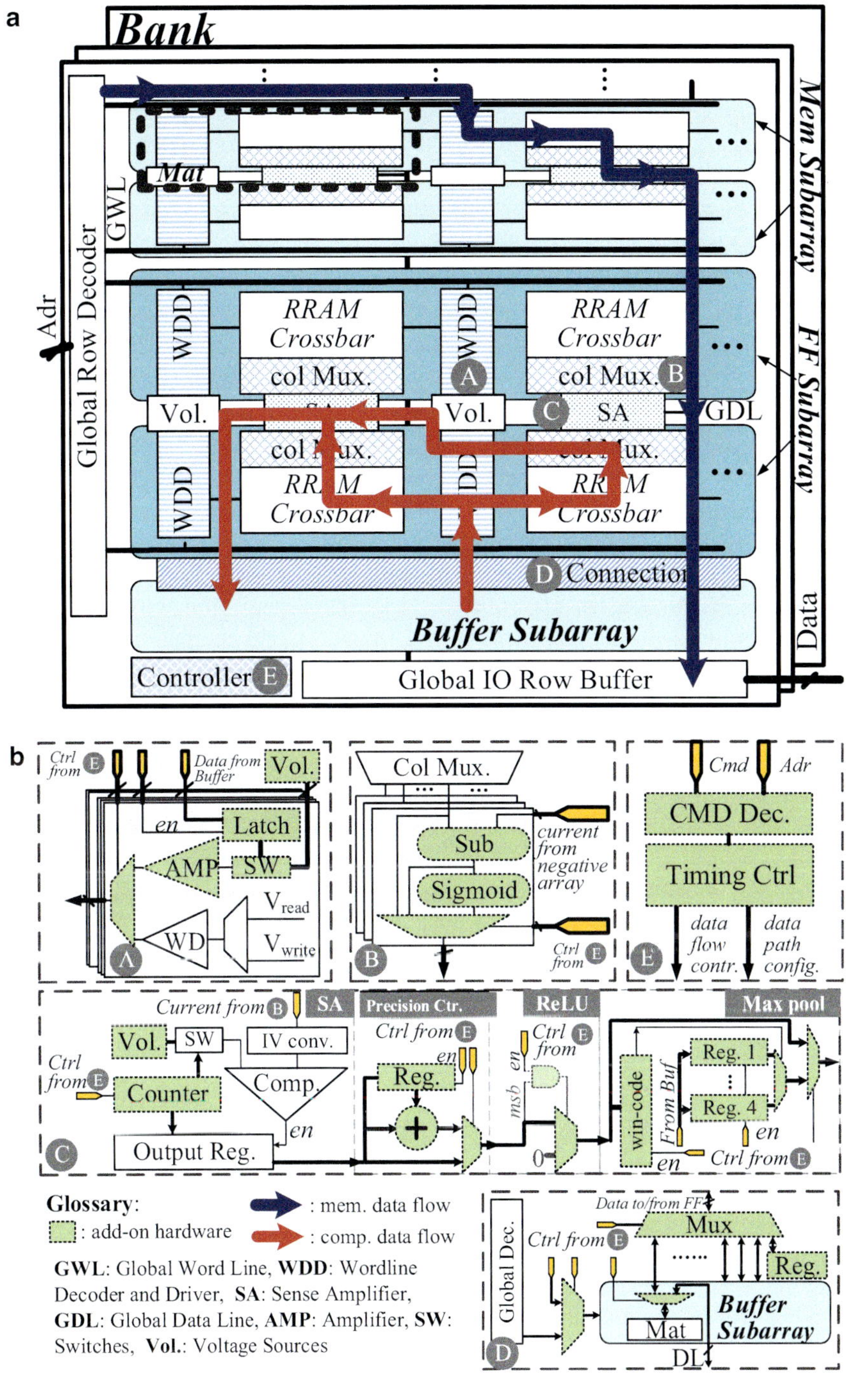

Fig. 10.2 The PRIME architecture. *Left*: bank structure. The *blue and red bold lines* represent the directions of the data flow for normal memory and for computation, respectively. *Right*: functional blocks modified/added in PRIME. (*A*) Wordline driver with multilevel voltage sources; (*B*) column multiplexer with analog subtraction and sigmoid circuitry; (*C*) reconfigurable SA with counters for multilevel outputs and added rectified linear unit (ReLU) and 4–1 max pooling function unit; (*D*) connection between the FF and Buffer subarrays; (*E*) PRIME controller

registers. Third, we allow low-precision RRAM cells to perform NN computation with a high-precision weight, by developing a precision control circuit that consists of a register and an adder. Fourth, we add a rectified linear unit (ReLU) to support the rectifier function, an activation function in the convolution layer of CNN. The ReLU circuit checks the sign bit of the result and outputs zero when the sign bit is negative and the result itself otherwise. Finally, a circuit to support 4–1 max pooling is included. More details are discussed in Sect. 10.8.

Buffer Connection Figure 10.2D shows the communication between the FF subarrays and the Buffer subarrays. We enable an FF subarray to access any physical location in a Buffer subarray to accommodate the random memory access pattern in NN computation (e.g., in the connection of two convolutional layers). To this end, extra decoders and multiplexers are employed in the buffer connection unit. Additionally, we allow the data transfer to bypass the Buffer subarray in certain scenarios, e.g., when the output of one mat is exactly the input of another. After bypassing the Buffer subarrays, we employ a register as an intermediate data storage.

Benefits of Our Design are twofold. First, our design efficiently utilizes the peripheral circuits by sharing them between memory and computation functions, which significantly reduces the area overhead. For example, in a typical RRAM-based neuromorphic computing system [29], DACs and ADCs are used for input and output signal conversions; in a RRAM-based memory system, SAs and write drivers are required for read and write operations. Yet, SAs and ADCs serve similar functions, while write drivers and DACs do similar functions. In PRIME, instead of using both, we reuse SAs and write drivers to serve ADC and DAC functions by slightly modifying the circuit design. Second, we enable the FF subarrays to flexibly and efficiently morph between memory and computation modes.

10.4.2 Morphing Between Two Modes

Figure 10.3 shows two FF subarrays that are configured into computation and memory modes, respectively. The black bold lines in the figure demonstrate the data flow in each configuration. As shown in Fig. 10.3a, in computation mode, the FF subarray fetches the input data of the NN from the Buffer subarray into the latch of the wordline decoder and driver. After the computation in the crossbar arrays that store positive and negative weights, their output signals are fed into the subtraction unit, and then the difference signal goes into the sigmoid unit. The analog output signal is converted to digital by the SA before it is written back to the Buffer subarray. As shown in Fig. 10.3b, in memory mode, the input comes from the read/write voltage selection (represented by an *m* box in the figure), and the output bypasses the subtraction and sigmoid units.

The morphing between memory and computation modes involves several steps. Before the FF subarrays switch from memory mode to computation mode, PRIME

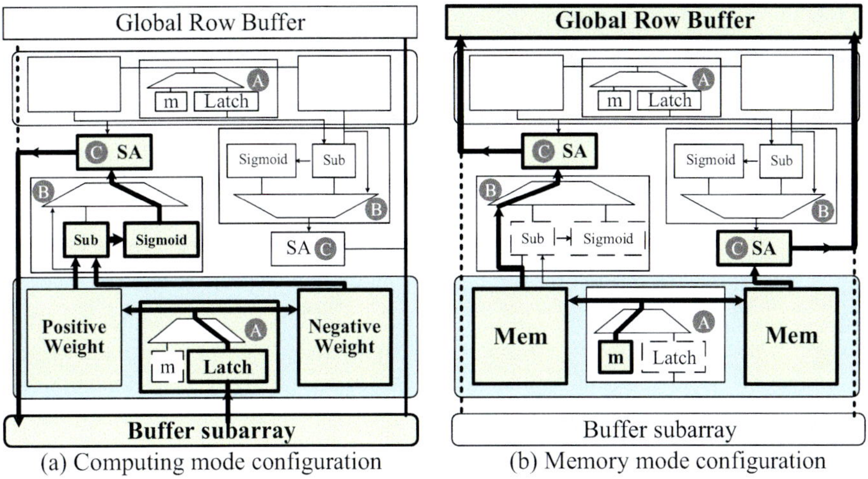

Fig. 10.3 An example of the configurations of FF subarrays. (**a**) Computation mode configuration; (**b**) memory mode configuration

migrates the data stored in the FF subarrays to certain allocated space in Mem subarrays and then writes the synaptic weights to be used by computation into the FF subarrays. When data preparations are ready, the peripheral circuits are reconfigured by the PRIME controller, and the FF subarrays are switched to computation mode and can start to execute the mapped NNs. After completing the computation tasks, the FF subarrays can be switched back to memory mode, by performing a wrap-up step that reconfigures the peripheral circuits.

10.5 Buffer Subarrays

The goal of the Buffer subarrays is twofold. First, they are used to cache the input and output data for the FF subarrays. Benefiting from the massive parallelism of weighted sum operations provided by RRAM crossbar structures, the computation itself takes a very short time. Moreover, the data input and output may be serial, and their latencies become potential bottlenecks. Therefore, it is necessary to cache the input and output data. Second, the FF subarrays can communicate with the Buffer subarrays directly without the involvement of the CPU, so that the CPU and the FF subarrays can work in parallel.

We choose to configure the adjacent memory subarray to the FF subarrays as the Buffer subarray, which is close to both the FF subarrays and the global row buffer so as to minimize the delay. We do not utilize the local row buffer because it is not large enough to serve typical NNs. We do not implement the buffer with low-latency SRAM due to its large area and cost overhead.

Table 10.1 PRIME controller commands

Datapath configure	Data flow control
prog/comp/mem [mat adr][0/1/2]	fetch [mem adr] to [buf adr]
bypass sigmod [mat adr] [0/1]	commit [buf adr] to [mem adr]
bypass SA [mat adr][0/1]	load [buf adr] to [FF adr]
input source [mat adr][0/1]	store [FF adr] to [buf adr]

As described in *Buffer Connection* in Sect. 10.4, the Buffer subarray and the FF subarrays are connected by the connection unit which enables the FF subarrays to access any data in the buffer. To fetch data for the FF subarrays, the data are first loaded from a Mem subarray to the global row buffer, and then they are written from the row buffer to the Buffer subarray. These two steps have to be done in serial due to the resource conflict, i.e., the global data lines (GDL). The communication between the Buffer subarray and the FF subarrays is independent with the communication between the Mem subarray and the globe row buffer. Therefore, when PRIME is accelerating NN computation, CPU can still access the memory and work in parallel. To write the data from the Buffer subarray back to memory, the data go through the global row buffer to the corresponding Mem subarray.

10.6 PRIME Controller

Figure 10.2E illustrates the PRIME controller that decodes instructions and provides control signals to all the peripheral circuits in the FF subarrays. A key role of the controller is to configure the FF subarrays in memory and computation modes. Table 10.1 lists the basic commands used by the controller. The left four commands generate control signals for the multiplexers in Fig. 10.2, including the function selection of each mat among programming synaptic weights, computation, and memory, and also the input source selection for computation, either from the Buffer subarray or from the output of the previous layer. These commands are performed once during each configuration of the FF subarrays. The right four commands in Table 10.1 control the data movement. They are applied during the whole computation phase.

10.7 Overcoming the Precision Challenge

The precision issue is one of the most critical challenges for RRAM-based NN computation. It contains several aspects: input precision, synaptic weight (or cell resistance) precision, output (or analog computation) precision, and their impacts on the results of NN applications (e.g., the classification accuracy of image recognition tasks).

Previous work has employed 1-bit to 12-bit synaptic weights for RRAM-based NN computation [30–32]. There has been active research going on with improving the resistance precision of MLC RRAM cells. With a simple feedback algorithm, the resistance of a RRAM device can be tuned with 1% precision (equivalent to 7-bit precision) for a single cell and about 3% for the cells in crossbar arrays [33, 34].

The latest results of the Dot-Product Engine project from HP Labs reported that, for a 256 × 256 crossbar array, given full-precision inputs (e.g., usually 8-bit for image data), 4-bit synaptic weights can achieve 6-bit output precision and 6-bit synaptic weights can achieve 7-bit output precision, when the impacts of noise on the computation precision of RRAM crossbar arrays are considered [35].

Our PRIME design can be adapted to different assumptions of input precision, synaptic weight precision, and output precision. According to the state-of-the-art technologies used in RRAM-based NN computation, one practical assumption is that: the input voltage has only 3-bit precision (i.e., eight voltage levels), and the RRAM cells can only represent 4-bit synaptic weights (i.e., 16 resistance levels), and the target output precision is 6-bit. The data format we use is dynamic fixed point [36]. To achieve high-computation accuracy with conservative assumptions, we propose an input and synapse composing scheme, which can use two 3-bit input signals to compose one 6-bit input signal and two 4-bit cells to represent one 8-bit synaptic weight.

10.7.1 Input and Synapse Composing Scheme

We present the input and synapse composing algorithm first and then present the hardware implementation. Table 10.2 lists the notations.

If the computation in a RRAM crossbar array has full accuracy, the weighted sum result should be

$$R_{\text{full}} = \sum_{i=1}^{2^{P_N}} \left(\sum_{k=1}^{P_{in}} I_k^i 2^{k-1} \cdot \sum_{k=1}^{P_w} W_k^i 2^{k-1} \right),$$

which has $(P_{\text{in}} + P_{\text{w}} + P_{\text{N}})$-bit full precision. Since the target output is P_{o}-bit, we will take the highest P_{o}-bit of R_{full}. Then, the target result is denoted as shifting R_{full} to the right by $(P_{\text{in}} + P_{\text{w}} + P_{\text{N}} - P_{\text{o}})$ bits:

Table 10.2 Notation description

$P_{\text{in}}, P_{\text{o}}, P_{\text{w}}$	The number of bits for input/output/synaptic weights
P_{N}	The number of inputs to a crossbar array is $2^{P_{\text{N}}}$
I_k^i, W_k^i	The kth bit of the ith input signal/synaptic weight
Ih_k^i, Il_k^i	The kth bit of HIGH/LOW-bit part of the ith input
Wh_k^i, Wl_k^i	The kth bit of HIGH/LOW-bit part of the ith weight

$$R_{target} = R_{full} \gg (P_{in} + P_W + P_N - P_0)$$

Now each input signal and synaptic weight are composed of two parts: high-bit part and low-bit part. We have

$$\text{input} : \sum_{k=1}^{P_{in}} I_k^i 2^{k-1} = \sum_{i=1}^{\frac{P_{in}}{2}} \left(Ih_k^i 2^{k-1} \cdot 2^{\frac{P_{in}}{2}} + Il_k^i 2^{k-1} \right),$$

$$\text{weight} : \sum_{k=1}^{P_w} W_k^i 2^{k-1} = \sum_{i=1}^{\frac{P_w}{2}} \left(Wh_k^i 2^{k-1} \cdot 2^{\frac{P_w}{2}} + Wl_k^i 2^{k-1} \right).$$

Then, R_{full} will contain four parts (i.e., HH-part, HL-part, LH-part, and LL-part),

$$\begin{aligned} R_{full} = \sum_{i=1}^{2^{P_N}} \{ & 2^{\frac{P_w + P_{in}}{2}} \cdot \left(\sum_{k=1}^{\frac{P_{in}}{2}} Ih_k^i 2^{k-1} \sum_{k=1}^{\frac{P_w}{2}} Wh_k^i 2^{k-1} \right)_{\mathrm{HH-part.}} \\ & + 2^{\frac{P_w}{2}} \cdot \left(\sum_{k=1}^{\frac{P_{in}}{2}} Il_k^i 2^{k-1} \sum_{k=1}^{\frac{P_w}{2}} Wh_k^i 2^{k-1} \right)_{\mathrm{HL-part}} \\ & + 2^{\frac{P_{in}}{2}} \cdot \left(\sum_{k=1}^{\frac{P_{in}}{2}} Ih_k^i 2^{k-1} \sum_{k=1}^{\frac{P_w}{2}} Wl_k^i 2^{k-1} \right)_{\mathrm{LH-part}} \\ & + \left(\sum_{k=1}^{\frac{P_{in}}{2}} Il_k^i 2^{k-1} \sum_{k=1}^{\frac{P_w}{2}} Wl_k^i 2^{k-1} \right)_{\mathrm{LL-part}} \quad \} \end{aligned}$$

.

Here, we rewrite R_{full} as

$$R_{full} = 2^{\frac{P_W + P_{in}}{2}} \cdot R_{full}^{HH} + 2^{\frac{P_W}{2}} \cdot R_{full}^{HL} + 2^{\frac{P_{in}}{2}} \cdot R_{full}^{LH} + R_{full}^{LL}. \tag{10.1}$$

We can also denote R_{target} with four parts

$$R_{target} = R_{tar}^{HH} + R_{tar}^{HL} + R_{tar}^{LH} + R_{tar}^{LL}. \tag{10.2}$$

In Eq. (10.1), if the output of each R_{full} part is only P_{o}-bit, then,

- R_{tar}^{HH}: take all the P_{o} bits of R_{full}^{HH} result.
- R_{tar}^{HL}: take the highest $P_{\mathrm{o}} - \frac{P_{in}}{2}$ bits of R_{full}^{HL} result.
- R_{tar}^{LH}: take the highest $P_{\mathrm{o}} - \frac{P_w}{2}$ bits of $R_{\mathrm{full}}^{\mathrm{LH}}$ result.

- R_{tar}^{LL}: take the highest $P_o - \frac{P_w+P_{in}}{2}$ bits of R_{full}^{LL} result.

According to our assumptions, we have $P_{in} = 6$ (composed of two 3-bit signals), $P_w = 8$ (composed of two 4-bit cells), and $P_o = 6$ (enabled by 6-bit precision reconfigurable sense amplifiers). The target result should be the summation of three components: all the 6 bits of R_{tar}^{HH} output, the highest 3 bits of R_{tar}^{HL} output, and the highest 2 bits of R_{tar}^{LH} output.

To implement synapse weight composing, P_{in} is loaded to the latch in the WL driver as shown in Fig. 10.2A. According to the control signal, the high-bit and low-bit parts of the input are fed to the corresponding crossbar array sequentially. To implement synapse composing, the high-bit and low-bit parts of the synaptic weights are stored in adjacent bitlines of the corresponding crossbar array. As shown in Eq. (10.2), R_{target} consists of four components. They are calculated one by one, and their results are accumulated with the adder in Fig. 10.2C. The right shift operation, i.e., taking the highest several bits of a result, can be implemented by the reconfigurable SA. To take the highest n-bit of a result, we simply configure the SA as an n-bit SA.

10.8 Implementing NN Algorithms

10.8.1 MLP/Fully Connected Layer: Weighted Sum

Weighted sum (matrix-vector multiplication) is one of the most important primitives in NN algorithms. The RRAM crossbar arrays are used to implement it: the weight matrix is preprogrammed in RRAM cells; the input vector is the voltages on the wordlines driven by the drivers (as shown in Fig. 10.2A); the output currents are accumulated at the bitlines. The synaptic weight matrix is separated into two matrices: one storing the positive weights and the other storing the negative weights. They are programmed into two crossbar arrays. A subtraction unit (as shown in Fig. 10.2B) is used to subtract the result of the negative part from that of the positive part.

10.8.2 MLP/Fully Connected Layer: Activation Function

Our circuit design supports two activation functions: sigmoid and rectifier. Sigmoid is implemented by the sigmoid unit in Fig. 10.2B, and rectifier is implemented by the ReLU in Fig. 10.2C. These two units can be configured to bypass in some scenarios.

10.8.3 Convolution Layer

The computation of the convolution layer is described as follows

$$f_i^{\text{out}} = \max\left(\sum_{j=1}^{n_{in}} f_j^{in} \otimes g_{i,j} + b_i, 0\right), 1 \leq i \leq n_{\text{out}},$$

where f_j^{in} is the j-th input feature map and f_i^{out} is the i-th output feature map, $g_{i,j}$ is the convolution kernel for f_j^{in} and f_i^{out}, b_i is the bias term, and n_{in} and n_{out} are the numbers of the input and output feature maps, respectively.

To implement the summation of n_{in} convolution operations $\left(f_j^{in} \otimes g_{i,j}\right)$ plus b_i, all the elements of j convolution kernels $g_{i,j}$ are preprogrammed in the RRAM cells of one BL or more BLs if they cannot fit in one, and the elements of f_j^{in} are performed as input voltages. We also write b_i in RRAM cells and regard the corresponding input as "1." Each BL will output the whole or part of the convolution result. If more BLs are used, it takes one more step to achieve the final result. Next, the max(x, 0) function is executed by ReLU in Fig. 10.2C.

10.8.4 Pooling Layer

To implement max pooling function, we adopt 4 : 1 max pooling hardware in Fig. 10.2C, which is able to support n : 1 max pooling with multiple steps for $n > 4$. For 4 : 1 max pooling, first, four inputs $\{a_i\}$ are stored in the registers, $i = 1, 2, 3, 4$; second, we execute the dot products of $\{a_i\}$ and six sets of weights [1,−1,0,0], [1,0,−1,0], [1,0,0,−1], [0,1,−1,0], [0,1,0,−1], and [0,0,1,−1] by using RRAM to obtain the results of $a_i - a_j$, $i \neq j$; next, the signs of their results are stored in the Winner Code register; finally, according to the code, the hardware determines the maximum and outputs it. Mean pooling is easier to implement than max pooling, because it can be done with RRAM and does not require extra hardware. To perform n : 1 mean pooling, we simply preprogram the weights $\left[\frac{1}{n}, \ldots, \frac{1}{n}\right]$ in RRAM cells and execute the dot product of the inputs and the weights to obtain the mean value of n inputs.

10.8.5 Local Response Normalization (LRN) Layer

Currently, PRIME does not support LRN acceleration. We did not add the hardware for LRN, because state-of-the-art CNNs do not contain LRN layers [37]. When LRN layers are applied, PRIME requires the help of CPU for LRN computation.

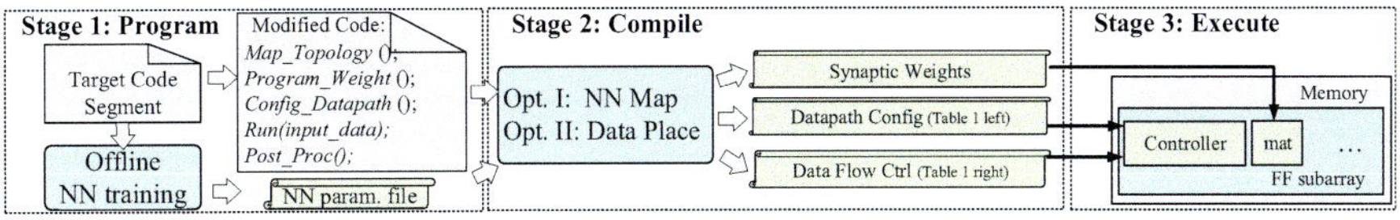

Fig. 10.4 The software perspective of PRIME

10.9 System-Level Design

In this section, we present the system-level design of PRIME. The software-hardware interface framework is described. Then, we focus on the optimization of NN mapping and data allocation during compile time. Next, we introduce the operating system (OS) support for switching FF subarrays between memory and computation modes at runtime.

10.9.1 Software-Hardware Interface

Figure 10.4 shows the stack of PRIME to support NN programming, which allows developers to easily configure the FF subarrays for NN applications. From software programming to hardware execution, there are three stages: programming (coding), compiling (code optimization), and code execution. In the programming stage, PRIME provides application programming interfaces (APIs) so that they allow developers to (1) map the topology of the NN to the FF subarrays, *Map_Topology*; (2) program the synaptic weights into mats, *Program_Weight*; (3) configure the data paths of the FF subarrays, *Config_Datapath*; (4) run computation, *Run*; and (5) post-process the result, *Post_Proc*. In our work, the training of NN is done offline so that the inputs of each API are already known (*NN param.file*). Prior work is explored to implement training with RRAM crossbar arrays [38–43], and we plan to further enhance PRIME with the training capability in future work.

In the compiling stage, the NN mapping to the FF subarrays and the input data allocation are optimized. The output of compiling is the metadata for synaptic weights mapping, data path configuration, and execution commands with data dependency and flow control. The metadata is also the input for the execution stage. In the execution stage, PRIME controller writes the synaptic weights to the mapped addresses in the FF subarrays; then it (re)configures the peripheral circuits according to the *Datapath Configure* commands (Table 10.1 left) to set up the data paths for computation; and finally, it executes *Data Flow Control* commands (Table 10.1 right) to manage data movement into or out of the FF subarrays at runtime.

10.9.2 Compile-Time Optimization

10.9.2.1 NN Mapping Optimization

The mapping of the NN topology to the physical RRAM cells is optimized during compile time. For different scales of NNs, we have different optimizations.

Small-Scale NN: Replication When an NN can be mapped to a single FF mat, it is small scale. Although we can simply map a small-scale NN to some cells in one mat, the other cells in this mat may be wasted. Moreover, the speedup for very small NNs is not obvious, because the latency of the peripheral circuits may overwhelm the latency of weighted sum operations on RRAM cells. Our optimization is to replicate the small NN to different independent portions of the mat. For example, to implement a $128-1$ NN, we duplicate it and map a $256-2$ NN to the target mat. This optimization can also be applied to convolution layers. Furthermore, if there is another FF mat available, we can also duplicate the mapping to the second mat, and then the two mats can work simultaneously, as long as the Buffer subarray has enough bandwidth.

Medium-Scale NN: Split-Merge When an NN cannot be mapped to a single FF mat, but can fit to the FF subarrays of one bank, it is medium scale. During the mapping at compile time, a medium-scale NN has to be split into small-scale NNs, and then their results are merged. For example, to implement a $512-512$ NN on PRIME with $256-256$ mats, it is split into four $256-256$ parts ($[M_{1,1}, M_{1,2}; M_{2,1}, M_{2,2}]$) and mapped to four different mats. After they finish computation, the results of $M_{1,1}$ and $M_{2,1}$ are added to get the first 256 elements of the final result, and the sum of the results of $M_{1,2}$ and $M_{2,2}$ forms the second 256 elements of the final result.

Large-Scale NN: Interbank Communication A large-scale NN is one NN that cannot be mapped to the FF subarrays in a single bank. Intuitively, we can divide it into several medium-scale trunks and map each trunk to the same bank serially in several stages. This naive solution requires reprogramming the FF subarrays at every stage, and the latency overhead of reprogramming may offset the speedup. Alternatively, PRIME allows to use multiple banks to implement a large-scale NN. These banks can transfer data to each other and run in a pipelined fashion to improve the throughput. Like prior work [44], the interbank data movement is implemented by exploiting the internal data bus shared by all the banks in a chip. PRIME controller manages the interbank communication and can handle arbitrary network connections. If all the banks are used to implement a single NN, PRIME can handle a maximal NN with ~2.7×10^8 synapses, which is larger than the largest NN that have been mapped to the existing NPUs (TrueNorth [4], 1.4×10^7 synapses). In Sect. 10.10, we implement an extremely large CNN on PRIME, *VGG-D* [37], which has 1.4×10^8 synapses.

Table 10.3 The benchmarks and topologies

MlBench		*MLP-S*	784–500–250-10
CNN-1	conv5x5-pool-720-70-10	*MLP-M*	784–1000–500-250-10
CNN-2	conv7x10-pool-1210-120-10	*MLP-L*	784–1500–1000-500-10
VGG-D	conv3x64-conv3x64-pool-conv3x128-conv3x128-pool-conv3x256-conv3x256-conv3x256-pool-conv3x512-conv3x512-conv3x512-pool-conv3x512-conv3x512-conv3x512-pool-25,088-4096-4096-1000		

10.9.2.2 Bank-Level Parallelism and Data Placement

Since FF subarrays reside in every bank, PRIME intrinsically inherits bank-level parallelism to speed up computation. For example, for a small-scale or medium-scale NN, since it can be fitted into one bank, the FF subarrays in all the banks can be configured the same and run in parallel. If the FF subarrays in each bank is regarded as an NPU, PRIME contains 64 NPUs in total (8 banks $\times$ 8 chips) so that 64 images can be processed in parallel. To take advantage of the bank-level parallelism, the OS is required to place one image in each bank and to evenly distribute images to all the banks. As current page placement strategies expose memory latency or bandwidth information to the OS [45, 46], PRIME exposes the bank ID information to the OS, so that each image can be mapped to a single bank. For large-scale NNs, they can still benefit from bank-level parallelism as long as we can map one replica or more to the spare banks.

10.9.2.3 A Mapping Case Study: CNN-1

We show how CNN-1 in Table 10.3 is mapped to PRIME in this section. We use 256×256 RRAM mats, and RRAM cells in FF subarrays can be programmed in 4-bit precision, and the synaptic weights of the NN are 8-bit in the application. Therefore, to present one weight, two cells are used. In CNN-1, the *convolutional layer* has five 5×5 kernels. Then, the weight matrix is of size $25(=5 \cdot 5) \times 10$ $(=5 \cdot 2)$.[1] In order to improve latency and RRAM's utility, the compile-time optimization may replicate the matrix ten times and map a 256×100 weight matrix to a 256×256 mat, as shown in Fig. 10.5a. Actually, the optimization can be better for a convolutional layer. Because two adjacent convolution operations share most of the inputs, e.g., 20 out of 25 inputs for a 5×5 kernel in this case, we are able to map more replicas of the 25×10 weight matrix in one mat in a smart way, as shown in Fig. 10.5b. In this way, we make 24 replicas in one mat, increasing the number of output data from 50 to 120 as well as reducing the number of input data from 250 to 140. In CNN-1, the *pooling layer* adopts 4 : 1 max pooling. Since each pooling requires a 4×6 weight matrix, we can map 42 replicas in one mat. In CNN-1, the

[1] For clarity, we only talk about the positive part of the weight matrix. The mapping optimization also applies to the negative part.

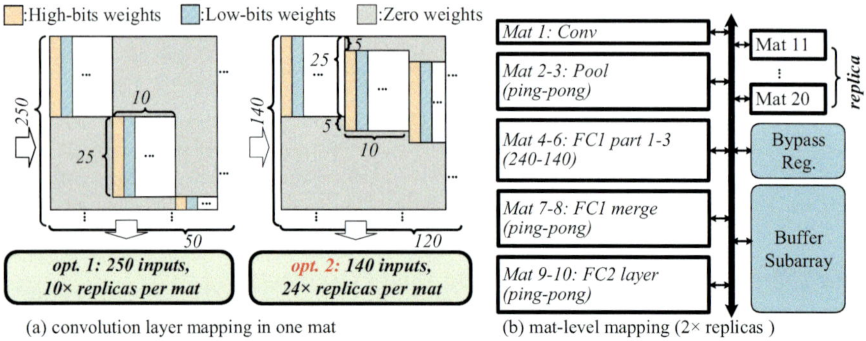

Fig. 10.5 A mapping example. (**a**) Optimization choice 1 and (**b**) optimization choice 2 for the convolutional layer; (**c**) mat-level mapping and optimization.

fully connected layers are 720 – 70 – 10. Since the input dimension is larger than 256, we apply the *split-merge* mapping. First, the $720 \times 140(=70 \cdot 2)$ weight matrix is mapped into three mats, each of size 240×140. Then, the merging of the results from those three mats is performed in another mat to achieve the final results. In the merging mat, since the weight matrix is 3×1, we can execute 85 merging operations in one mat at the same time.

In each mat, computation and data input/output can work in a pipelined way, thanks to input latches and output registers in Fig. 10.2A, C. All mats can work in parallel, and the data communication among them either goes through the Buffer subarray or bypasses it with the help of the bypassing registers (as shown in Fig. 10.2C). When a mat has bypassed inputs (not from the Buffer subarray but from the bypassing registers) and is dedicated for a whole layer (not one of a set of mats for split-merge mapping), in order to improve the throughput, we can duplicate the mat and make the two mats work in a ping-pong mode, in which one mat receives its inputs from the previous layer while the other does the computation. As shown in Fig. 10.5c, the pooling layer, the merging layer of the fully connected layer 1, and the fully connected layer 2 are configured to work in ping-pong mode. The whole system takes ten mats, mats 1–10, and we duplicate it using mats 11–20. We do not replicate more because with two copies the data communication latency totally hides the computation latency. More copies cannot further improve the throughput because the latency-dominated data communication is serial among mats through the bus-based Buffer subarray.

10.9.3 Runtime Optimization

When FF subarrays are configured for NN applications, the memory space is reserved and supervised by the OS so that it is invisible to other user applications. However, during the runtime, if none or few of their crossbar arrays are used for computation, and the page miss rate is higher than the predefined threshold (which indicates the memory capacity is insufficient), the OS is able to release the reserved memory addresses as normal memory. It was observed that the memory requirement varies among workloads, and prior work has proposed to dynamically adjust the memory capacity by switching between SLC and MLC modes in PCM-based memory [47]. The page miss rate curve can be tracked dynamically by using either hardware or software approaches [48]. In our design, the granularity to flexibly configure a range of memory addresses for either computation or memory is crossbar array (mat): when an array is configured for computation, it stores multi-bit synaptic weights; when an array is used as normal memory, it stores data as single-bit cells. The OS works with the memory management unit (MMU) to keep all the mapping information of the FF subarrays and decides when and how much reserved memory space should be released, based on the combination of the page miss rate and the utilization of the FF subarrays for computation.

10.10 Evaluation

In this section, we evaluate our PRIME design. We first describe the experiment setup and then present the performance and energy results and estimate the area overhead.

10.10.1 Experiment Setup

Benchmark The benchmarks we use (*MlBench*) comprise six NN designs for machine learning applications, as listed in Table 10.3. *CNN-1* and *CNN-2* are two CNNs, and *MLP-S*/*M*/*L* are three multilayer perceptrons (MLPs) with different network scales: small, medium, and large. Those five NNs are evaluated on the widely used *MNIST* database of handwritten digits [49]. The sixth NN, *VGG-D*, is well known for ImageNet ILSVRC [37]. It is an extremely large CNN, containing 16 weight layers and 1.4×10^8 synapses and requiring $\sim 1.6 \times 10^{10}$ operations.

PRIME Configurations There are two FF subarrays and one Buffer subarray per bank (totally 64 subarrays). In FF subarrays, for each mat, there are 256×256 RRAM cells and eight 6-bit reconfigurable SAs; for each RRAM cell, we assume 4-bit MLC for computation while SLC for memory; the input voltage has eight levels (3-bit) for computation while two levels (1-bit) for memory. With our input and synapse composing scheme, for computation, the input and output data are 6-bit dynamic fixed point, and the weights are 8-bit.

Methodology We compare PRIME with several counterparts. The baseline is a CPU-only solution. The configurations of CPU and RRAM main memory are shown in Table 10.4, including key memory timing parameters for simulation.

Table 10.4 Configurations of CPU and memory

Processor	4 cores; 3GHz; out of order
L1 I&D cache	Private; 32 KB; 4-way; 2 cycles access
L2 cache	Private; 2 MB; 8-way; 10 cycles access
RRAM-based main memory	16GB RRAM; 533 MHz IO bus; 8 chips/rank; 8 banks/chip; tRCD-tCL-tRP-tWR 22.5–9.8-0.5-41.4 (ns)

Table 10.5 The configurations of comparatives

Description		Data path	Buffer
pNPU-co	Parallel NPU as coprocessor, similar to DianNao [1]	16×16 multiplier, $256 - 1$ adder tree	2 KB in/out, 32 KB weight
pNPU-pim	PIM version of the parallel NPU, 3D stacked to each bank		

We also evaluate two different NPU solutions: using a complex parallel NPU [1] as a coprocessor (pNPU-co) and using the NPU as a PIM processor through 3D stacking (pNPU-pim). The configurations of these comparatives are described in Table 10.5.

We model the above NPU designs using Synopsys Design Compiler and PrimeTime with 65 *nm* TSMC CMOS library. We also model RRAM main memory and our PRIME system with modified NVSim [50], CACTI-3DD [51], and CACTI-IO [52]. We adopt Pt/TiO_{2-x}/Pt devices [34] with $R_{on}/R_{off} = 1k\Omega/20k\Omega$ and 2 V SET/RESET voltage. The FF subarray is modeled by heavily modified NVSim, according to the peripheral circuit modifications, i.e., write driver [53], sigmoid [27], and sense amplifier [28] circuits. We built a trace-based in-house simulator to evaluate different systems, including CPU-only, PRIME, NPU coprocessor, and NPU PIM processor.

10.10.2 Performance Results

The performance results for *MlBench* are presented in Fig. 10.6. *MlBench* benchmarks use large NNs and require high-memory bandwidth, and therefore, they can benefit from PIM. To demonstrate the PIM advantages, we evaluate two pNPU-pim solutions: pNPU-pim-x1 is a PIM processor with a single parallel NPU stacked on top of memory and pNPU-pim-x64 with 64 NPUs, for comparison with PRIME which takes advantages of bank-level parallelism (64 banks). By comparing the speedups of pNPU-co and pNPU-pim-x1, we find that the PIM solution has a $9.1\times$ speedup on average over a coprocessor solution. Among all the solutions, PRIME achieves the highest speedup over the CPU-only solution, about $4.1\times$ of pNPU-pim-x64's. PRIME achieves a smaller speedup in *VGG-D* than other benchmarks, because it has to map the extremely large *VGG-D* across eight chips where the data

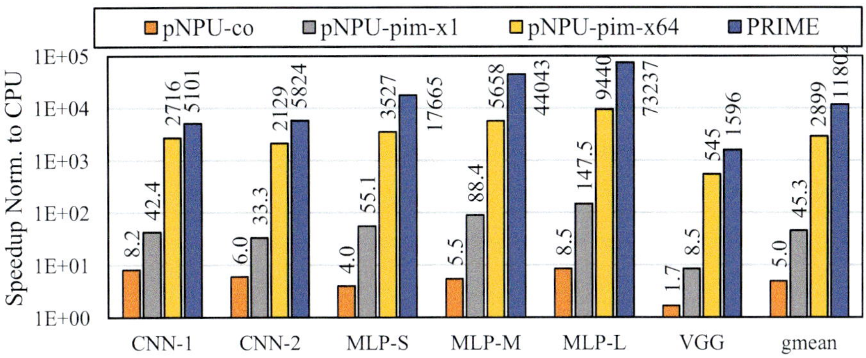

Fig. 10.6 The performance speedups (vs. CPU) and the geometric mean (gmean)

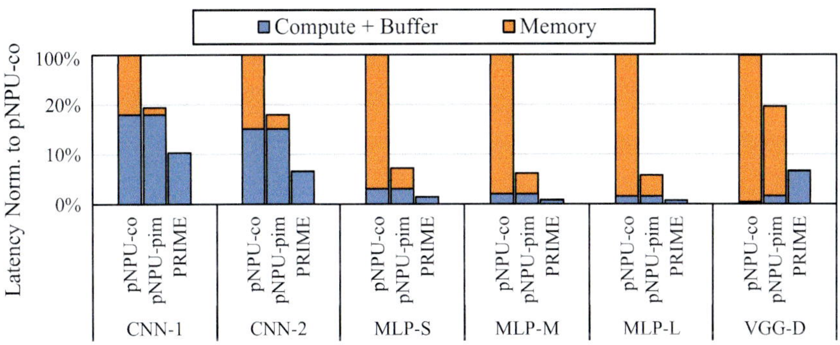

Fig. 10.7 The execution time breakdown (vs. pNPU-co)

communication between banks/chips is costly. The performance advantage of PRIME over the 3D-stacking PIM solution (pNPU-pim-x64) for NN applications comes from the efficiency of using RRAM for NN computation, because the synaptic weights have already been preprogrammed in RRAM cells and do not require data fetches from the main memory during computation. In our performance and energy evaluations of PRIME, we do not include the latency and energy consumption of configuring RRAM for computation, because we assume that once the configuration is done, the NNs will be executed for tens of thousands of times to process different input data.

Figure 10.7 presents the breakdown of the execution time normalized to pNPU-co. To clearly show the breakdown, we evaluate the results of pNPU-pim with one NPU and PRIME without leveraging bank parallelism for computation. The execution time is divided into two parts, computation and memory access. The computation part also includes the time spent on the buffers of NPUs or the Buffer subarrays of PRIME in managing data movement. We find that pNPU-pim reduces the memory access time a lot, and PRIME further reduces it to zero. Zero memory

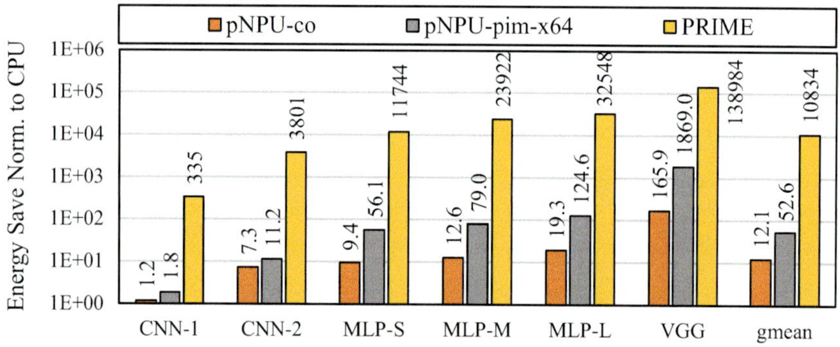

Fig. 10.8 The energy-saving results (vs. CPU) and the geometric mean (gmean)

access time does not imply that there is no memory access, but it means that the memory access time can be hidden by the Buffer subarrays.

10.10.3 Energy Results

The energy-saving results for *MlBench* are presented in Fig. 10.8. Figure 10.8 does not show the results of pNPU-pim-x1, because they are the same with those of pNPU-pim-x64. From Fig. 10.8, PRIME shows its superior energy efficiency to other solutions. pNPU-pim-x64 is several times more energy efficient than pNPU-co, because the PIM architecture reduces memory accesses and saves energy. The energy advantage of PRIME over the 3D-stacking PIM solution (pNPU-pim-x64) for NN applications comes from the energy efficiency of using RRAM for NN computation.

Figure 10.9 provides the breakdown of the energy consumption normalized to pNPU-co. The total energy consumptions are divided into three parts, computation energy, buffer energy, and memory energy. From Fig. 10.9, pNPU-pim-x64 consumes almost the same energy in computation and buffer with pNUP-co, but saves the memory energy by 93.9% on average by decreasing the memory accesses and reducing memory bus and I/O energy. PRIME reduces all the three parts of energy consumption significantly. For computation, RRAM-based analog computing is very energy efficient. Moreover, since each RRAM mat can store 256×256 synaptic weights, the cache and memory accesses to fetch the synaptic weights are eliminated. Furthermore, since each RRAM mat can execute as large as a $256 - 256$ NN at one time, PRIME also saves a lot of buffer and memory accesses to the temporary data. From Fig. 10.9, CNN benchmarks consume more energy in buffer and less energy in memory than MLP benchmarks. The reason is that the convolution layers and pooling layers of CNN usually have a small number of input data, synaptic weights, and output data, and buffers are effective to reduce memory accesses.

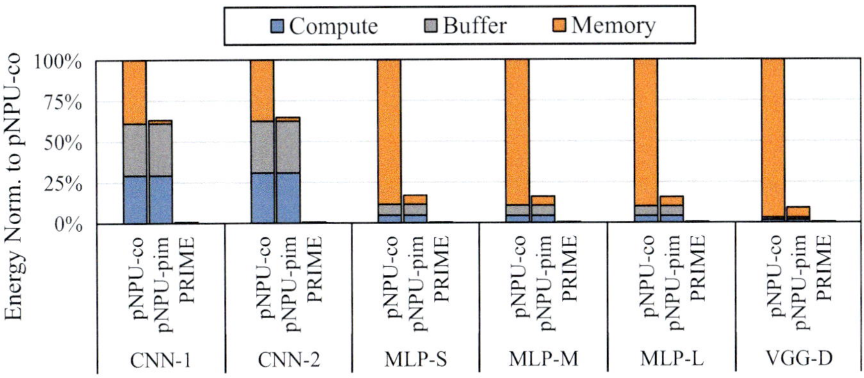

Fig. 10.9 The energy breakdown (vs. pNPU-co)

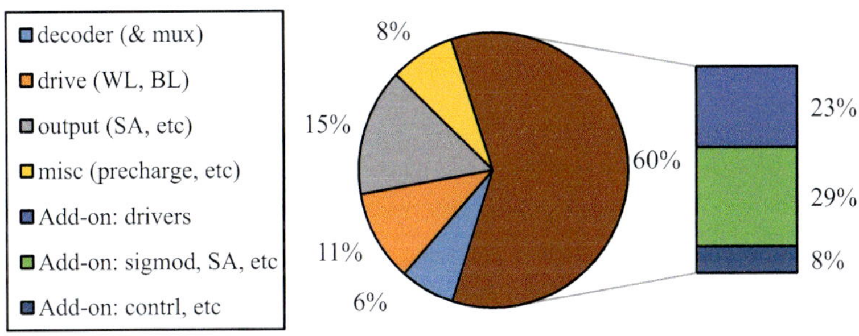

Fig. 10.10 The area overhead of PRIME

10.10.4 Area Overhead

Given two FF subarrays and one Buffer subarray per bank (64 subarrays in total), PRIME only incurs 5.76% area overhead. The choice of the number of FF subarrays is a trade-off between peak giga operations per second (GOPS) and area overhead. Our experimental results on *Mlbench* (except *VGG-D*) show that the utilities of FF subarrays are 39.8% and 75.9% on average before and after replication, respectively. For *VGG-D*, the utilities of FF subarrays are 53.9% and 73.6% before and after replication, respectively. Figure 10.10 shows the breakdown of the area overhead in a mat of an FF subarray. There is 60% area increase to support computation: the added driver takes 23%, the subtraction and sigmoid circuits take 29%, and the control, the multiplexer, etc. cost 8%.

10.11 Summary

This chapter introduced a novel processing in RRAM-based main memory design, PRIME, which substantially improves the performance and energy efficiency for neural network (NN) applications, benefiting from both the PIM architecture and the efficiency of RRAM-based NN computation. In PRIME, part of the RRAM memory arrays is enabled with NN computation capability. They can either perform computation to accelerate NN applications or serve as memory to provide a larger working memory space. The designs are presented from circuit level to system level. With circuit reuse, PRIME incurs an insignificant area overhead to the original RRAM chips. The experimental results show that PRIME can achieve a high speedup and significant energy saving for various NN applications using MLP and CNN.

References

1. T. Chen, Z. Du, N. Sun, J. Wang, C. Wu, Y. Chen, O. Temam, DianNao: a small-footprint high-throughput accelerator for ubiquitous machine-learning, in *ACM International Conference on Architectural Support for Programming Languages and Operating Systems (ASPLOS)* (2014)
2. Y. Chen, T. Luo, S. Liu, S. Zhang, L. He, J. Wang, L. Li, T. Chen, Z. Xu, N. Sun, O. Temam, DaDianNao: a machine-learning supercomputer, in *IEEE/ACM International Symposium on Microarchitecture (MICRO)* (2014)
3. P. Merolla, J. Arthur, F. Akopyan, N. Imam, R. Manohar, D.S. Modha, A digital neurosynaptic core using embedded crossbar memory with 45pj per spike in 45 nm, in *IEEE Custom Integrated Circuits Conference (CICC)* (2011)
4. S. K. Esser, A. Andreopoulos, R. Appuswamy, P. Datta, D. Barch, A. Amir, J. Arthur, A. Cassidy, M. Flickner, P. Merolla, S. Chandra, N. Basilico, S. Carpin, T. Zimmerman, F. Zee, R. Alvarez-Icaza, J. Kusnitz, T. Wong, W. Risk, E. McQuinn, T. Nayak, Cognitive computing systems: algorithms and applications for networks of neurosynaptic cores, in *International Joint Conference on Neural Networks (IJCNN)* (2013)
5. J. Seo, B. Brezzo, Y. Liu, B. Parker, S. Esser, R. Montoye, B. Rajendran, J. Tierno, L. Chang, D. Modha, D. Friedman, A 45 nm cmos neuromorphic chip with a scalable architecture for learning in networks of spiking neurons, in *IEEE Custom Integrated Circuits Conference (CICC)* (2011)
6. P. Chi, S. Li, C. Xu, T. Zhang, J. Zhao, Y. Liu, Y. Wang, Y. Xie, PRIME: a novel processing-in-memory architecture for neural network computation in ReRAM-based main memory, in *ACM/IEEE International Symposium on Computer Architecture (ISCA)* (2016)
7. D. Patterson, T. Anderson, N. Cardwell, R. Fromm, K. Keeton, C. Kozyrakis, R. Thomas, K. Yelick, A case for intelligent RAM. IEEE Micro. **17**(2), 34–44 (1997)
8. C.E. Kozyrakis, S. Perissakis, D. Patterson, T. Anderson, K. Asanovic, N. Cardwell, R. Fromm, J. Golbus, B. Gribstad, K. Keeton, R. Thomas, N. Treuhaft, K. Yelick, Scalable processors in the billion-transistor era: IRAM. Computer **30**(9), 75–78 (1997)
9. D. Patterson, K. Asanovic, A. Brown, R. Fromm, J. Golbus, B. Gribstad, K. Keeton, C. Kozyrakis, D. Martin, S. Perissakis, R. Thomas, N. Treuhaft, K. Yelick, Intelligent RAM (IRAM): the industrial setting, applications, and architectures, in *IEEE International Conference on Computer Design (ICCD)* (1997)

10. J. Draper, J. Chame, M. Hall, C. Steele, T. Barrett, J. LaCoss, J. Granacki, J. Shin, C. Chen, C. W. Kang, I. Kim, G. Daglikoca, The architecture of the DIVA processing-in-memory chip, in *ACM/IEEE International Conference on Supercomputing* (2002)
11. M. Gokhale, B. Holmes, K. Iobst, Processing in memory: the terasys massively parallel PIM array. Computer **28**(4), 23–31 (1995)
12. D. Elliott, M. Stumm, M. Snelgrove, Computational RAM: the case for simd computing in memory, in *Workshop on Mixing Logic and DRAM: Chips that Compute and Remember at ACM/IEEE International Symposium on Computer Architecture (ISCA)* (1997)
13. T. Yamauchi, L. Hammond, K. Olukotun, A single chip multiprocessor integrated with DRAM, in *Workshop on Mixing Logic and DRAM: Chips that Compute and Remember at ACM/IEEE International Symposium on Computer Architecture (ISCA)* (1997)
14. M. Oskin, F.T. Chong, T. Sherwood, Active pages: a computation model for intelligent memory. SIGARCH Comput Arch News **26**(3), 192–203 (1998)
15. R. Balasubramonian, J. Chang, T. Manning, J.H. Moreno, R. Murphy, R. Nair, S. Swanson, Near-data processing: insights from a micro-46 workshop. IEEE Micro. **34**(4), 36–42 (2014)
16. R. Nair, S.F. Antao, C. Bertolli, P. Bose, J.R. Brunheroto, T. Chen, C.Y. Cher, C.H.A. Costa, J. Doi, C. Evangelinos, B.M. Fleischer, T.W. Fox, D.S. Gallo, L. Grinberg, J.A. Gunnels, A.C. Jacob, P. Jacob, H.M. Jacobson, T. Karkhanis, C. Kim, J. Moreno, Active memory cube: a processing-in-memory architecture for exascale systems. IBM J. Res. Dev. **59**(2/3), 17:1–17:14 (2015)
17. D. Zhang, N. Jayasena, A. Lyashevsky, J.L. Greathouse, L. Xu, M. Ignatowski, TOP-PIM: throughput-oriented programmable processing in memory, in *ACM/IEEE International Symposium on High-performance Parallel and Distributed Computing (HPDC)* (2014)
18. Z. Guz, M. Awasthi, V. Balakrishnan, M. Ghosh, A. Shayesteh, T. Suri, Real-time analytics as the killer application for processing-in-memory, in *2nd Workshop on Near-Data Processing at IEEE International Symposium on Microarchitecture* (2014)
19. B. Akin, F. Franchetti, J.C. Hoe, Data reorganization in memory using 3D–stacked DRAM, in *ACM/IEEE International Symposium on Computer Architecture (ISCA)* (2015)
20. J. Ahn, S. Hong, S. Yoo, O. Mutlu, K. Choi, A Scalable Processing-in-memory Accelerator for Parallel Graph Processing, in *ACM/IEEE International Symposium on Computer Architecture (ISCA)* (2015)
21. S. H. Pugsley, J. Jestes, H. Zhang, R. Balasubramonian, V. Srinivasan, A. Buyuktosunoglu, A. Davis, F. Li, NDC: analyzing the impact of 3d–stacked memory + logic devices on mapreduce workloads, in *International Symposium on Performance Analysis of Systems and Software* (2014)
22. N. S. Mirzadeh, O. Kocberber, B. Falsafi, B. Grot, Sort vs. hash join revisited for near-memory execution, in *5th Workshop on Architectures and Systems for Big Data (ASBD) at IEEE International Symposium on Computer Architecture* (2015)
23. F. Alibart, T. Sherwood, D.B. Strukov, Hybrid CMOS/nanodevice circuits for high throughput pattern matching applications, in *NASA/ESA Conference on Adaptive Hardware and Systems (AHS)* (2011)
24. Q. Guo, X. Guo, Y. Bai, E. Ipek, A resistive TCAM accelerator for data-intensive computing, in *IEEE/ACM International Symposium on Microarchitecture (MICRO)* (2011)
25. Q. Guo, X. Guo, R. Patel, E. Ipek, E.G. Friedman, AC-DIMM: associative computing with STT-MRAM, in *ACM/IEEE International Symposium on Computer Architecture (ISCA)* (2013)
26. C. Xu, D. Niu, N. Muralimanohar, R. Balasubramonian, T. Zhang, S. Yu, Y. Xie, Overcoming the challenges of crossbar resistive memory architectures, in *IEEE International Symposium on High Performance Computer Architecture (HPCA)* (2015)
27. B. Li, P. Gu, Y. Shan, Y. Wang, Y. Chen, H. Yang, RRAM-based analog approximate computing. IEEE Trans. Comput. Aided Des. Integr. Circuits Syst. **34**(12), 1905–1917 (2015)
28. J. Li, C. I. Wu, S. C. Lewis, J. Morrish, T. Y. Wang, R. Jordan, T. Maffitt, M. Breitwisch, A. Schrott, R. Cheek, H. L. Lung, C. Lam, A novel reconfigurable sensing scheme for variable

level storage in phase change memory, in *IEEE International Memory Workshop (IMW)* (2011)
29. M. Hu, H. Li, Q. Wu, G.S. Rose, Hardware realization of BSB recall function using memristor crossbar arrays, in *ACM/IEEE Design Automation Conference (DAC)* (2012)
30. B. Li, Y. Shan, M. Hu, Y. Wang, Y. Chen, H. Yang, Memristor-based approximated computation, in *IEEE/ACM International Symposium on Low Power Electronics and Design (ISLPED)* (2013)
31. M. Prezioso, F. Merrikh-Bayat, B. Hoskins, G. Adam, K.K. Likharev, D.B. Strukov, Training and operation of an integrated neuromorphic network based on metal-oxide memristors. Nature **521**, 61–64 (2015)
32. Y. Kim, Y. Zhang, P. Li, A reconfigurable digital neuromorphic processor with memristive synaptic crossbar for cognitive computing. ACM J. Emerg. Technol Comput. Syst. **11**(4), 38:1–38:25 (2015)
33. F. Alibart, L. Gao, B.D. Hoskins, D.B. Strukov, High precision tuning of state for memristive devices by adaptable variation-tolerant algorithm. Nanotechnology **23**(7), 075201 (2012)
34. L. Gao, F. Alibart, D.B. Strukov, A high resolution nonvolatile analog memory ionic devices, in *4th Non-Volatile Memories Workshop (NVMW)* (2013)
35. M. Hu, J.P. Strachan, E. Merced-Grafals, Z. Li, R.S. Williams, Dot-product engine: programming memristor crossbar arrays for efficient vector-matrix multiplication, in *Workshop on "Towards Efficient Computing in the Dark Silicon Era" at IEEE/ACM International Conference on Computer-Aided Design* (2015)
36. M. Courbariaux, Y. Bengio, J.-P. David, Low precision storage for deep learning, in *arXiv:1412.7024* (2015)
37. K. Simonyan, A. Zisserman, Very deep convolutional networks for large-scale image recognition, in *International Conference on Learning Representations (ICLR)* (2015)
38. F. Alibart, E. Zamanidoost, D.B. Strukov, Pattern classification by memristive crossbar circuits using ex situ and in situ training. Nat. Commun. **4**, 2013 (2072)
39. M. Hu, H. Li, Y. Chen, Q. Wu, G. S. Rose, BSB training scheme implementation on memristor-based circuit, in *IEEE Symposium on Computational Intelligence for Security and Defense Applications (CISDA)* (2013)
40. B. Li, Y. Wang, Y. Wang, Y. Chen, H. Yang, Training itself: mixed-signal training acceleration for memristor-based neural network, in *ACM/IEEE Asia and South Pacific Design Automation Conference (ASP-DAC)* (2014)
41. B. Liu, M. Hu, H. Li, Z.-H. Mao, Y. Chen, T. Huang, W. Zhang, Digital-assisted noise-eliminating training for memristor crossbar-based analog neuromorphic computing engine, in *ACM/IEEE Design Automation Conference (DAC)* (2013)
42. B. Liu, H. Li, Y. Chen, X. Li, T. Huang, Q. Wu, M. Barnell, Reduction and IR-drop compensations techniques for reliable neuromorphic computing systems, in *IEEE/ACM International Conference on Computer-Aided Design (ICCAD)* (2014)
43. M. Prezioso, F. Merrikh-Bayat, B. Hoskins, G. Adam, K.K. Likharev, D.B. Strukov, Training and operation of an integrated neuromorphic network based on metal-oxide memristors. Nature **521**, 61–64 (2014)
44. V. Seshadri, Y. Kim, C. Fallin, D. Lee, R. Ausavarungnirun, G. Pekhimenko, Y. Luo, O. Mutlu, P.B. Gibbons, M.A. Kozuch, T.C. Mowry, RowClone: fast and energy-efficient in-DRAM bulk data copy and initialization, in *IEEE/ACM International Symposium on Microarchitecture (MICRO)* (2013)
45. B. Verghese, S. Devine, A. Gupta, M. Rosenblum, Operating system support for improving data locality on CC-NUMA compute servers, in *ACM International Conference on Architectural Support for Programming Languages and Operating Systems (ASPLOS)* (1996)
46. N. Agarwal, D. Nellans, M. Stephenson, M. O'Connor, S.W. Keckler, Page placement strategies for GPUs within heterogeneous memory systems, in *ACM International Conference on Architectural Support for Programming Languages and Operating Systems (ASPLOS)* (2015)

47. M.K. Qureshi, M.M. Franceschini, L.A. Lastras-Montano, J.P. Karidis, Morphable memory system: a robust architecture for exploiting multi-level phase change memories, in *ACM/IEEE International Symposium on Computer Architecture (ISCA)* (2010).
48. P. Zhou, V. Pandey, J. Sundaresan, A. Raghuraman, Y. Zhou, S. Kumar, Dynamic tracking of page miss ratio curve for memory management, in *ACM International Conference on Architectural Support for Programming Languages and Operating Systems (ASPLOS)* (2004)
49. Y. Lecun, L. Bottou, Y. Bengio, P. Haffner, Gradient-based learning applied to document recognition. Proc. IEEE **86**(11), 2278–2324 (1998)
50. X. Dong, C. Xu, Y. Xie, N.P. Jouppi, Nvsim: a circuit-level performance, energy, and area model for emerging nonvolatile memory. IEEE Trans. Comput. Aided Des. Integr. Circuits Syst. **31**(7), 994–1007 (2012)
51. K. Chen, S. Li, N. Muralimanohar, J.H. Ahn, J.B. Brockman, N.P. Jouppi, CACTI-3DD: architecture-level modeling for 3D die-stacked DRAM main memory, in *ACM/IEEE Design, Automation Test in Europe Conference (DATE)* (2012)
52. N. P. Jouppi, R.B. Kahng, N. Muralimanohar, V. Srinivas, CACTI-IO: CACTI with off-chip power-area-timing models, in *ACM/IEEE International Conference on Computer-Aided Design (ICCAD)* (2012)
53. C. Xu, D. Niu, N. Muralimanohar, N.P. Jouppi, Y. Xie, Understanding the trade-offs in multi-level cell ReRAM memory design, in *ACM/IEEE Design Automation Conference (DAC)* (2013)

Chapter 11
Multilayer Perceptron Algorithm: Impact of Nonideal Conductance and Area-Efficient Peripheral Circuits

Lucas L. Sanches, Alessandro Fumarola, Severin Sidler, Pritish Narayanan, Irem Boybat, Junwoo Jang, Kibong Moon, Robert M. Shelby, Yusuf Leblebici, Hyunsang Hwang, and Geoffrey W. Burr

11.1 Introduction

By performing computation at the location of data, non-Von Neumann (non-VN) computing *ought* to provide significant power and speed benefits (Fig. 11.1) on specific and assumedly important tasks. For one such non-VN approach – on-chip training of large-scale ANN using NVM-based synapses [1–4] – viability will require several things. First, despite the inherent imperfections of NVM devices such as phase-change memory (PCM) [1, 2] or resistive RAM (RRAM) [4], such NVM-based networks *must* achieve competitive performance levels (e.g., classification accuracies) when compared to ANN trained using CPUs or GPUs. Second, the benefits of performing computation at the data (Fig. 11.2) must confer a decided advantage in either training power or training speed (or preferably both). And finally, any on-chip accelerator should be applicable toward networks of different

L.L. Sanches • P. Narayanan • R.M. Shelby • G.W. Burr (✉)
IBM Research – Almaden, 650 Harry Road, San Jose, CA 95120, USA
e-mail: gwburr@us.ibm.com

A. Fumarola • S. Sidler • I. Boybat
IBM Research – Almaden, 650 Harry Road, San Jose, CA 95120, USA

EPFL, ELB111, Station 11, Lausanne CH–1015, Switzerland

J. Jang
IBM Research – Almaden, 650 Harry Road, San Jose, CA 95120, USA

Department of Materials Science and Engineering, Pohang University of Science and Technology, 77 Cheongamro, Nam-gu, Pohang, Gyeongbuk 37673, South Korea

K. Moon • H. Hwang
Department of Materials Science and Engineering, Pohang University of Science and Technology, 77 Cheongamro, Nam-gu, Pohang, Gyeongbuk 37673, South Korea

Y. Leblebici
EPFL, ELB111, Station 11, Lausanne CH–1015, Switzerland

S. Yu (ed.), *Neuro-inspired Computing Using Resistive Synaptic Devices*,
DOI 10.1007/978-3-319-54313-0_11

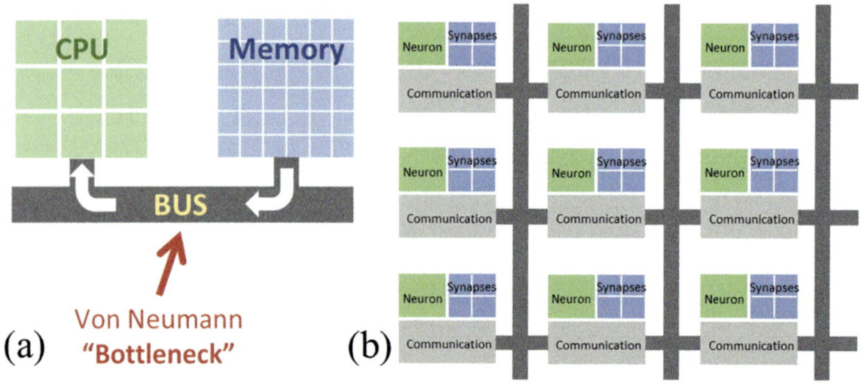

Fig. 11.1 In the Von Neumann architecture (**a**), data (both operations and operands) must move to and from the dedicated central processing unit (CPU) along a bus. In contrast, in a non-Von Neumann architecture (**b**), distributed computations take place at the location of the data, reducing the time and energy spent moving data around (Adapted from Burr et al. [1])

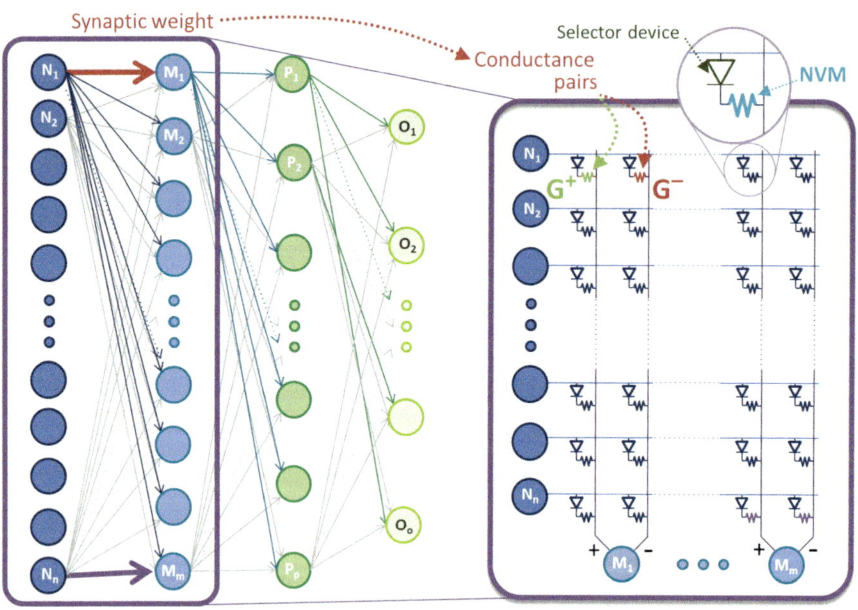

Fig. 11.2 Neuro-inspired non-Von Neumann computing [1–4], in which neurons activate each other through dense networks of programmable synaptic weights, can be implemented using dense crossbar arrays of nonvolatile memory (*NVM*) and selector device pairs (Adapted from Burr et al. [1])

types (fully connected "deep" neural network or convolutional neural network) and be reconfigurable for networks of different shapes (wide, with many neurons, or deep, with many layers).

We briefly review our work [1–4] in assessing the accuracy, speed, and power potential of on-chip NVM-based machine learning.

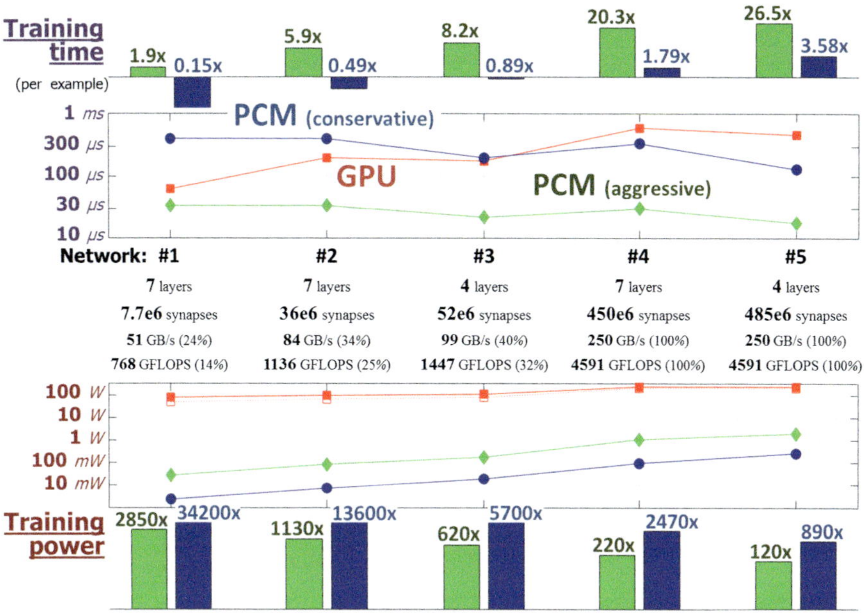

Fig. 11.3 Predicted training time (per ANN example) and power for five ANNs, ranging from 0.2 GB to nearly 6 GB [3]. Under moderately aggressive assumptions for parallel-read and parallel-write speed, PCM-based on-chip machine learning can offer lower-power and faster training for both large and small networks (Adapted from Burr et al. [3])

11.1.1 Comparative Analysis of Speed and Power

We have previously assessed the potential advantages, in terms of speed and power, of on-chip machine learning (ML) of large-scale artificial neural networks (ANN) using nonvolatile memory (NVM)-based synapses, in comparison to conventional GPU-based hardware [3].

Under moderately aggressive assumptions for parallel-read and parallel-write speed, *PCM-based on-chip machine learning can potentially offer lower-power and faster training (per ANN example) than GPU-based training* for both large and small networks (Fig. 11.3), even with the time and energy required for occasional RESET (forced by the large asymmetry between gentle partial SET and abrupt RESET in PCM). Critical here is the design of area-efficient read/write circuitry so that many copies of this circuitry operate in parallel (each handling a small number of columns (rows), c_s).

11.1.2 Potential for Competitive Classification Accuracies

Using two phase-change memory (PCM) devices per synapse, we demonstrated a three-layer perceptron (fully connected ANN) with 164,885 synapses [1], trained

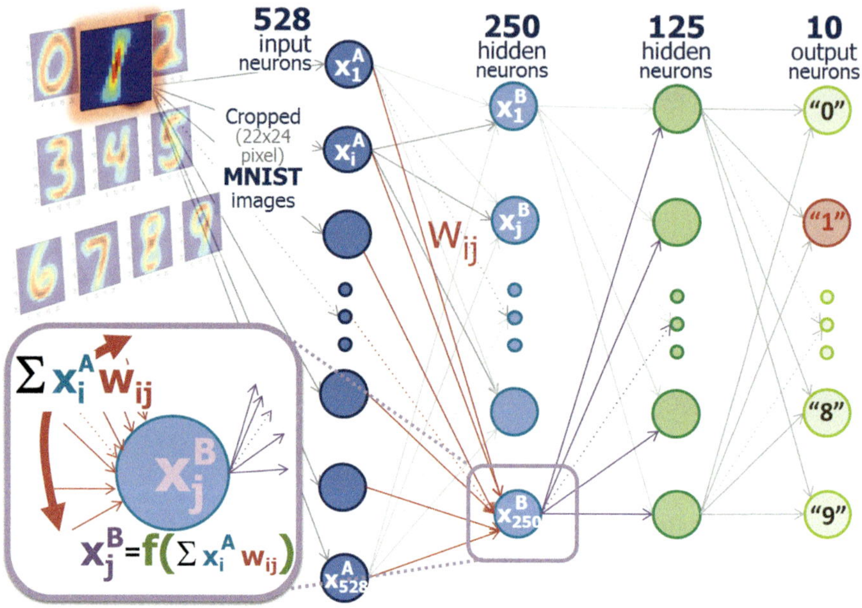

Fig. 11.4 In forward evaluation of a multilayer perceptron, each layer's neurons drive the next layer through weights w_{ij} and a nonlinearity $f()$. Input neurons are driven by input (for instance, pixels from successive MNIST images (cropped to 22×24)); the ten output neurons classify which digit was presented (Adapted from Burr et al. [1])

with backpropagation [5] on a subset (5000 examples) of the MNIST database of handwritten digits [6] (Fig. 11.4), using a modified weight-update rule compatible with NVM+selector crossbar arrays [1]. We proved that this weight-update modification does not degrade the high "test" (generalization) accuracies that such a three-layer network inherently delivers on this problem when trained in software [1]. However, nonlinearity and asymmetry in PCM conductance response limited both "training" and "test" accuracy in our original, mixed hardware-software experiments to 82–83% [1] (Fig. 11.5).

Asymmetry (between the gentle conductance increases of PCM partial SET and the abrupt conductance decrease of a PCM RESET operation) was mitigated by an occasional RESET strategy, which could be both infrequent and inaccurate [1]. While in these initial experiments, network parameters such as learning rate η had to be tuned very carefully, a modified "local gains" algorithm offered wider tolerance to η, higher classification accuracies, and lower training energy [3] (Fig. 11.6).

Tolerancing results showed that all NVM-based ANN can be expected to be *highly resilient to random effects* (NVM variability, yield, and stochasticity), but *highly sensitive to "gradient" effects that act to steer all synaptic weights* [1]. We showed that a bidirectional NVM with a symmetric, linear conductance response of

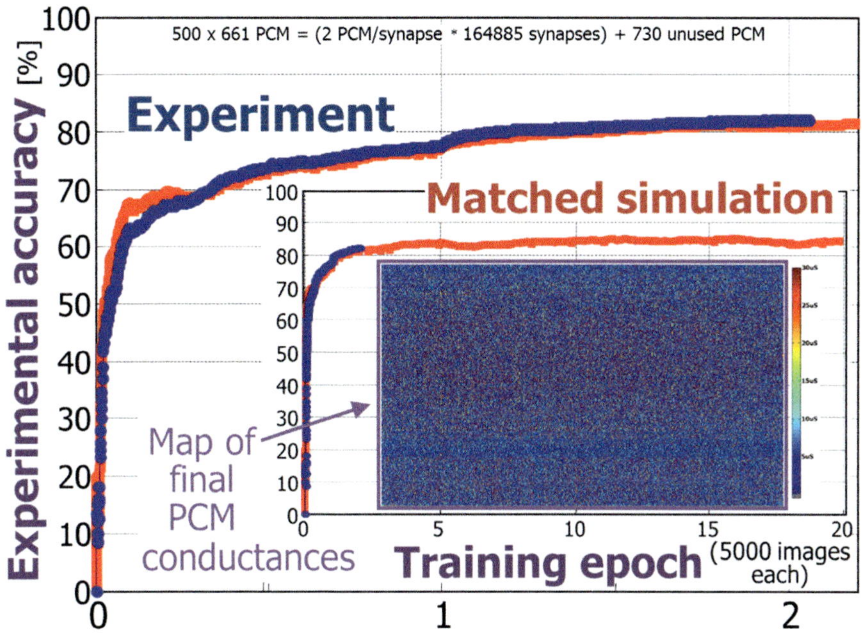

Fig. 11.5 Training accuracy for a three-layer perceptron of 164,885 hardware synapses [1], with all weight operations taking place on a 500×661 array of mushroom-cell PCM devices. Also shown is a matched computer simulation of this ANN, using parameters extracted from the experiment (Adapted from Burr et al. [1])

finite but large dynamic range (e.g., each conductance step is relatively small) can deliver the same high classification accuracies on the MNIST digits as a conventional, software-based implementation (Fig. 11.7). One key observation is the importance of avoiding constraints on weight magnitude that arise when the two conductances are either both small or both large – e.g., synapses should remain in the center stripe of the "G-diamond" [2].

In this chapter, we extend upon these observations and address several different yet useful topics. We assess the impact of undesired, time-varying conductance change, including drift in phase-change memory (PCM) and leakage of analog CMOS capacitors. We investigate the use of nonfilamentary, bidirectional RRAM devices based on $PrCaMnO_3$ (PCMO), with an eye to develop material variants that provide sufficiently linear conductance change. And finally, we explore trade-offs in designing peripheral circuitry, balancing simplicity and area efficiency against the impact on ANN performance.

11.1.3 Jump Table Concept

A highly useful concept in modeling the behavior of real NVM devices for neuromorphic applications is the concept of a "jump table." For backpropagation

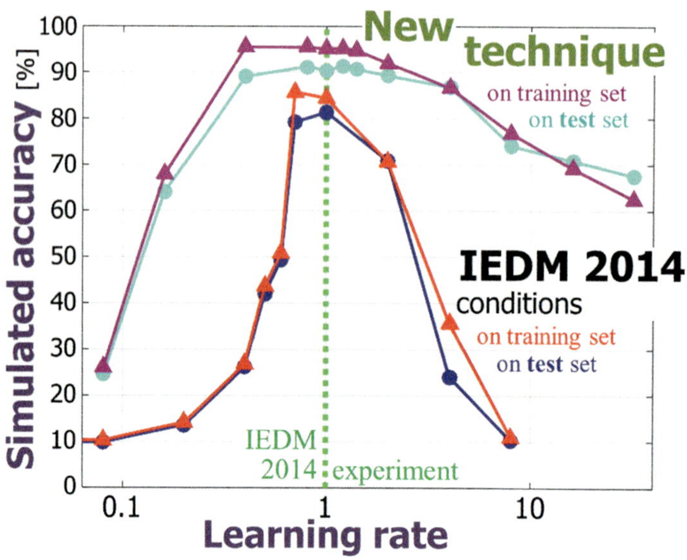

Fig. 11.6 A large number of synapses tend to "dither" in place, with frequent updates whose aggregate effect *ought* to be zero (but which is nonzero due to the nonlinearity and asymmetry of NVM-based synapses). By suppressing update of such synapses, ANN performance can be improved and training energy reduced while reducing the need to tune the learning rate precisely (Adapted from Burr et al. [3])

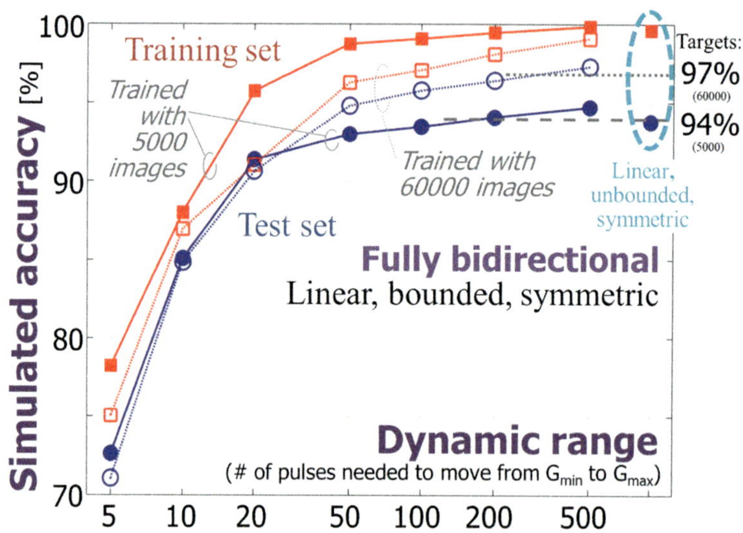

Fig. 11.7 When the dynamic range of the linear response is large, the classification accuracy can now reach the peak accuracy supported by the original neural network (a *test* accuracy of 94% when trained with 5000 images and of 97% when trained with all 60,000 images) (Adapted from Burr et al. [2])

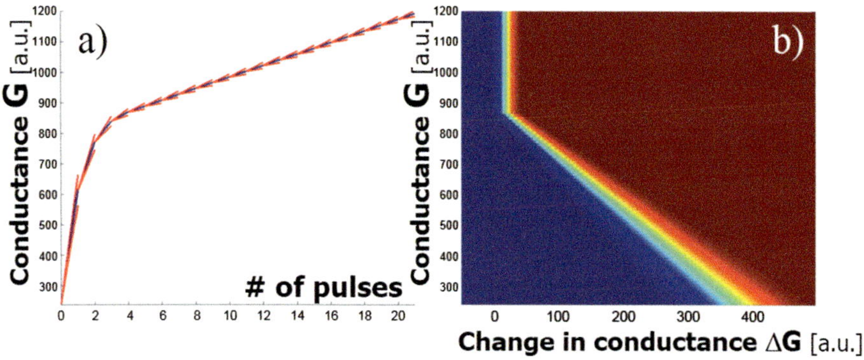

Fig. 11.8 (**a**) Example median (*blue*) and $\pm 1\sigma$ (*red*) conductance response for potentiation. (**b**) Associated jump table that fully captures this (artificially constructed in this case) conductance response, with cumulative probability plotted in color (from 0% to 100%) of any conductance change ΔG at any given initial conductance G (Adapted from Sidler et al. [7])

training, where one or more copies of the same programming pulse are applied to the NVM for adjusting the weights [1], we simply need one jump table for potentiation (SET) and one for depression (RESET).

With a pair of such jump tables, we can capture the nonlinearity of conductance response as a function of conductance (e.g., the same pulse might create a large "jump" at low conductance, but a much smaller jump at high conductance), the asymmetry between positive (SET) and negative (RESET) conductance changes, and the inherent stochastic nature of each jump. Figure 11.8a plots median conductance change for potentiation (blue) together with the $\pm 1\sigma$ stochastic variation about this median change (red). Figure 11.8b shows the jump table that fully captures this conductance response, plotting the cumulative probability (in color, from 0 to 100%) of any conductance change ΔG at any given initial conductance G. This table is ideal for computer simulation because a random number r (uniform deviate, between 0.0 and 1.0) can be converted to a resulting ΔG produced by a single pulse by scanning along the row associated with the conductance G (of the device before the pulse is applied) to find the point at which the table entry just exceeds r.

11.2 Impact of Nonlinear Conductance Response

We have previously used a measured jump table to simulate the SET response of PCM devices [1] and are currently exploring the use of similarly measured jump tables for PCMO. In order to develop an intuitive understanding of the impact that various features of such jump tables have on the classification performance in the ANN application, we have studied various artificially constructed jump tables

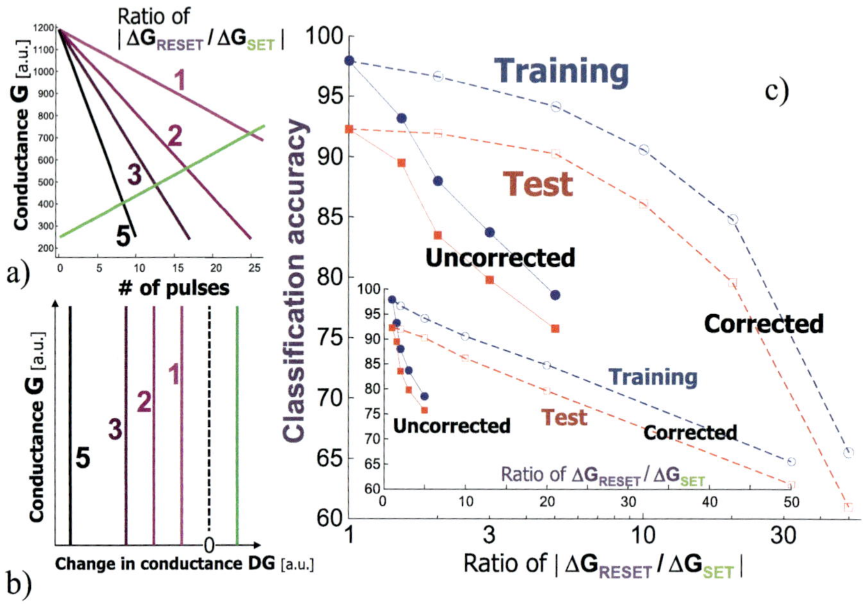

Fig. 11.9 (**a**) For a set of constructed linear conductance responses where the depression (RESET, *magenta*) response is steeper than the base potentiation (SET, *green*) response, the (**b**) resulting jump table shows larger (but constant) steps for RESET. (For clarity, only median response is shown.) (**c**) Although even a small SET/RESET asymmetry causes performance to fall off steeply (*solid curves with filled symbols*), the downstream neuron can partially compensate for this asymmetry by firing fewer RESET pulses (or more SET pulses). Inset shows same data plotted on a linear horizontal scale (Adapted from Sidler et al. [7])

[7]. Except for the specific jump tables, these simulations were identical to those performed in Ref. [1], spanning 20 epochs.

The first question we addressed was the impact of asymmetry in conductance response. We assumed both conductance responses are linear (Fig. 11.9a) but assumed that RESET conductance response was much steeper than SET, corresponding to an increased step size for the depression (RESET) jump table (Fig. 11.9b). As shown by the solid curves with filled symbols in Fig. 11.9c, even a small degree of asymmetry can cause classification accuracy to fall steeply. However, each downstream neuron has knowledge of the sign of the backpropagated correction, δ, and thus knows whether it is attempting a SET or RESET. This implies that asymmetry can be partly offset by "correcting" a steeper RESET response by firing commensurately fewer RESET pulses (or more SET pulses) [7]. As shown by the dotted curves with open symbols in Fig. 11.9c, this markedly expands the asymmetry that could potentially be accommodated.

Figure 11.10 examines jump tables that incorporate some degree of initial nonlinearity in the SET conductance response (Fig. 11.10a). The relative extent of the linear region was varied from 100% (fully linear) down to near 0% (fully nonlinear). For this and all subsequent studies, we assume that RESET operations

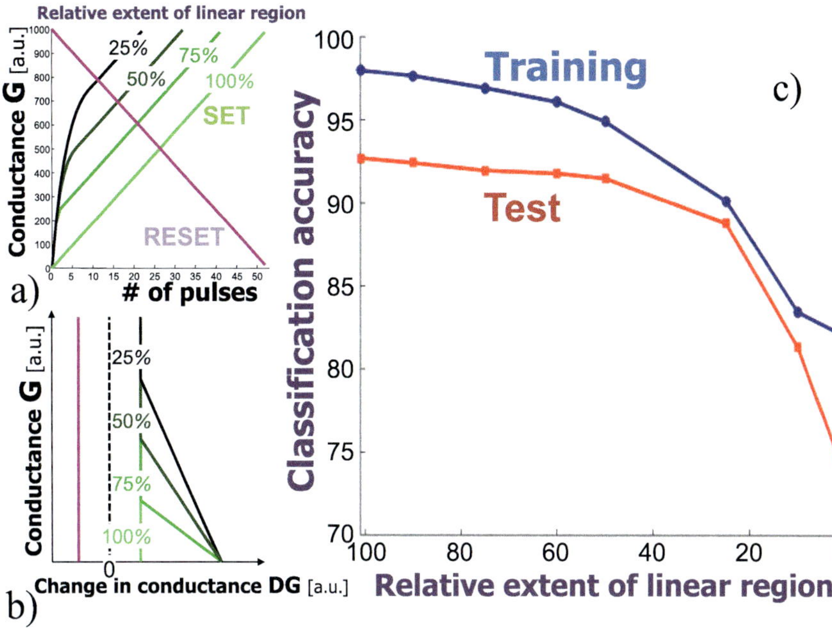

Fig. 11.10 Impact of relative extent of linear region on neural network performance. (RESET conductance response remains linear at all times.) (**a**) Conductance vs. number of pulses, (**b**) hypothetical jump tables studied, and (**c**) impact on training and test accuracy. A substantial nonlinear conductance region (up to ~50%) can be accommodated without loss in application performance (Adapted from Sidler et al. [7])

remain perfectly linear and symmetric to SET (Fig. 11.10b). We find that a substantial nonlinear conductance region (up to ~50%) can be accommodated without a significant drop-off in the neural network performance (Fig. 11.10c) [7].

Figure 11.11 examines the impact of the strength of this initial nonlinearity on the neural network performance. In these experiments, a stronger (weaker) nonlinearity implies fewer (more) steps to traverse the extent of the nonlinear region (representing 25% of the total conductance range, Fig. 11.11a). The strength is defined as the ratio between the size of the final (minimum) conductance jump and the initial (maximum) conductance jump (Fig. 11.11b). Again, we find that the strength of the nonlinearity has little impact on the test accuracy (Fig. 11.11c), so long as the linear region is sufficiently large [7].

We also investigate fully nonlinear conductance responses of varying strengths (Fig. 11.12a, b). We find that it is still possible to achieve high classification accuracies (Fig. 11.12c), so long as the ratio of the minimum to maximum conductance jumps is >0.5. However, larger nonlinearities cause a marked drop-off in network performance, as a large portion of the dynamic range can be used up by just a few training pulses [7].

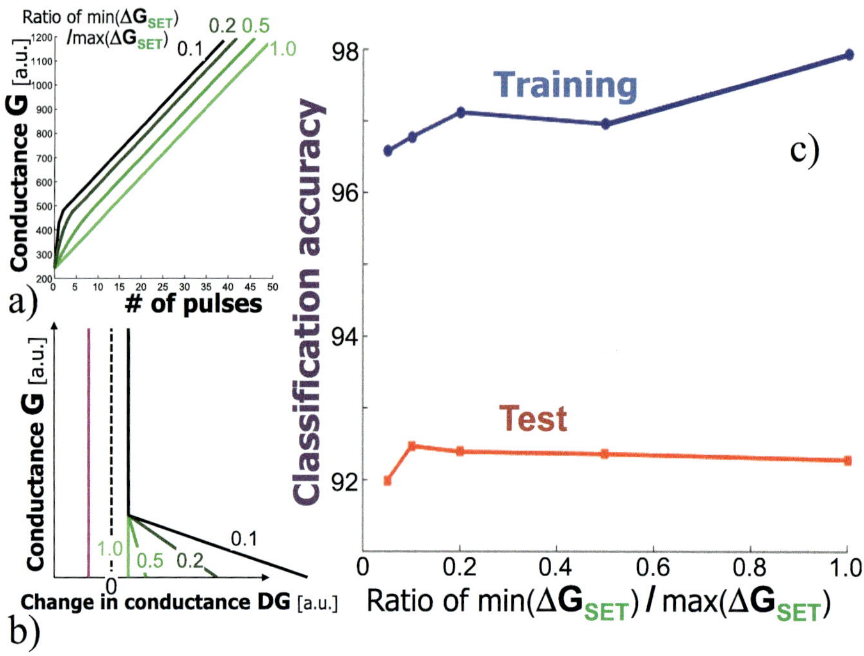

Fig. 11.11 Impact of the strength of an initial nonlinearity on neural network performance. (**a**) Conductance vs. number of pulses, (**b**) hypothetical jump tables studied, and (**c**) impact on training and test accuracy. Strength of an initial nonlinearity does not impact test classification accuracy, so long as a sufficiently large linear region is available (Adapted from Sidler et al. [7])

11.3 Time-Dependent Conductance Response

One aspect of phase-change memory that we did not address in our original tolerancing paper [1] was the role of resistance drift [8], also known as amorphous relaxation. As shown in Fig. 11.13, after a RESET operation, amorphous relaxation causes conductances to decrease, rapidly at first but then more and more slowly. We have modeled this in our neural network simulator for the network of Fig. 11.4, for an otherwise near-perfect PCM device, in which partial-SET conductance increases are gentle and linear (each ~0.5% of the conductance extent) and occasional RESET is performed fairly frequently (every 100 examples) with high precision [9]. The time response for drift starts upon RESET operations, with partial-SET operations assumed only to shift the conductance states without affecting the underlying time response of the amorphous relaxation.

As expected, as drift coefficients increase dramatically (to the values of $\nu \sim 0.1$ observed for fully amorphous (strong RESET) states), then accuracy is eventually affected (Fig. 11.14). However, for the much lower ν values ($\nu \sim 0.005$–0.010) associated with SET and the near-SET states relevant to PCM-based implementations of neural networks, accuracy is only minimally affected.

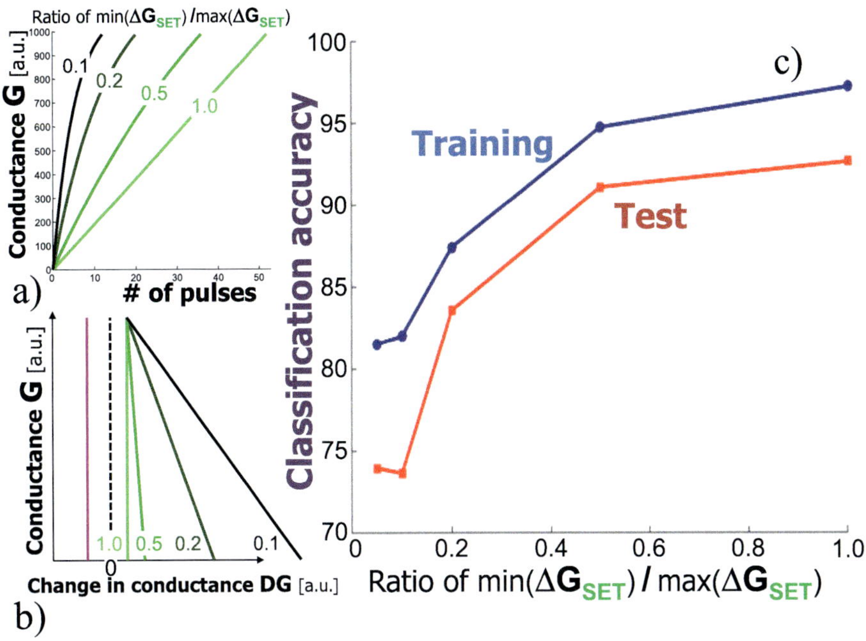

Fig. 11.12 Impact of fully nonlinear conductance response. (**a**) Conductance vs. number of pulses, (**b**) hypothetical jump tables studied, and (**c**) impact on training and test accuracy. Even in the absence of a linear region, it is possible to achieve high performance – however, the ratio of minimum to maximum conductance change needs to be sufficiently large (>0.5) (Adapted from Sidler et al. [7])

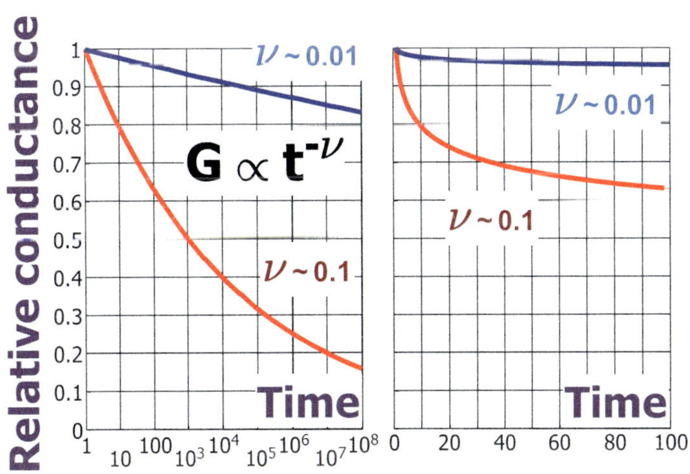

Fig. 11.13 After a RESET operation, amorphous relaxation causes conductances to decrease, rapidly at first but then more and more slowly. Plots show the same evolution of linear conductance as a function of time on log (*left*) and linear (*right*) scales, for two different values of drift coeffcient ν (Adapted from Fumarola et al. [9])

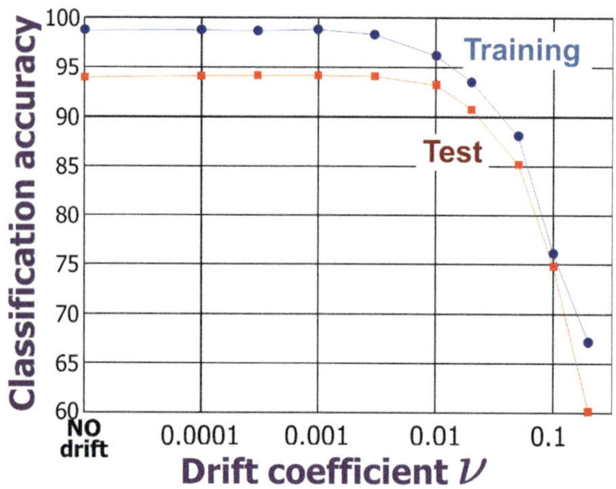

Fig. 11.14 At the large drift coefficients associated with fully amorphous RESET phase-change memory devices ($\nu \sim 0.1$), neural network accuracy is significantly degraded. At the values of $\nu \sim 0.005$–0.010 relevant to the SET and the near-SET states that dominate the PCM-based implementations of neural networks, accuracy is only slightly degraded (Results shown for ten epochs of simulated training on 5000 examples from the MNIST dataset) (Adapted from Fumarola et al. [9])

We performed a similar study for the case of fully volatile analog memory elements, such as CMOS capacitors, in which any conductance state immediately begins to decay toward zero after a programming operation. This study was performed with perfectly linear bidirectional conductances with ~0.5% conductance change per pulse, and thus with no decay (e.g., at large RC time constants), the network is identical to the right-hand side of Fig. 11.7, where accuracy becomes extremely high for high synaptic dynamic range.

In this study, we quantify the effective decay constant (the "RC time constant") not in absolute units but relative to the time required for training of a single data example (e.g., forward evaluation, reverse propagation, and weight update). As shown in Fig. 11.15, accuracy is strongly affected as soon as the ratio between the RC time constant and the time per example falls below 10,000. However, these initial results revealed an extremely interesting dependence on the choice of learning rate, implying that some further optimization may be possible. Figure 11.16 shows that the same global learning rate which is optimal for a truly nonvolatile conductance (infinite RC time constant) is decidedly suboptimal when the RC time constant becomes lower. This implies that it is better to either update so many weights that one can counteract the loss of conductance by retraining those weights or so few that the number of weights being touched (and thus placed into a mode where they will decay rapidly) is much lower.

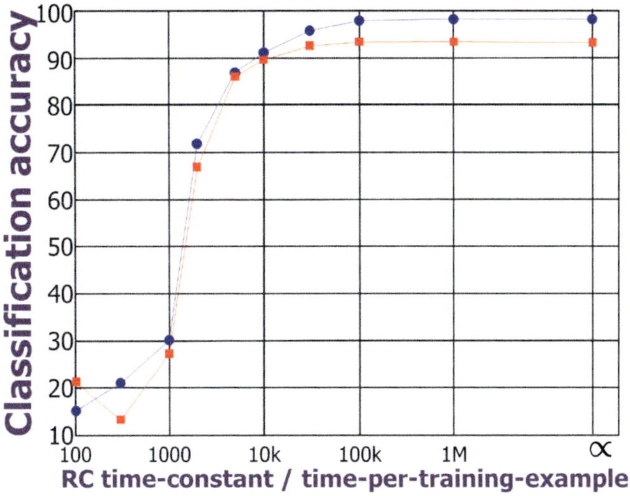

Fig. 11.15 Neural network accuracy is strongly affected as soon as the RC time constant becomes less than 10,000× larger than the time needed for each training example (Results shown for ten epochs of simulated training on 5000 examples from the MNIST dataset, all at the same global learning rate, $\eta \sim 1$) (Adapted from Fumarola et al. [9])

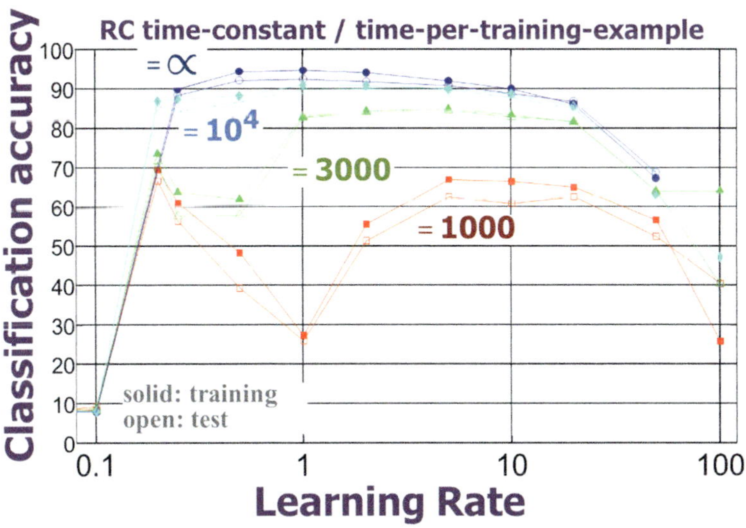

Fig. 11.16 For truly nonvolatile weights (infinite RC time constant), neural network accuracy is optimized by using a global learning rate that is large enough to affect a moderate number of weights, but not so many that chaos ensues. However, as the RC time constant decreases, the volatility of the conductance states favors either a larger learning rate (e.g., we adjust for the decaying weights by retraining many more of them) or, curiously, *lower* learning rates (assumably reducing the number of recently touched weights that cannot be trusted not to move without being actively programmed) (Adapted from Fumarola et al. [9])

11.4 Impact of Measured PCMO Conductance Response

We have previously studied the impact of the conductance response of PCMO material by fitting a set of functions to the average conductance response [4]. However, this approach is limited by the discrepancy between the real conductance response and the function chosen, and it does not include any stochastic aspect of the conductance response, for scenarios where the conductance response can vary significantly from the average conductance response.

We have studied the use of measured jump tables for the nonfilamentary RRAM material $Pr_xCa_{1-x}MnO_3$, also known as PCMO [9].

11.4.1 Analog Bidirectional Switching

Resistive switching in PCMO-based devices is caused by slow and gradual drift of oxygen ions and vacancies in the polycrystalline PCMO layer. Injection (removal) of oxygen ions takes place at the PCMO-oxide (-metal) interface through oxidation (reduction) reactions. Asymmetry in the device structure and the oxidation-reduction reactions contribute to the asymmetry in the switching characteristics, but PCMO-based NVMs show gradual SET and RESET characteristics. Thus, unlike phase-change memory (PCM) materials, there is no need to stop training and perform an occasional RESET operation [1]. Both the average conductance response and its statistical behavior can be described by a measured jump table (Fig. 11.17). (Note that unlike nonfilamentary RRAM such as PCMO, a filament-based RRAM such as HfO_x, TaO_x, or TiO_x exhibits only gradual RESET characteristics, meaning that such filamentary RRAM devices will likely still require an "occasional SET" step, analogous to "occasional RESET" with PCM.)

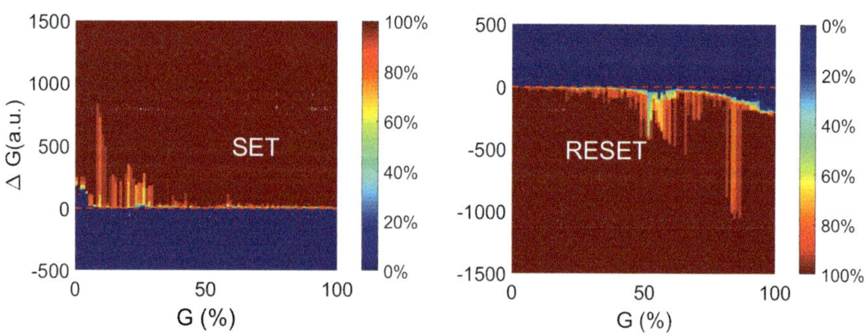

Fig. 11.17 Jump table of Al/Mo:PCMO-based RRAM devices for positive (SET) and negative (RESET) conductance changes. Unlike phase-change memory (PCM) devices, these materials provide both gradual RESET and gradual SET, enabling truly bidirectional programming. 50,000 total SET pulses (4.0 V, 10 ms) and RESET pulses (3.5 V, 10 ms) followed by 1.0 V read pulses were used on three identically sized (200 nm) devices (Adapted from Fumarola et al. [9])

11.4.2 Fabrication Process

A 30-nm PCMO polycrystalline layer was deposited on a 50-nm-thick Pt layer, which served as bottom electrode. Next, an 80-nm-thick SiN_x layer was deposited by chemical vapor deposition, and via holes (from 0.15 to 1.0 μm) were formed by conventional lithography and reactive-ion etching. The Al and Mo layers (5 nm each) and an 80-nm-thick Pt layer (top electrode) were deposited and patterned by conventional lithography. Electrical characteristics of the Al/Mo:PCMO-based resistive memory devices were measured using an Agilent B1500A [9].

11.4.3 Simulated Performance

A three-layer perceptron with two PCMO-based devices per synapse was simulated performing a classification task on the MNIST database (same network shown in Fig. 11.4) [9]. Figure 11.17 plots the modeled conductance response of the resistive switching elements. For average values of conductance G (e.g., the central region of the plot), the response is mostly linear, although somewhat asymmetric, with different average jump values for SET and RESET. In contrast, for extreme values of the conductances (left and right edges of each jump table), a high degree of nonlinearity is observed. However, we have previously observed that when the extent of the nonlinear region is sufficiently small, high classification accuracies can still be achieved [7].

The network parameters were tuned to achieve a good performance, with particular focus given to the ratio of $\eta_{\mathrm{SET}}/\eta_{\mathrm{RESET}}$, used to compensate the asymmetry of the jump table [9]. Figure 11.18 shows a schematic version of the crossbar-compatible weight-update rule for backpropagation, in which upstream neurons fire a set of pulses (shown in red) along the horizontal word lines, based solely on their knowledge of x_i and the global learning rate ($\eta = \eta_{\mathrm{SET}}$) [1]. Simultaneously, the downstream neuron fires pulses (shown in magenta) along the vertical bit lines connected to a large number of G^+ and G^- conductances. These pulses are based only on the downstream neuron's knowledge of δ_j and the global learning rate.

Because these pulses affect all the devices along the shared word lines and bit lines, their amplitude and duration cannot be tuned to optimize the programming of any one particular conductance value. This leads to significant problems when conductance response is nonlinear, since the same pulse can cause small conductances to increase much more significantly than conductances that are already large.

However, the downstream neuron can easily fire *different* pulse trains on the separate G^+ and G^- bit lines, and knowledge of δ_j can be sufficient to identify whether SET or RESET will occur (x_i need only be constrained to be nonnegative). Thus it is straightforward to apply a different global learning rate for RESET and for SET, thus leading to more or fewer pulses and providing a way to compensate

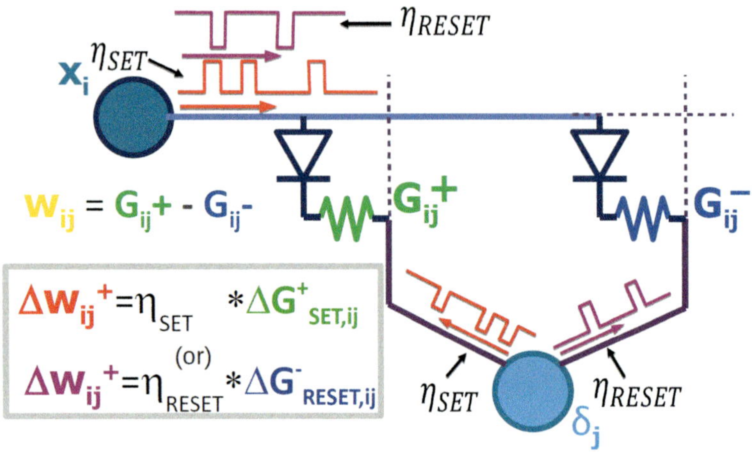

Fig. 11.18 Schematic showing crossbar-compatible [1] weight-update rule for analog bidirectional NVMs. Weight increases (decreases) can be implemented either as a SET operation on G^+ (G^-) or a RESET operation on G^- (G^+) devices. Asymmetry in the partial SET and RESET operation is compensated by applying a different learning rate parameter (η_{SET}, η_{RESET}) that modulates the number of pulses fired from the neurons into the array (Adapted from Fumarola et al. [9])

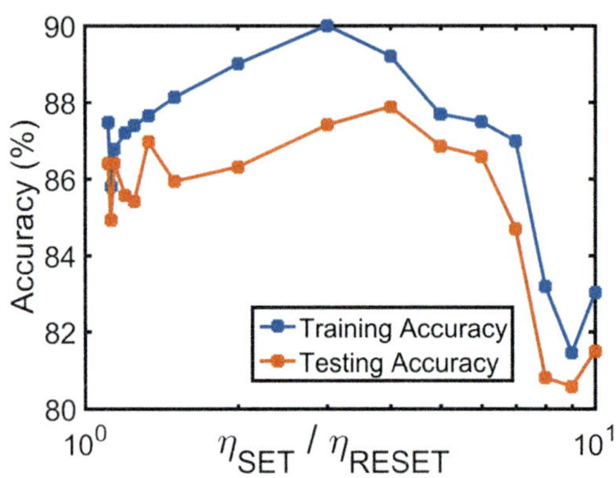

Fig. 11.19 Simulated training and test accuracy for a three-layer perceptron using PCMO-based devices as synaptic weights. The asymmetry between positive and negative jumps can be compensated by tuning individually the learning rates for SET and RESET (see Fig. 11.18). The classification accuracy of the network improves as the ratio of SET to RESET learning rate (η_{SET}/ η_{RESET}) increases (Adapted from Fumarola et al. [9])

for jump-table asymmetry [9]. Figure 11.19 shows that classification accuracy can be improved for the Al/Mo:PCMO jump tables shown in Fig. 11.17, with an optimal ratio of η_{SET}/η_{RESET} of approximately three to four.

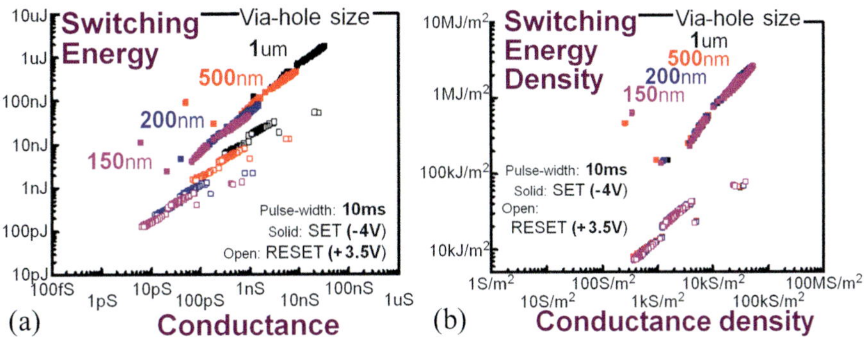

Fig. 11.20 Switching (**a**) energy as a function of conductance and (**b**) energy density as a function of conductance density, measured for Al/Mo:PCMO-based devices with 1.0 V reading voltage (Adapted from Fumarola et al. [9])

11.4.4 Switching Energy

The switching energy of the devices was measured by integrating the product between the voltage and the current for the duration of a programming pulse (10 ms) [9]. The conductance was measured with read pulses of 1.0 V. PCMO-based memory devices (like other nonfilamentary switching elements) show a dependence of the programming energy on the active area. Switching energy ranging from sub-nJ to tens of μJ was measured on devices with hole sizes from 0.15 to 1.0 μm (Fig. 11.20a). The switching energy was then normalized with respect to the active device area (Fig. 11.20b) to show a good linear dependence between switching current and device hole size. It is expected that one could achieve femto joule switching energy by aggressive scaling of the device and reduction of the switching time, enabling highly parallelized weight-update operations.

11.5 Circuit Needs

A crossbar array-based neural network implements the multiply-accumulate operations at the heart of most neural network algorithms extremely efficiently, through Ohm's law followed by current summation (Kirchhoff's current law). However, an important consideration is the design of highly area-efficient neuron circuits that reside at the edges of these arrays enabling read and write of many synaptic rows or columns in parallel. Such high parallelism is essential if we wish to achieve orders of magnitude performance and power benefits over conventional CPU/GPU approaches [3]. Given this need for a large number of distinct copies of neural circuits that can be executed in parallel, it is critical to embrace approximate functionality (e.g., nonlinear squashing functions, calculating and multiplying derivatives, etc.) rather than rigorously precise yet highly area-inefficient functionality.

In this section, we present examples of design choices that simplify the underlying hardware by leveraging the inherent tolerance of ANN algorithms to error. We discuss circuit needs for the forward- and reverse-evaluate operations, including precision/range of the computed neuron activations and backpropagated errors, using piecewise linear (PWL) approximations of nonlinear squashing functions and simplifying the derivatives included during reverse propagation to avoid complex floating-point arithmetic operations. We then demonstrate that these approximations do not significantly degrade classification accuracies as compared to neuron implementations with rigorously precise functionality [9].

11.5.1 Circuit Needs for Forward and Reverse Propagate

Forward propagation (Fig. 11.21) in a fully connected neural network involves the calculation of the neuron activations of a hidden/output layer, based on the neuron activations of the previous layer and the intervening synaptic weights. This is a two-stage process, with the multiply-accumulate operation occurring in the crossbar array and the nonlinear squashing function applied at the periphery. One commonly used function in software implementations is *tanh*() (the hyperbolic tangent function), which is difficult to implement exactly unless a large number of

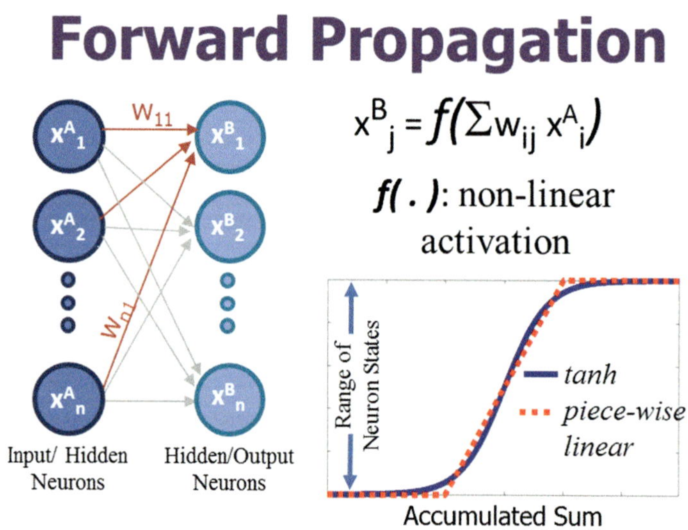

Fig. 11.21 Forward propagation operation in a deep neural network. The multiply-accumulate operation occurs on the crossbar array. Neuron circuitry must handle the nonlinear squashing function (Adapted from Fumarola et al. [9])

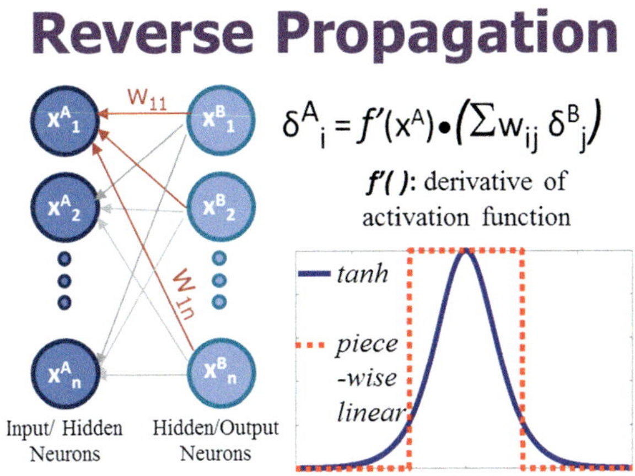

Fig. 11.22 Reverse propagation operation in a deep neural network. Multiply-accumulate operation on δ occurs on the crossbar array. Neuron circuitry must handle generation and multiplication of the derivative of the squashing function (Adapted from Fumarola et al. [9])

transistors are included. However, a piecewise linear implementation of this squashing function would be fairly straightforward to implement (Fig. 11.21).

A second design choice is the range of distinct neuron activation values that need to be supported by the hardware. In a digital implementation, this translates into the number of bits, which would have area implications depending on the amount of local storage required, as well as the resolution of any analog to digital conversion circuits used to convert signals from the crossbar array into those bits. In an analog implementation, this would directly translate into the resolution between analog voltage levels and/or time steps.

Reverse propagation (Fig. 11.22) is similar to forward propagation, but from output/hidden neurons to preceding hidden neurons. The quantity δ, known as the correction or error, together with the forward-propagated neuron activations, controls the weight updates for neural network training (see Fig. 11.18). An important distinction from forward propagation is that the nonlinear squashing function is not applied. Instead, the multiply-accumulate sum (integrated on the crossbar array, but in a direction orthogonal to the integration performed during the forward propagation step) needs to be scaled by the derivative of the activation function, as evaluated at the neuron activation value. Again, an exact *tanh*() derivative is not efficient to compute and multiply.

Instead, a step-function derivative with two distinct states can be used. Multiplication by derivative values of zero and one is fairly straightforward to implement in hardware. This corresponds to simply enabling or disabling the transmission of an accumulated sum of deltas from any neuron stage to the preceding stage. However, multiplication by arbitrary scale factors may be difficult to achieve since floating-point multipliers are not readily available. The impact of such approximations on neural network training is studied in the next subsection.

11.5.2 Results: Circuit Approximations

We explored the impact of the aforementioned circuit approximations on the training and test performance of the MNIST dataset of handwritten digits through simulations. A subset of only 5000 training images from the original dataset of 60,000 images is used. Images were cropped to 24×22 pixels. The same three-layer neural network (528-250-125-10) was used (Fig. 11.4), with a crossbar-compatible weight-update rule [1] emulating how weight updates would be done on a real crossbar array. The baseline training and test accuracies assuming 20 epochs of training, 256 neuron activation states, a *tanh*() activation function, and exact derivatives were found to be 99.7% and 93.6%, respectively (blue curve and star, Fig. 11.23). Note that, as per Fig. 11.7, both training and test accuracy increase (~100% and ~97–98%) when all 60,000 examples are used for training.

Figure 11.23 also shows the training and test accuracies using a piecewise linear (PWL) activation function. On MNIST, one observes that the test accuracy obtained (92.7%) is already comparable to the full *tanh*() implementation [9]. Further improvements in test accuracy can be obtained by optimizing the low value of the derivative. This is akin to the intentional implementation of "leaky" derivatives in some conventional machine learning techniques, especially in the case of rectified linear units (ReLU). A leaky derivative ensures that some contribution from the downstream neuron gets passed on to earlier stages, thereby participating in the programming of those weights.

Figure 11.24 shows that the test accuracy can be further improved to 93.2% when the derivative of the piecewise linear squashing function at extreme values is made nonzero. However, the multiplication operation is nontrivial. In a digital implementation, one might be able to do bit-shift operations (restricting derivative values to powers of two). An analog implementation can offer more freedom, since

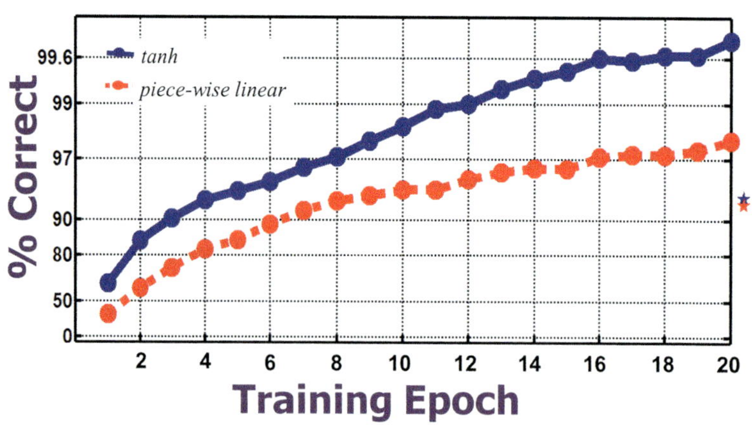

Fig. 11.23 Training and test accuracies obtained on MNIST with *tanh*() and piecewise linear activation functions. PWL achieves test accuracy comparable to tanh() (Adapted from Fumarola et al. [9])

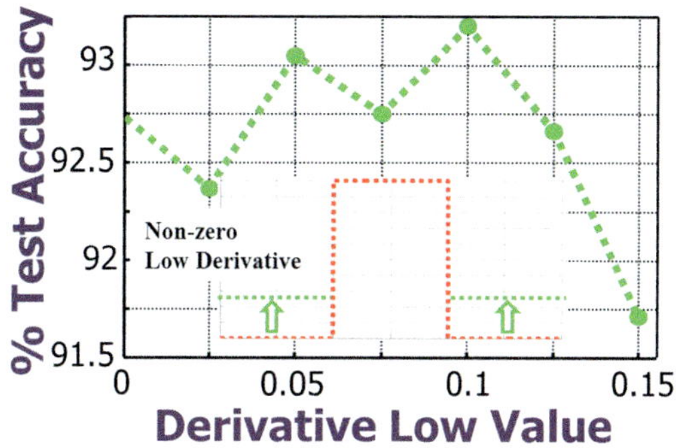

Fig. 11.24 Optimizing the low derivative value enables further improvements in test accuracy yet requires some circuit complexity to implement an approximate multiplication function (Adapted from Fumarola et al. [9])

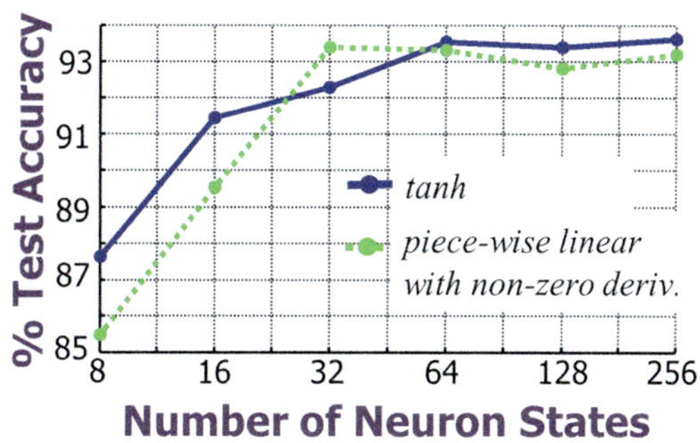

Fig. 11.25 If the number of distinct neuron activation and error states is lower than 32, then test accuracy degrades. However, reducing the total number of neuron states can help enable significantly more area-efficient peripheral circuitry (Adapted from Fumarola et al. [9])

we need to only enable one of two nonzero scale factors when transmitting accumulated analog voltages to preceding stages.

In addition to the squashing function and its derivative, the impact of the number of distinct neuron activation and error states on the test accuracy was analyzed [9]. Values from 8 to 256 were considered (Fig. 11.25). High test accuracies are maintained down to 32 distinct neuron states for both the *tanh*() and piecewise linear implementations. Reducing the total number of neuron states can be extremely beneficial in area-efficient circuit design. In a digital implementation, this allows a reduction in the total number of latches or flip-flops. In an analog implementation, it permits a wider separation of analog voltage levels, relaxing noise constraints and enabling simpler circuits.

11.6 Conclusion

We have studied several aspects of system design when nonvolatile memory (NVM) devices are employed as the synaptic weight element for on-chip acceleration of the backpropagation training of large-scale artificial neural networks (ANN).

In order to develop an intuitive understanding of the impact that various features of such jump tables have on the classification performance in the ANN application, we have studied various artificially constructed jump tables [7]. We have also assessed the impact of undesired, time-varying conductance change, including drift in phase-change memory (PCM) devices and leakage of analog CMOS capacitors [9]. We have investigated the use of nonfilamentary, bidirectional RRAM devices based on PCMO, which can be considered a promising material variant that could potentially provide both gradual conductance increase *and* conductance decrease [9]. And finally, we have explored some of the trade-offs in designing peripheral circuitry, balancing simplicity and area efficiency against the impact on ANN performance for the nonlinear squashing function, the evaluation of its derivation, and the number of resolvable levels when integrating both x (forward propagate) and δ (reverse propagate) values [9].

We briefly reviewed our previous work toward achieving competitive performance (classification accuracies) for such ANN with both phase-change memory [1, 2] and nonfilamentary RRAM based on PCMO [4] and toward assessing the potential advantages for ML training over GPU-based hardware in terms of speed (up to 25$\times$ faster) and power (from 120–2850$\times$ lower power) [3]. We discussed the "jump table" concept, previously introduced to model real-world NVM such as PCM [1] or PCMO, to describe the full cumulative distribution function (CDF) of resulting conductance change at each possible conductance value, for both potentiation (SET) and depression (RESET).

While the "local gains" algorithm, together with other approaches, should help a nonlinear, asymmetric NVM (such as PCM) that acts more like an ideal linear, bidirectional NVM, the identification of NVM devices and/or pulse schemes that can offer a conductance response that is at least partly linear, using circuitry that can be highly area efficient (and thus massively parallel), will help significantly in achieving equally high classification accuracies while offering faster and lower-power training than conventional GPUs and CPUs.

References

1. G.W. Burr, R.M. Shelby, C. di Nolfo, J.W. Jang, R.S. Shenoy, P. Narayanan, K. Virwani, E.U. Giacometti, B. Kurdi, H. Hwang, Experimental demonstration and tolerancing of a large-scale neural network (165,000 synapses), using phase-change memory as the synaptic weight element, in *IEEE International Electron Devices Meeting (IEDM)* (IEEE, San Francisco, 2014)
2. G.W. Burr, R.M. Shelby, S. Sidler, C. di Nolfo, J. Jang, I. Boybat, R.S. Shenoy, P. Narayanan, K. Virwani, E.U. Giacometti, B. Kurdi, H. Hwang, Experimental demonstration and tolerancing of a large-scale neural network (165 000 synapses), using phase change memory as the synaptic weight element. IEEE Trans. Electron Devices **62**(11), 3498–3507 (2015)
3. G.W. Burr, P. Narayanan, R.M. Shelby, S. Sidler, I. Boybat, C. di Nolfo, Y. Leblebici, Large-scale neural networks implemented with non-volatile memory as the synaptic weight element: comparative performance analysis (accuracy, speed, and power), in *IEEE International Electron Devices Meeting (IEDM)* (IEEE, Washington, DC, 2015)

4. J.-W. Jang, S. Park, G.W. Burr, H. Hwang, Y.-H. Jeong, Optimization of conductance change in Pr1-xCaxMnO3–based synaptic devices for neuromorphic systems. IEEE Electron Device Lett. **36**(5), 457–459 (2015)
5. D. Rumelhart, G.E. Hinton, J.L. McClelland, in *Parallel Distributed Processing*. A general framework for parallel distributing processing (MIT Press, Cambridge, MA, 1986)
6. Y. LeCun, L. Bottou, Y. Bengio, P. Haffner, Gradient-based learning applied to document recognition. Proc. IEEE **86**(11), 2278–2324 (1998)
7. S. Sidler, I. Boybat, R.M. Shelby, P. Narayanan, J. Jang, A. Fumarola, K. Moon, Y. Leblebici, H. Hwang, G.W. Burr, Large-scale neural networks implemented with non-volatile memory as the synaptic weight element: impact of conductance response, in *IEEE European Solid State Device Research Conference (ESSDERC)* (IEEE, Lausanne, 2016)
8. A. Pirovano, A.L. Lacaita, F. Pellizzer, S.A. Kostylev, A. Benvenuti, R. Bez, Low–field amorphous state resistance and threshold voltage drift in chalcogenide materials. IEEE Trans. Electron Devices **51**(5), 714–719 (2004)
9. A. Fumarola, P. Narayanan, L.L. Sanches, S. Sidler, J. Jang, K. Moon, R.M. Shelby, H. Hwang, G.W. Burr, Accelerating machine learning with non-volatile memory: exploring device and circuit tradeoffs, in *IEEE International Conference on Rebooting Computing (ICRC)* (IEEE, San Diego, 2016)

Chapter 12
Impact of Nonideal Resistive Synaptic Device Behaviors on Implementation of Sparse Coding Algorithm

Pai-Yu Chen and Shimeng Yu

12.1 Introduction

The neuro-inspired computing is attractive due to the bottleneck of bandwidth between the processor and memory in conventional von Neumann computation. The primary goal of neuro-inspired computing is to develop biological artificial systems that enable better interaction with the natural environment for problems, such as image and speech recognition, while achieving significant computation speedup and low power consumption. Several large-scale neuro-inspired hardware platforms have been developed in the recent years, e.g., FACETs [1], Caviar [2], SpiNNaker [3], TrueNorth [4], etc. However, to build these neuro-inspired systems with a massive number of synapses, a tremendous hardware cost is unavoidable due to a large volume of CMOS-based memories for implementing these synapses. Low-cost and ultra-high memory arrays are the current target of memory technology. Synapses based on the mature floating-gate memory have been successfully implemented on a single chip for neuromorphic computation [5, 6]. To achieve even higher integration density, faster speed, and lower programming voltage, compact synaptic devices based on emerging nonvolatile memory have been proposed for the neuromorphic systems [7]. Resistive synaptic devices in this book chapter refer to a special subset of the resistive memory devices that can continuously tune its conductance into multilevel states [8]. For on-chip implementation of the learning algorithm, the conductance of a synaptic device represents a weight element. Prior works suggest that the cross-point array architecture can efficiently map the weight matrix and can perform the weighted sum and weight update in a parallel fashion [9–11]. At device level, the nonideal characteristics of

P.-Y. Chen • S. Yu (✉)
School of Electrical, Computer, and Energy Engineering, Arizona State University, 781 E Terrace Rd, ISTB4 room 591, Tempe, AZ 85287, USA
e-mail: shimengy@asu.edu

S. Yu (ed.), *Neuro-inspired Computing Using Resistive Synaptic Devices*,
DOI 10.1007/978-3-319-54313-0_12

phase-change memory-based synaptic devices have been experimentally calibrated to evaluate the learning performance for the supervised multilayer perceptron algorithm [12, 13]. In this book chapter, we will focus on the impact of nonideal resistive synaptic device properties on the unsupervised sparse coding algorithm, and we will propose potential solutions to mitigate the accuracy loss.

The nonideal resistive synaptic device properties that are considered in this book chapter include (1) the nonlinearity in the weight update; (2) the limited ON/OFF weight ratio; (3) the device variation in weight update and read noise; and (4) at the array level, the IR drop along interconnect resistance. An ideal synaptic device behavior assumes a linear update of the weight with the input stimulus, e.g., the number of voltage pulses. However, this assumption may not hold in reality. Figure 12.1 shows the experimental data of the multilevel conductance tuning by applying positive (negative) pulses on the resistive synaptic devices reported in the literature. Long-term potentiation (LTP) (long-term depression (LTD)) means the increase (decrease) of the device conductance in the weight update. Figure 12.1a–c show the representative LTP/LTD behaviors in the TaO_x/TiO_2-based [14], PCMO-based [12], and Ag:a-Si-based [15] synaptic devices, respectively. It is found that today's synaptic devices commonly have the nonlinear behavior of LTP and LTD, possibly due to the inherent drift and diffusion dynamics of the ions/vacancies in these materials. Besides the nonlinearity in weight update, another realistic characteristic that deviates from an ideal device is the limited ON/OFF weight ratio. The off-state conductance is not perfectly zero in the examples of Fig. 12.1a–c, while the ideal synaptic device assumes that the minimum weight is zero. Moreover, the device variation is generally a concern for the emerging devices at the nanoscale. The synaptic device variation can be categorized into two aspects: one is the spatial variation from device to device, and the other is the temporal variation from pulse to pulse. The device variation occurs not only in the weight update, but also the read

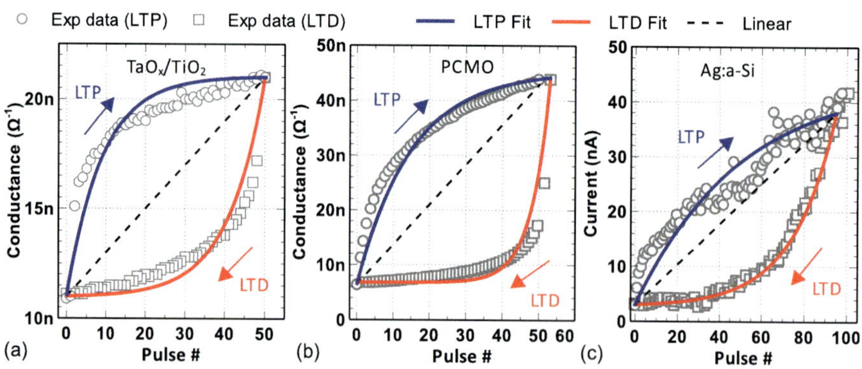

Fig. 12.1 The measured experimental data of conductance modulation for the weight update reported in the literature: (**a**) TaO_x/TiO_2 device [14], (**b**) PCMO device [12], and (**c**) Ag:a-Si device [15]. Long-term potentiation (LTP) (long-term depression (LTD)) means the increase (decrease) of the device conductance in the weight update. Exponential function of pulse number is used to fit the nonlinear LTP and LTD curves

noise may show up in the weighted sum operation. Finally, the effect of IR drop problem along interconnect resistance may become remarkable when the array size is large, as it may distort the weighted sum results in the crossbar array. All of these realistic device characteristics can potentially degrade the learning accuracy of neuromorphic systems. In this work, we propose array design methodologies for the co-optimization of the realistic synaptic device properties and the array architecture to mitigate these undesirable effects.

12.2 Sparse Coding Algorithm for On-Chip Feature Extraction

12.2.1 Sparse Coding (SC) Algorithm

Sparse coding algorithm [16] is an important class of unsupervised learning algorithms that can efficiently extract the features from the input data pattern. The sparse coding aims to project the input data pattern (possibly dense in the information intensity) to a sparse feature domain; here sparse means most of the features are zero. Therefore, such sparse feature representation makes it easier for the next stage classifier to cluster the input data pattern into different classes. Although the sparse coding is only a single-layer neural network of input neurons and output neurons and one weight synaptic matrix in between, it can still achieve reasonably high learning accuracy with invariance for data pattern's spatial shift and rotation. The sparse coding has been found to be a bio-physiological plausible model: neurons in mammalian primary visual cortex can form a sparse representation of natural scenes [17, 18], which is believed to emerge from an unsupervised learning algorithm that looks for a factorial code of independent features such as lines, edges, and corners. For real-world applications, the sparse coding algorithm has demonstrated its power in numerous domains such as audio processing, text mining, and image recognition. In this work, our goal is to evaluate and optimize the synaptic device properties and cross-point array architecture for fast and compact on-chip sparse feature learning as a case study.

Figure 12.2a shows the process flow of the sparse coding algorithm (SC module), which is obtained from [19] with optimized algorithm parameters. In the training phase, with a given input vector set $\{X\}$ (braces mean a collection of objects), the corresponding feature vector set $\{Z\}$ and the dictionary matrix (D) are trained iteratively by minimizing the objective error function (E):

$$E = \min \sum \left(\|DZ - X\|^2 + \lambda \|Z\|_1 \right) \tag{12.1}$$

As each X is a sparse linear combination of Z via D, the first term of Eq. (12.1) generally measures how well the dictionary reconstructs the input X. The second term of Eq. (12.1) imposes constraint of the sparsity of the feature vector by the L1

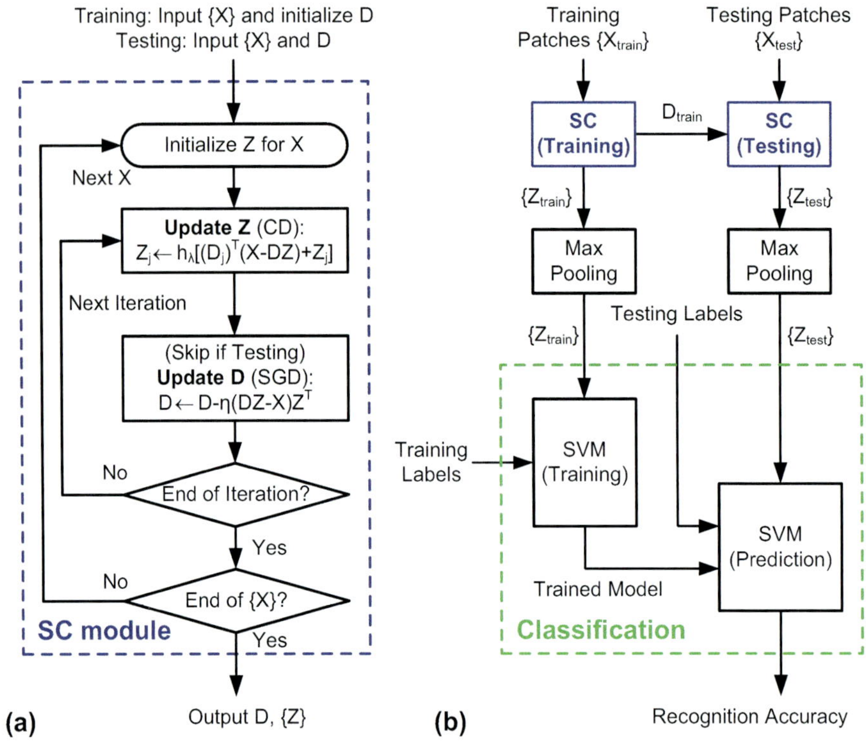

Fig. 12.2 (**a**) Process flow of the sparse coding (*SC*) module and (**b**) the entire process consisting of the unsupervised feature extraction by sparse coding (*SC*) and the supervised classification by support vector machine (*SVM*)

norm, where λ is a constant. The objective error function (E) is minimized when the reconstructed pattern DZ approaches the original X with minimal number of Z feature vector elements. The above optimization problem is a non-convex problem given both D and Z are unknown. It is proposed to iteratively optimize Z with fixed D by the coordinate descent (CD) method and optimize D with fixed Z by the stochastic gradient descent (SGD) method, which converts the problem into a convex optimization problem. Compared to conventional full-gradient descent, SGD is more computation efficient with large-scale dataset [19]. Using SGD, the weight update process of D can be expressed as:

$$D \leftarrow D - \eta R Z^{\mathrm{T}} (R = DZ - X) \tag{12.2}$$

It can be seen that D is updated by the product of $\eta R Z^{\mathrm{T}}$, where η is the learning rate and R is the reconstruction error, which is essentially the delta rule of a single-layer neural network, a special case of the backpropagation in multilayer neural network. For the algorithm ideally implemented in software, the exact value of $\eta R Z^{\mathrm{T}}$ can be calculated for the update of D. However, the D update has to be translated

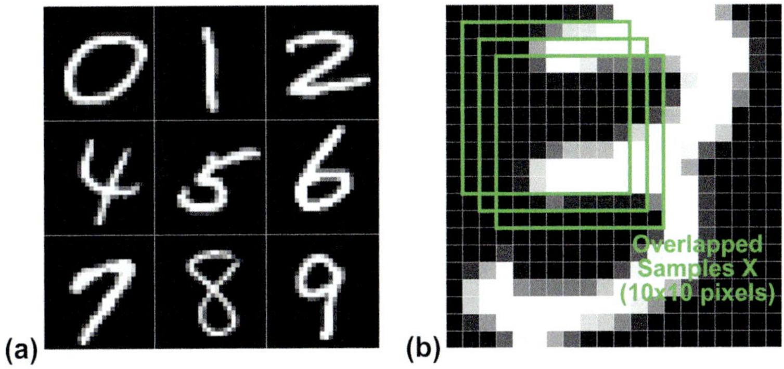

Fig. 12.3 (**a**) Examples of MNIST handwritten digits [20]. (**b**) Image patches (10 × 10 pixels) are extracted as the input vector X

to the number of pulses applied on the synaptic devices for on-chip implementation; thus the effect of the programming pulses on the conductance of the devices may not represent the exact value of ηRZ^{T} due to the nonideal properties of synaptic devices as mentioned earlier. In this work, we model the weight update curve and incorporate this device behavior model in the D update code in the SC algorithm.

Figure 12.2b describes the entire process flow that consists of dictionary learning (training phase) and classification (testing phase). In this work, the MNIST handwritten digits [20] are used as the training and testing dataset, where the raw images are densely sampled into small patches with 10 × 10 pixels; thus the X input vector has a dimension of 100, as shown in Fig. 12.3. In the later analyses, a set of 40,000 images is used for training, and a different set of 5,000 images is used for testing, as we have found that using the entire 60,000 training images does not have noticeable increase on the accuracy (only 1%), and it will greatly slow down the simulation speed.

After the training process, the trained dictionary D_{train} is fixed and used in the testing phase to generate the testing features $\{Z_{\mathrm{test}}\}$. Before the classification process, a simple maximum pooling operation is employed on both the trained and testing features of each image to select the most active neuron of each feature node:

$$Z_{\mathrm{i}} = \max\left(Z_i^1, \ldots, Z_i^k\right) \tag{12.3}$$

where $Z_i^1, \ldots, Z_i^k$ are the i_{th} elements of the feature vectors of total k image patches in one image. In the maximum pooling operation, all the feature vectors of small image patches are merged into one feature vector per image using the maximum value of each i_{th} element. Finally, the supervised support vector machine (SVM) algorithm [21] is used to classify the 10 digits 0–9. Based on the input of testing labels, SVM performs classification and gives out the recognition accuracy.

Figure 12.4 shows the recognition accuracy of the MNIST dataset as a function of Z vector dimension. The learning accuracy does not increase much beyond a

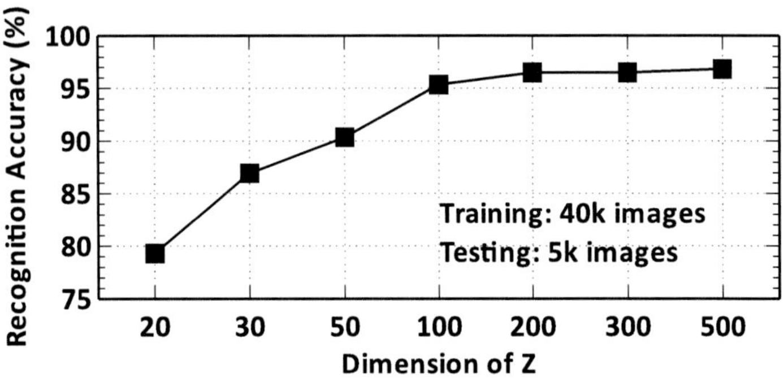

Fig. 12.4 Learning accuracy as a function of Z dimension. The accuracy does not increase much for Z dimension >200

dimension of 200. In this work, the Z dimension is fixed to be 300 in this work; thus the size of the D matrix is 100×300 ($X \times Z$).

12.2.2 Limited On-Chip Precision of SC

For on-chip implementation of the SC algorithm, the precision reduction of D and Z is needed in the algorithm as the chip cannot afford the floating-point computation. In the cross-point architecture, the values of the Z vector are stored on local memories in the peripheral circuitry, and the synaptic weights in the array represent the values in the D matrix. Figure 12.5 shows the learning accuracy as a function of the reduced precision bits of D and Z in the SC algorithm. It indicates that a 4-bit Z (neuron value) is sufficient while D (weight matrix) should be at least 6 bits to achieve an accuracy >95%. This requirement of a high precision in the weight update for the learning (in the backpropagation) is also reported in other works [22, 23]. High precision is required to preserve the small error information since the training of the SC algorithm is based on the error-driven backpropagation. As the number of bits in D matrix means how many levels of conductance that the synaptic device should achieve, a 6-bit D means that 64 levels are needed for the synaptic devices, which may be available in today's synaptic devices (see Fig. 12.1).

12.3 Operation Scheme of Weighted Sum and Weight Update in Cross-Point Array for On-Chip Learning

Figure 12.6a shows the schematic of the proposed resistive cross-point array architecture. At each cross-point, there is one selector in series with one synaptic device to form a synaptic cell. The selector introduces the nonlinear I–V

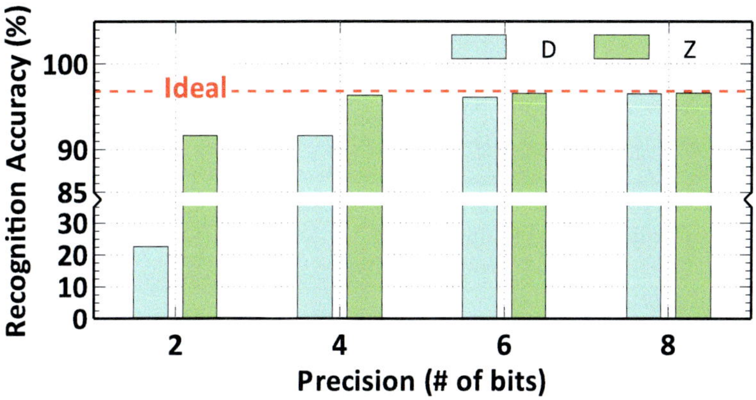

Fig. 12.5 Learning accuracy with different number of precision bits of *D* matrix and *Z* vector in the SC algorithm

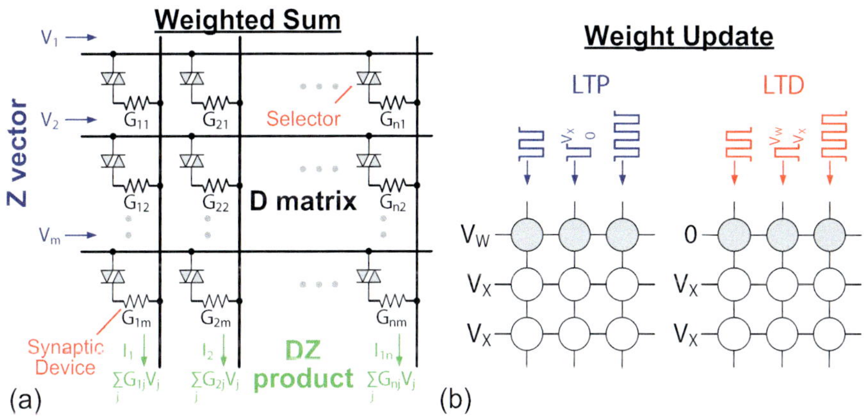

Fig. 12.6 (**a**) Proposed resistive cross-point array architecture. Read scheme performs the weighted sum (matrix-vector product *DZ* as an example) in a fully parallel fashion. The read voltage (*Z*) multiplied by the conductance (*G*) of the synaptic device at each cross-point along the column generates weighted sum current at the end of all columns. (**b**) Weight update operation consists of the LTP and LTD weight update phase. The selected row (*gray circles*) is biased at V_W and 0 V, with V_X-0-V_X negative and V_X-V_W-V_X positive write pulses applied at all columns to perform LTP and LTD weight update in a row-by-row fashion, respectively

characteristics to the synaptic cell and is helpful for both weight update and weighted sum operations, which will be discussed later in this section. To compute the weighted sum (*DZ*) in the read operation, the read voltage (V_R) is applied in parallel to each row for every nonzero element of *Z*. The weighted sum current will sink to the end of all columns because V_R is multiplied by the conductance of synaptic device at each cross-point. The read peripheral circuitry at each column then converts the analog weighted sum current to the digital output. It should be noted that the sneak path problem due to the leakage of unselected cells in the conventional memory application [24] does not exist in the weighted sum operation. This is because a bit or a row of data are read out in conventional memory application, while the data in the entire array are read out in the weighted sum operation; thus all the cells in the array are participating in the computation according to the Kirchhoff's law. It is preferred that the value of Z elements is encoded by the number of V_R pulses (4 bits mean 16 pulses), which has less distortion on the DZ product compared to the analog encoding scheme with varying voltage amplitude [11]. In the analog encoding scheme, it is also challenging to split V_R (typically <0.5 V) into 16 voltage levels due to noise consideration and practical bias circuit design constraints.

To accelerate the weight update operation, a fully parallel write scheme was developed in prior work to update the entire array using the product of Z and R Eq. (12.2) [25]; however, it requires a complex peripheral circuit design; thus there will be a tremendous hardware cost especially when the array size is large. As shown in Fig. 12.6b, we propose a simpler write scheme to perform the LTP (LTD) weight update in the write operation to select one row at a time with the write voltage (V_W) (0 V) applied at the edge, while other rows are biased at an intermediate voltage V_X to prevent the write disturbance. Then, the V_X-0-V_X negative (V_X-V_W-V_X positive) write pulses are then applied at all columns to perform the weight update. Similar to the fully parallel one, a write cycle of each row also has two phases (the LTP and LTD weight update), considering they have different voltage polarities. This write scheme is essentially a row-by-row operation with less speedup, but the hardware cost is also much lower compared to the fully parallel one.

Unlike the weighted sum operation, the proposed weight update scheme suffers from the sneak path problem, as the cross-point array is partially selected and the half-selected cells on other unselected rows or columns can form sneak paths. The half-selected cells see a voltage drop of V_X (LTP) or V_W-V_X (LTD) during the weight update. Therefore, the selector is placed in series with the synaptic device to suppress the leakage current at these voltages. Figure 12.7 shows the I–V characteristics of a TaO_x/TiO_2-based synaptic device in ON state, the selector, and the series of these two devices. In this work, we use the mixed-ionic-electronic-conduction (MIEC)-based selector that has high nonlinearity (~85 mV/dec) [26] and set the original $V_W = 2$ V and $V_R = 1$ V for a single synaptic device. Without the selector, V_X is designed to be 1 V and is just the $V/2$ write scheme in conventional memory application [27]. With the selector, the overall cell resistance increases, which reduces the IR drop along interconnects in weighted sum with only little impact on the mapping from device conductance to weight values because the conductance of selector is relatively higher than the conductance range of synaptic device at 1 V. Also, the selector can suppress the leakage on the half-selected cells during weight update, and it does not affect the weight update because it is already turned on at sufficient large voltage. In this case, V_W should be as high as 3 V, and the V_X for LTP and LTD weight update are then 1 V and 2 V, respectively. It makes the voltage drop on the selected cells to be 2 V, which is the same as the original write condition for a single synaptic device. In addition, the voltage drop on the

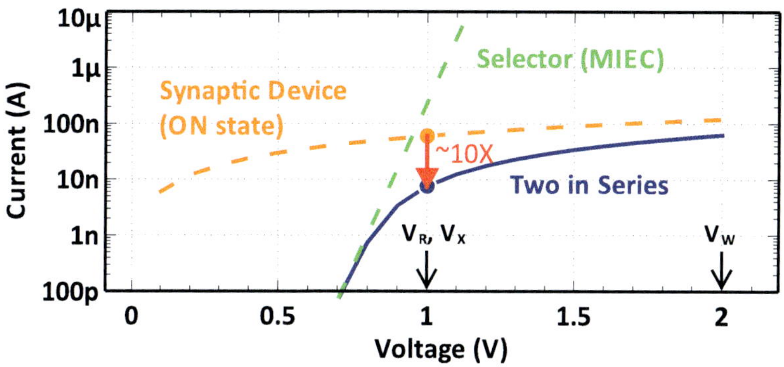

Fig. 12.7 The I–V characteristics of resistive synaptic device, MIEC selector [26], and the series of the above two devices. With selector, the cell resistance increases to mitigate the IR drop along interconnects in the weighted sum and reduce the leakage (~10X) on the unselected cells, without affecting the weight update. V_W, V_R, and V_X labeled here indicate the voltages for the original synaptic array without the selector

half-selected cells will then be 1 V, where the leakage reduction is ~10X as shown in Fig. 12.7. Since most of the cells during the weight update are half-selected cells, the energy consumption is significantly reduced compared to the conventional $V/2$ write scheme where the voltage drop on the half-selected cells are $V_W/2 = 1.5$ V.

12.4 Impact of Realistic Device Properties on Learning Accuracy

As previously shown in Fig. 12.1, the realistic synaptic behaviors include (1) the nonlinearity and (2) device variations in weight update and (3) the read noise and (4) limited ON/OFF weight ratio in weighted sum. We also considered the circuit model of synaptic device including the interconnect resistance in the array-level analysis. In this section, we incorporated these realistic properties individually into the sparse coding algorithm and investigated their impact on the learning accuracy. For the limited precision of synaptic devices, we use 6 bits (64 levels) as a baseline.

12.4.1 *Nonlinear Weight Update*

To analyze the impact of nonlinear weight update on the learning, a general behavior for the LTP and LTD conductance (G_{LTP} and G_{LTD}) modulation with the number of voltage pulses (P) is modeled using the following exponential equations:

$$G_{\mathrm{LTP}} = B\left(1 - e^{\left(-\frac{P}{A}\right)}\right) + G_{\min} \tag{12.4}$$

$$G_{\mathrm{LTD}} = -B\left(1 - e^{\left(\frac{P-P_{\max}}{A}\right)}\right) + G_{\max} \tag{12.5}$$

$$B = \frac{G_{\max} - G_{\min}}{1 - e^{\frac{-P_{\max}}{A}}} \tag{12.6}$$

where $G_{\max}$, $G_{\min}$, and $P_{\max}$ can be directly fitted with the experimental data, which represents the maximum conductance, minimum conductance, and the maximum number of pulses required to switch the device between the minimum and maximum conductance states. A is the parameter that controls the nonlinearity of the weight update, and B is simply a function of A that fits the functions within the range of $G_{\max}$, $G_{\min}$, and $P_{\max}$. A and B may be different in (12.4) and (12.5). A set of nonlinear LTP and LTD behavior can be obtained with different values of A as illustrated in Fig. 12.8a, where the nonlinearity value is ranged from +6 to −6. Here the plus and minus are merely the signs to label LTP and LTD, respectively.

Then, these nonlinear functions are applied into the weight update operation in the SC algorithm. Figure 12.8b shows that the learning accuracy slightly decreases in the high nonlinearity region of LTP and LTD, and a relatively larger drop from ~96% to ~92% occurs at maximum nonlinearities (+6/−6 curves). For today's synaptic devices (in Fig. 12.1), the nonlinearities of LTP and LTD are also labeled in the Fig. 12.8b. It is observed that the nonlinearity in the weight update has a moderate impact on the learning performance. To improve the nonlinearity, smart programming schemes have been designed in prior work [28]; however, it requires extra circuitry and thereby results in the overhead of area, latency, and energy consumption. Given that the accuracy loss at the maximum nonlinearities is only ~4%, it may not be crucial to apply the smart programming schemes; thus those overheads can be saved.

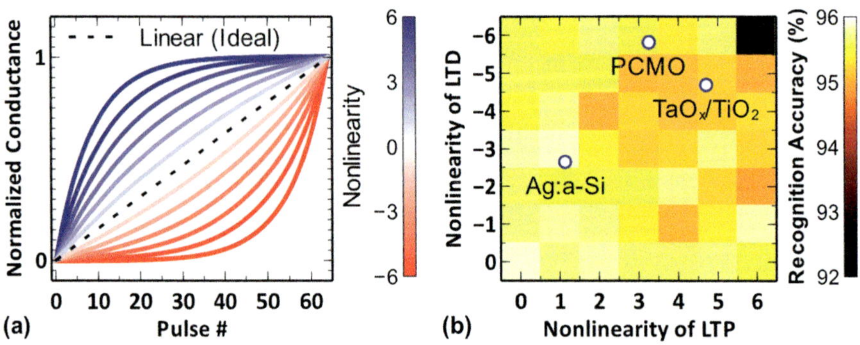

Fig. 12.8 (**a**) LTP and LTD weight update curves at nonlinearities from +6 to −6. (**b**) Learning accuracy with different LTP and LTD nonlinearities. Nonlinearities of the three realistic synaptic devices from Fig. 12.1 are also labeled

12.4.2 Device Variations

It is well known that the synaptic devices show considerable variation from device to device and even from pulse to pulse within one device due to the drift and diffusion's stochastic dynamics of the ions/vacancies. The device-to-device weight update variation leads to different nonlinearity baselines in different devices of the array. On the other hand, the cycle-to-cycle weight update variation can be interpreted as the pulse-to-pulse noise on top of the nonlinearity baseline. For the read noise, there will be some temporal fluctuation in the readout current of a weight.

The impact of device-to-device variation can be analyzed by applying the variation to the nonlinearity baseline for each synaptic device, as illustrated in Fig. 12.9a. For example, if a synaptic device has a +100% device-to-device variation, there will be a +1 deviation of the nonlinearity. As shown in Fig. 12.9b, the device-to-device variation in the weight update only has little impact on the learning accuracy even with 30% standard deviation from the baseline. The reason may be that the SC algorithm can partially tolerate the device-to-device variation as the solution to the optimized D is not unique as this is a non-convex problem. With the different weight update curves in different devices, the optimized D matrix may have converged at another local minima. Therefore, the neural network has certain resilience to the variations in weight update.

The cycle-to-cycle variation of the conductance occurs whenever a write pulse is applied to the synaptic device, as illustrated in Fig. 12.10a. In this work, the cycle-to cycle weight update variation is defined as the variation of weight change with one applied write pulse. As shown in Fig. 12.10b, high learning accuracy is still maintained at large cycle-to-cycle weight update variation. Instead, at nonlinearity baseline (LTP, LTD) = (6, −6), the learning accuracy somehow slightly improves

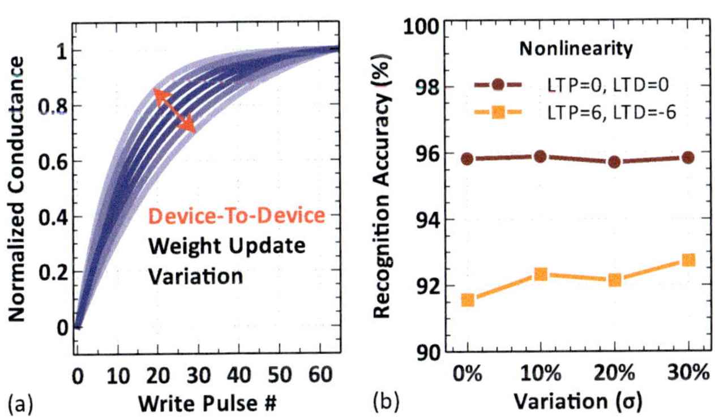

Fig. 12.9 (**a**) Illustration of the device-to-device weight update variation. The nonlinearity baselines will be different for different devices in the array. (**b**) Learning accuracy with different standard deviations of device-to-device weight update variation. The accuracy does not degrade at high variation

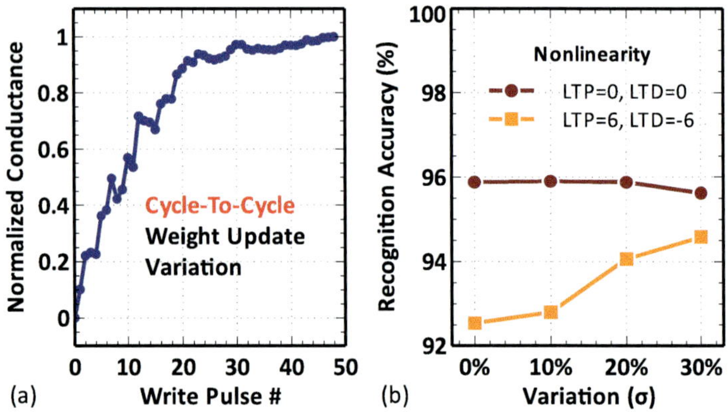

Fig. 12.10 (**a**) Illustration of the cycle-to-cycle weight update variation. There is pulse-to-pulse noise on top of the nonlinearity baseline. (**b**) Learning accuracy with different standard deviations of cycle-to-cycle weight update variation, which has almost no impact but somehow slightly improves the learning accuracy at nonlinearity baseline $(LTP, LTD) = (6, -6)$

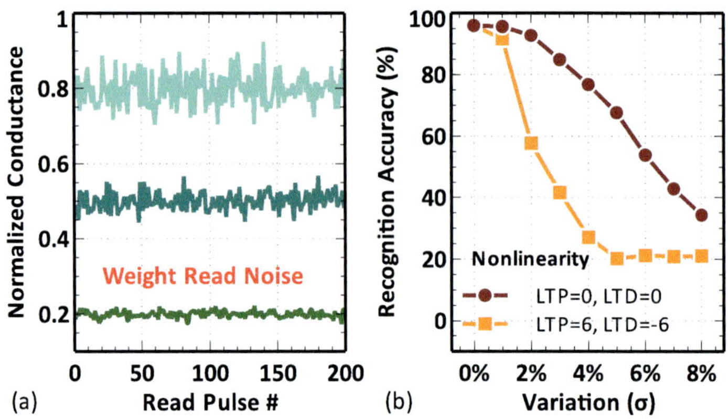

Fig. 12.11 (**a**) Illustration of the weight read noise on top of different conductance states. (**b**) Learning accuracy with different standard deviations of weight read noise. Even small variation of read noise can cause significant degradation on the learning accuracy and is even severer at nonlinearity baseline $(LTP, LTD) = (6, -6)$

with larger variation possibly because the randomness in the pulse amplitude may partially compensate the high nonlinearity.

Similar to the cycle-to-cycle variation in weight update, the read noise occurs at every read access to the synaptic device, but it does not disturb the average conductance. As illustrated in Fig. 12.11a, the readout current fluctuates at different conductance states with different number of read pulses, but the average conductance does not change. Figure 12.11b shows significant degradation of learning accuracy due to the read noise. Since the read noise occurs not in the weight update process but in the weighted sum process, the neural network may be susceptible to

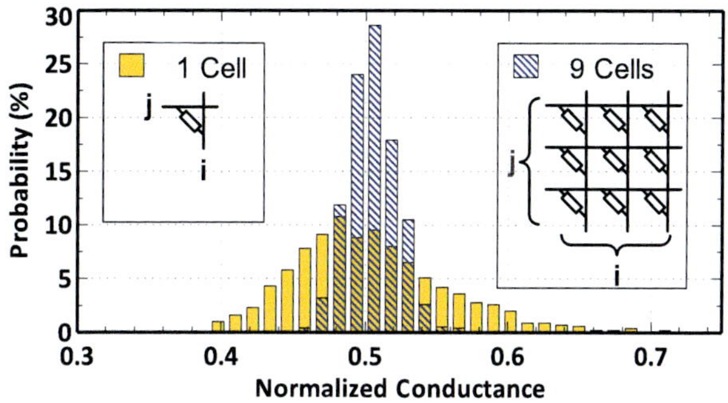

Fig. 12.12 Illustration of multiple cells as one weight element to average out the device variations or the read noise

the read noise as the weighted sum results may deviate from what is learned in the weight update process. The impact is even more critical with nonlinearity baseline (LTP, LTD) = (6, −6). From our measurement, the read noise in the TaO_x/TiO_2-based synaptic device has a variation of ~2.89%, which could cause the accuracy drop below 90% with this read noise effect only.

To alleviate the impact of device variations (especially the read noise), we propose to use multiple cells as one *D* weight element. This approach statistically averages out all the conductance variation of synaptic devices. If there are *n* cells as one weight element, the standard deviation of the conductance variation will be reduced by a factor of $1/\sqrt{n}$ given that variations are normally distributed. Figure 12.12 illustrates an example of the reduction on the variation using nine cells compared to that using only one cell. This strategy is believed to have great improvement on the accuracy loss due to the read noise, and it does not have a large overhead in the array area as this area is typically determined by the pitch of the peripheral circuits in the logic design rule. For example, the array cell height should be aligned with the standard cell height of the array row driver. At 65 nm technology node and 200 nm wire width, we estimate that the chip area of 9 resistive synaptic cells per weight increases by ~20% compared to that of 1 cell per weight. It should be noted that part of the peripheral circuitry can be placed underneath the synaptic array to save the total area since the synaptic devices are integrated on top of the CMOS circuits at the interconnect level. However, the energy consumption inevitably becomes *n* times larger when *n* cells are used to represent one *D* weight element.

12.4.3 Limited ON/OFF Weight Ratio

In the SC algorithm, the range of *D* value is from 0 to 1; thus we use the normalized conductance of synaptic devices to represent the *D* values. However, the minimum conductance can be considered as $D = 0$ only when the ratio between the maximum and minimum conductance (ON/OFF ratio) approaches infinity, which is not feasible in today's synaptic devices. For example, with an ON/OFF ratio = 10,

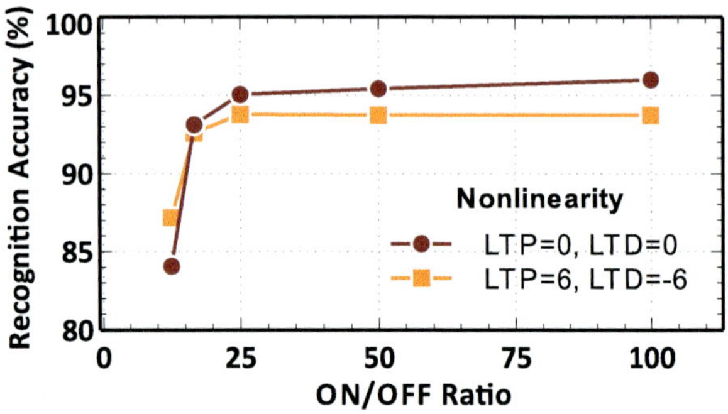

Fig. 12.13 Learning accuracy with different ON/OFF ratios at nonlinearity baselines (*LTP*, *LTD*) = (0, 0) and (6, −6). The Ag:a-Si device in Fig. 12.1 has the maximum ON/OFF ratio about 15, which results in >10% accuracy loss

the minimal weight that the device can represent is 0.1, and the maximum weight that the device can represent is 1. Figure 12.13 shows the learning accuracy with different ON/OFF ratios. When the ON/OFF ratio shrinks below 25, the learning accuracy dramatically decreases because the calculations involved with small values of D in the algorithm will be significantly distorted. The Ag:a-Si device has the largest ON/OFF ratio of ~15 among the devices in Fig. 12.1, while other devices show even smaller ON/OFF ratio. For on-chip implementation of sparse learning, this means that none of these synaptic devices can achieve high learning accuracy without any optimization.

To remedy this situation, one approach is to eliminate the effect of the off-state current in every weight element with the aid of a dummy column. The cross-point array architecture with a dummy column is illustrated in Fig. 12.14. The dummy column consists of the synaptic devices in the minimum conductance state, such that the readout value at the output of dummy column is the weighted sum of the Z vector and the off-state conductance. In the peripheral circuitry, the off-state weighted sum is subtracted from all the partial weighted sums, D_iZ, at all the columns. This virtually eliminates the effect of off-state current in the learning except for the case that there is spatial variation between the synaptic devices in the same row. An extra column will give 1% overhead on the array area as there are totally 100 columns ($X = 100$), and the area of subtractors is estimated to be ~7.84% of the array area with 9 cells as one weight element at 65 nm technology node and 200 nm wire width. However, since the array can partially hide the subtractors, its area overhead can be further reduced.

12.4.4 Synaptic Device Model in Cross-Point Array

To perform array-level simulation of weighted sum operation, we model the synaptic device in SPICE as a resistor in parallel with a capacitor. The synaptic device is in series with a selector as mentioned earlier. We also considered the wire resistances and parasitic capacitances. The wire resistance is calculated using the parameters from the ITRS Table [29]. Figure 12.15a shows a sub-circuit module of

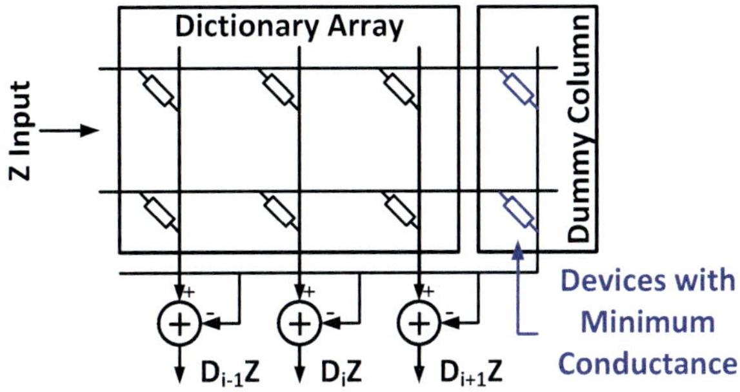

Fig. 12.14 Cross-point array architecture with the dummy column and subtractors to eliminate the common off-state current in the weighted sum operation

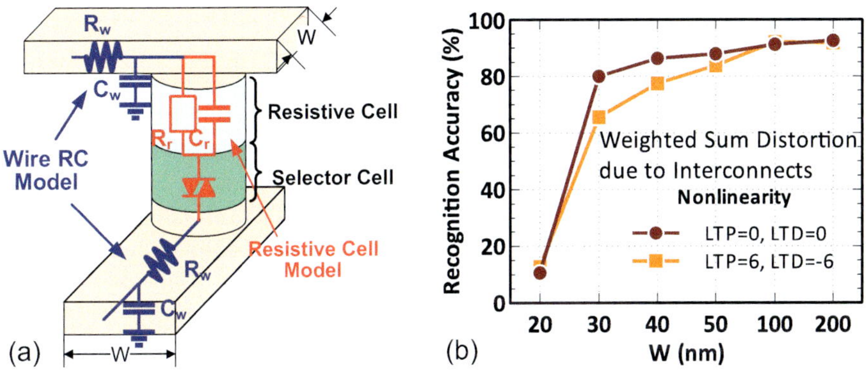

Fig. 12.15 (**a**) Sub-circuit module of a synaptic device cell (*W* wire width). The cell consists of a resistive synaptic device and a selector. The resistive cell has capacitor (C_r) in parallel with the cell resistor (R_r). There are also wire resistors (R_W) and capacitors (C_W) for top and bottom interconnect. Sub-circuit is duplicated for the entire array to perform SPICE simulation. (**b**) Learning accuracy with different wire widths. Smaller wire width will degrade the learning accuracy due to the IR drop along the interconnects

a cross-point, and such module is duplicated for the entire array in SPICE. We extract statistical *D*, *Z*, and *R* data at different learning stages at the runtime of SC algorithm by software and use these values to simulate the weighted sum *DZ* and *DR* (in the CD method in Fig. 12.2a) by SPICE using the read scheme described in Sect. 12.3. We then calculated the deviation of weighted sum by SPICE and incorporated back into the SC algorithm to evaluate its impact on the learning accuracy. We find that the RC delay (<ns) for the crossbar array (e.g., with a size of 100 × 300) is not critical, while the interconnect wire resistance may cause a sufficient IR drop. Figure 12.15b shows the learning accuracy with different array wire widths. Wires with smaller width have larger wire resistance; thus the weighted sum result becomes inaccurate, and the learning accuracy is greatly reduced. To alleviate this inaccuracy, reverse scaling on the wire's geometrical dimension is proposed. The result suggests that the wire width should be larger than 100 nm to achieve a learning accuracy higher than 90%. Such reverse scaling plus the redundant cells for reduction of device variations dramatically increase the

array area, but this may be acceptable considering the size of peripheral neuron circuits is large, and it may be comparable to the cell pitch of a synaptic cell (with the reverse scaling design rule) in the array design.

12.5 Accuracy Improvement by Proposed Circuit-Level Strategies

If we consider all the nonideal device properties and array parasitics mentioned above in the simulation, the learning accuracy will drop terribly low to ~30%. Therefore, it is necessary that we should apply the proposed mitigation strategies in the SC algorithm. Specifically, we assume that the following improvements on the realistic properties are achieved: (1) the ON/OFF weight ratio increases by 4X from 12.5 (within the range of the Ag:a-Si device) to 50 with the aid of dummy column but assuming that the off-state current is not completely removed due to device-to-device variation, and (2) 9 cells as 1 weight element are used to reduce the variation of read noise from ~2.89% to ~0.96%. It is also assumed that the nonlinearity remains the same, which is (LTP, LTD) = (4.7, −4.7) for the TaO_x/TiO_2-based synaptic device, and the array wire width is relaxed to be 200 nm. As shown in Fig. 12.16, the recognition accuracy of synaptic devices can achieve an accuracy improvement of >65% and closely approaches that of the accuracy in the ideal algorithm.

However, the proposed mitigation strategies will have some overhead on the chip area, latency, and energy. Compared to the design without mitigation strategies, the area overhead mainly comes from the redundant cells with relaxed wire width (~20% for 9 cells as 1 weight element and 200 nm wire width). The area overhead of the subtractors can be smaller (<7.84%) if they are partially hidden underneath the array. The total latency of weighted sum operation will be similar if the readout of weighted sum current is based on the principle of integrate-and-fire neuron model [25], where both the weighted sum current and parasitic column

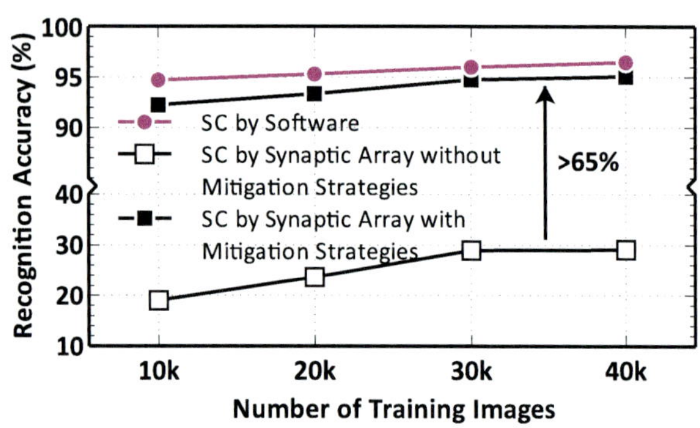

Fig. 12.16 Comparison of the recognition accuracy of the MNIST handwritten digits using traditional software approach and on-chip hardware implementation of the sparse coding algorithm. With the proposed mitigation strategies, the recognition accuracy by hardware can approach the ideal value that uses software approach

capacitance increase by 9X (nine cells as one weight element); thus these two effects cancel out each other. The total latency of the weight update will also be similar since the nine cells are physically wired together and being programmed at the same time. However, the energy consumption for both the weighted sum and weight update operation will increase by ~9X because 9 cells are used.

12.6 Conclusion

As the core elements of a neuromorphic system, synapses are responsible for establishing communication between groups of neurons. Several nonideal device properties are identified in the synaptic devices available today, e.g., the nonlinearity in weight update, device variations, read noise, and limited ON/OFF weight ratio. The wire parasitics in nanoscale cross-point array architecture also cannot be neglected. Sparse coding algorithm is used as a case study algorithm for unsupervised learning in the image recognition application, and we have built a platform based on this algorithm to evaluate the performance of learning accuracy using realistic models of synaptic devices and arrays. It is observed that the nonideal synaptic device properties and the wire parasitics can greatly reduce the image recognition accuracy from ~96% to ~30%. Therefore, we proposed the mitigation strategies to remedy this issue, including (1) the use of multiple cells as one weight element to alleviate the impact of device variations and read noise; (2) a dummy column and subtractors to eliminate the off-state current; and (3) the use of selector and larger wire width to reduce the IR drop along interconnects; thus the accuracy of weighted sum is increased. By applying these strategies with affordable overhead on-chip area, latency, and energy, the synaptic behavior is greatly improved, and the recognition accuracy comes back to ~95%, enabling the synaptic devices for practical on-chip hardware implementation of the sparse learning algorithm. We believe that the device-algorithm co-design methodologies presented in this work can also be applied to other neuro-inspired learning algorithms in general.

References

1. J. Schemmel, D. Bruderle, A. Grubl, M. Hock, K. Meier, S. Millner, A wafer-scale neuromorphic hardware system for large-scale neural modeling, in *IEEE International Symposium on Circuits and Systems (ISCAS)* (2010), pp. 1947–1950
2. R. Serrano-Gotarredona, M. Oster, P. Lichtsteiner, A. Linares-Barranco, R. Paz-Vicente, F. Gómez-Rodríguez, L. Camuñas-Mesa, R. Berner, M. Rivas-Pérez, T. Delbruck, CAVIAR: a 45 k neuron, 5 M synapse, 12G connects/s AER hardware sensory–processing–learning–actuating system for high-speed visual object recognition and tracking. IEEE Trans. Neural Netw. **20**(9), 1417–1438 (2009)

3. X. Jin, M. Lujan, L.A. Plana, S. Davies, S. Temple, S. Furber, Modeling spiking neural networks on SpiNNaker. Comput. Sci. Eng. **12**(5), 91–97 (2010)
4. P.A. Merolla, J.V. Arthur, R. Alvarez-Icaza, A.S. Cassidy, J. Sawada, F. Akopyan, B.L. Jackson, N. Imam, C. Guo, Y. Nakamura, A million spiking-neuron integrated circuit with a scalable communication network and interface. Science **345**(6197), 668–673 (2014)
5. S. Brink, S. Nease, P. Hasler, S. Ramakrishnan, R. Wunderlich, A. Basu, B. Degnan, A learning-enabled neuron array IC based upon transistor channel models of biological phenomena. IEEE Trans. Biomed. Circuits Sys. **7**(1), 71–81 (2013)
6. J. Hasler, B. Marr, Finding a roadmap to achieve large neuromorphic hardware systems. Front. Neurosci. **7**(118), 1–29 (2013)
7. S. Yu, P.-Y. Chen, Y. Cao, L. Xia, Y. Wang, H. Wu, Scaling-up resistive synaptic arrays for neuro-inspired architecture: challenges and prospect, in *IEEE International Electron Devices Meeting (IEDM)* (2015), pp. 451–454
8. D. Kuzum, S. Yu, H.-S.P. Wong, Synaptic electronics: materials, devices and applications. *Nanotechnology* **24**(38), 382001 (2013)
9. M. Hu, H. Li, Q. Wu, G. S. Rose, Y. Chen, Memristor crossbar based hardware realization of BSB recall function, in *International Joint Conference on Neural Networks (IJCNN)* (2012), pp. 1–7
10. P. Gu, B. Li, T. Tang, S. Yu, Y. Cao, Y. Wang, H. Yang, Technological exploration of rram crossbar array for matrix-vector multiplication, in *Asia and South Pacific Design Automation Conference (ASP-DAC)* (2015), pp. 106–111
11. P.-Y. Chen, D. Kadetotad, Z. Xu, A. Mohanty, B. Lin, J. Ye, S. Vrudhula, J.-s. Seo, Y. Cao, S. Yu, Technology-design co-optimization of resistive cross-point array for accelerating learning algorithms on chip, in *IEEE/ACM Design, Automation & Test in Europe Conference & Exhibition (DATE)* (2015), pp. 854–859
12. S. Park, A. Sheri, J. Kim, J. Noh, J. Jang, M. Jeon, B. Lee, B. Lee, B. Lee, H.-J. Hwang, Neuromorphic speech systems using advanced ReRAM-based synapse, in *IEEE International Electron Device Meeting (IEDM)* (2013), pp. 625–628
13. G.W. Burr, R.M. Shelby, S. Sidler, C. Di Nolfo, J. Jang, I. Boybat, R.S. Shenoy, P. Narayanan, K. Virwani, E.U. Giacometti, Experimental demonstration and tolerancing of a large-scale neural network (165 000 synapses) using phase-change memory as the synaptic weight element. IEEE Trans. Electron Devices **62**(11), 3498–3507 (2015)
14. I.-T. Wang, Y.-C. Lin, Y.-F. Wang, C.-W. Hsu, T.-H. Hou, 3D synaptic architecture with ultralow sub-10 f. energy per spike for neuromorphic computation, in *IEEE International Electron Devices Meeting (IEDM)* (2014), pp. 665–668
15. S.H. Jo, T. Chang, I. Ebong, B.B. Bhadviya, P. Mazumder, W. Lu, Nanoscale memristor device as synapse in neuromorphic systems. Nano Lett. **10**(4), 1297–1301 (2010)
16. H. Lee, A. Battle, R. Raina, A. Y. Ng, Efficient sparse coding algorithms, in *Advances in Neural Information Processing Systems (NIPS)* (2006), pp. 801–808
17. D.H. Hubel, T.N. Wiesel, Receptive fields, binocular interaction and functional architecture in the cat's visual cortex. J. Physiol. **160**(1), 106–154 (1962)
18. B.A. Olshausen, Emergence of simple-cell receptive field properties by learning a sparse code for natural images. Nature **381**(6583), 607–609 (1996)
19. B. Lin, Q. Li, Q. Sun, M.-J. Lai, I. Davidson, W. Fan, J. Ye, Stochastic coordinate coding and its application for drosophila gene expression pattern annotation, *CoRR abs/1407.8147* (2014)
20. Y. LeCun, C. Cortes, The MNIST database of handwritten digits, 1998
21. C.-C. Chang, C.-J. Lin, LIBSVM: a library for support vector machines. ACM Trans. Intell. Syst. Technol. **2**(3), 27 (2011)
22. E. Stromatias, D. Neil, M. Pfeiffer, F. Galluppi, S.B. Furber, S.-C. Liu, Robustness of spiking Deep Belief Networks to noise and reduced bit precision of neuro-inspired hardware platforms. Front. Neurosci. **9**, 222 (2015)

23. M. Courbariaux, I. Hubara, D. Soudry, R. El-Yaniv, Y. Bengio, Binarized neural networks: training deep neural networks with weights and activations constrained to +1 or −1, *arXiv preprint arXiv:1602.02830v3* (2016)
24. J. Liang, H.-S.P. Wong, Cross-point memory array without cell selectors – device characteristics and data storage pattern dependencies. IEEE Trans. Electron Devices **57**(10), 2531–2538 (2010)
25. D. Kadetotad, Z. Xu, A. Mohanty, P.-Y. Chen, B. Lin, J. Ye, S. Vrudhula, S. Yu, Y. Cao, J.-s. Seo, Parallel architecture with resistive crosspoint array for dictionary learning acceleration. IEEE J. Emer. Sel. Top. Circuits Syst. **5**(2), 194–204 (2015)
26. P. Narayanan, G. Burr, R. Shenoy, K. Virwani, B. Kurdi, Circuit-level benchmarking of access devices for resistive nonvolatile memory arrays, in *IEEE International Electron Devices Meeting (IEDM)* (2014), pp. 705–708
27. A. Chen, Comprehensive methodology for the design and assessment of crossbar memory array with nonlinear and asymmetric selector devices, in *IEEE International Electron Devices Meeting (IEDM)* (2013), pp. 746–749
28. P.-Y. Chen, B. Lin, I. Wang, T.-H. Hou, J. Ye, S. Vrudhula, J.-s. Seo, Y. Cao, S. Yu, Mitigating effects of non-ideal synaptic device characteristics for on-chip learning, in *IEEE/ACM International Conference on Computer-Aided Design (ICCAD)* (2015), pp. 194–199
29. International Technology Roadmap for Semiconductors (ITRS), http://www.itrs2.net/

Chapter 13
Binary OxRAM/CBRAM Memories for Efficient Implementations of Embedded Neuromorphic Circuits

Elisa Vianello, Thilo Werner, Giuseppe Piccolboni, Daniele Garbin, Olivier Bichler, Gabriel Molas, Jean Michel Portal, Blaise Yvert, Barbara De Salvo, and Luca Perniola

13.1 Synaptic Array Implementation Using Multiple Binary OxRAM/CBRAM Devices

In the domain of disruptive technologies for nonvolatile memory applications, since more than a decade, different technologies have been blossoming. Among them, resistive RAMs, including both oxide-based resistive RAM (OxRAM) and conductive-bridge RAM (CBRAM), appear as leading candidates to replace conventional flash memories, because of their simple fabrication (three additional masks), fast programming time (tens of ns), BEOL integration, and scalability. In this work, we explore the possibility to exploit these technologies to implement synapse-like devices.

In most OxRAM and CBRAM, both directions of programming are noncumulative. They are intrinsically binary devices: they switch between two distinct resistance states, a low-resistance state (LRS) and a high-resistance state (HRS), when appropriate identical SET/RESET pulses are applied. It is possible to provide multilevel memory by carefully choosing the programming current or the reset voltage [1], but applying the same programming pulse

The original version of this chapter was revised. An erratum to the chapter can be found at DOI: 10.1007/978-3-319-54313-0_14

E. Vianello (✉) • T. Werner • G. Piccolboni • D. Garbin • G. Molas • B. De Salvo • L. Perniola
CEA-Leti, Minatec Campus, F-38054 Grenoble, France
e-mail: elisa.vianello@cea.fr

O. Bichler
CEA-List, F-91191 Gif-sur-Yvette, France

J.M. Portal
IM2NP, UMR CNRS, 7334 Marseille, France

B. Yvert
INSER, Clinatec UA01, 38000 Grenoble, France

S. Yu (ed.), *Neuro-inspired Computing Using Resistive Synaptic Devices*,
DOI 10.1007/978-3-319-54313-0_13

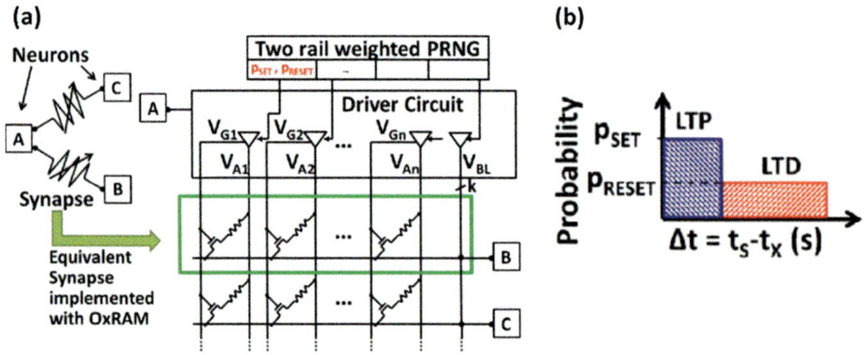

Fig. 13.1 (**a**) Schematic of an *anode connected matrix* scheme. All the OxRAM devices on the same row build one equivalent synapse (*n*T*n*R structure). Driver circuit is used to individually program OxRAM devices and propagate spikes to the next neuron layer. The weighted PRNG is used for online learning, to implement extrinsic stochasticity in STDP learning rule. (**b**) Simplified probabilistic STDP used in our work (Adapted from Garbin et al. [3])

multiple times has no further effect. However, the use of only two resistance levels per synapse, with respect to the multilevel approach, can be insufficient to achieve good performances in neuromorphic systems designed for some complex applications, as for example, image recognition [2]. Consequently, to emulate synaptic conductance, a single synapse composed of *n* multiple binary OxRAM cells operating in parallel is proposed (Fig. 13.1) [3]. Since parallel conductances sum up, the conductance of the equivalent synapse ranges from the sum of the *n* conductance values in the HRS to the sum of all the *n* conductance values in the LRS.

The simplest way to form a synaptic array is to organize devices in crossbar structure; the RRAM elements are located at the intersections of perpendicular wires. Free from any driving elements, the crossbar structure can be potentially very dense. However, the major issue of this topology is the sneak paths and parasitic resistance introduced by the connection network. To reduce sneak paths, we work with an active matrix, which has a selection transistor in series with each synaptic device. We adopted an *anode connected matrix* scheme; the cells in the active matrix are gate connected together following the anode lines (Fig. 13.1a). This configuration is ideal in reading mode since each input spike gets integrated into every output neuron automatically, but it requires sequential programming [4].

In order to define the resistance state (LRS or HRS) of each RRAM device needed to obtain the desired equivalent synaptic conductance, two alternative approaches can be used: supervised or unsupervised learning. *Supervised learning* is obtained using classical back-propagation algorithm [5], where the LRS/HRS status of each RRAM device is determined with computer simulations (offline learning), then discretized and imported in the memory array with a one-time programming operation. In *unsupervised learning*, the LRS/HRS status of the devices is learned in situ (online learning) with a simplified spike-time-dependent plasticity (STDP) learning rule. In this work, we investigate the benefit of spiking resistive neural networks programmed with both supervised (Sect. 13.2) and unsupervised learning (Sect. 13.3).

Concerning unsupervised learning, we adopted a simplified probabilistic STDP (Fig. 13.1b). First, the analog time dependence of STDP is ignored; when a programming event occurs, each RRAM cell of the synapse only has a given probability to have a SET (synaptic potentiation) or RESET (synaptic depression) operation. Second, the synaptic weight changes only when a post-spike occurs. If the presynaptic neuron was active recently, the synapse is potentiated; in any other case, it is depressed. The major benefit of this approach is that only simple square voltage pulses are adopted simplifying the programming circuitry. The switching probability can be governed by the RRAM itself (internal switching probability): SET and RESET conditions can be tuned to control the probability to switch the memory [6, 7]. Another possibility that allows a finer tuning of the switching probability, at the expense of a small increase of the circuit complexity, consists in using stronger programming conditions that do not show intrinsic stochasticity, i.e., switching probability is equal to one. Extrinsic stochasticity is thus obtained using an external pseudorandom number generator (PRNG) circuit block, which provides tunable switching probabilities p_{SET} and p_{RESET}.

In order to validate the functionality of the proposed synapse design and learning rule, we carried out simulations of synaptic potentiation and depression operations on RRAM synapses composed by 10 HfO_2-based OxRAM devices connected in parallel. Figures 13.2a and b show the typical binary behavior during an endurance test of a single OxRAM cell, and the corresponding evolution of the conductance when 100 potentiation events (SET) followed by 100 depression events (RESET) were applied with $p_{\mathrm{SET}} = 0.02$ and $p_{\mathrm{RESET}} = 0.04$.

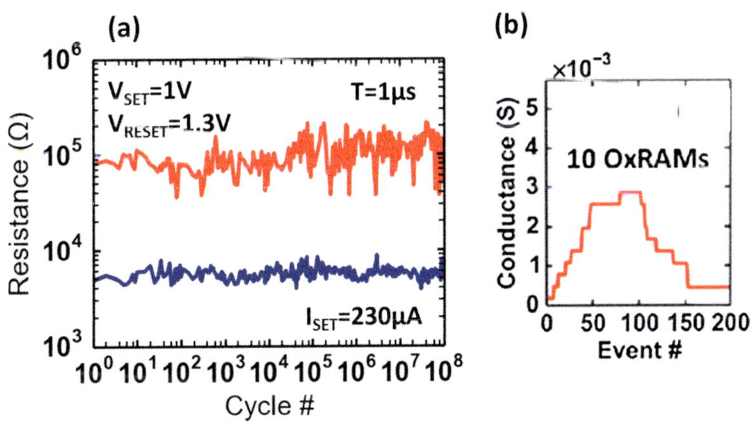

Fig. 13.2 (**a**) Endurance test on a single HfO_2-based cell, showing the typical binary behavior. The device can be successfully switched for more than 10^8 cycles. (**b**) 100 consecutive potentiation (*SET*) and depression (*RESET*) events, with $p_{\mathrm{SET}} = 0.02$ and $p_{\mathrm{RESET}} = 0.04$ on a synapse composed of 10 OxRAM devices. The use of multiple devices allows to implement a multilevel equivalent synapse (Adapted from Garbin et al. [3])

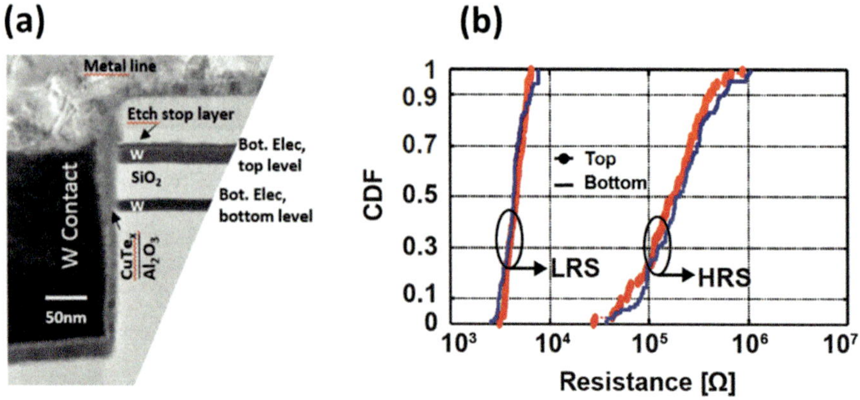

Fig. 13.3 (**a**) TEM cross section of a vertical CBRAM. (**b**) LRS and HRS distributions for top and bottom vertical CBRAM (Adapted from Piccolboni et al. [8])

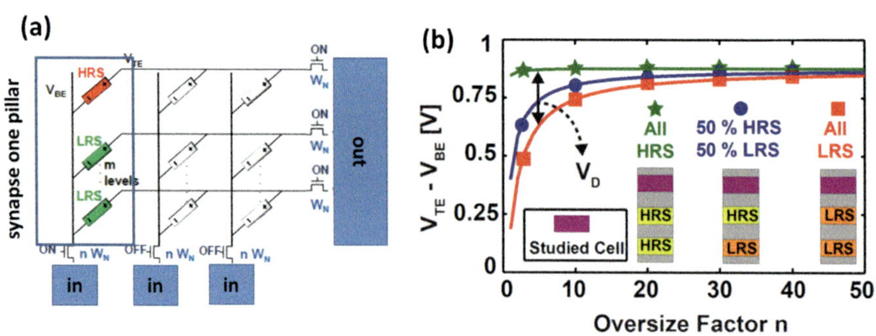

Fig. 13.4 (**a**) Matrix structure adopting one RRAM pillar composed of m levels per synapse. All levels share a common transistor. Each pillar has a transistor, with a channel n times wider compared to the plane select transistors. (**b**) Evolution of the voltage drop on a studied cell, violet cell, for a three-level synapse ($m = 3$) as a function of the oversize factor n (Adapted from Piccolboni et al. [8])

The weakest point of the proposed approach is the silicon area consumption for each synapse that is proportional to the number of devices needed to obtain a multilevel behavior. A possible solution to overcome this problem is the adoption of the vertical RRAM (VRRAM) technology [6, 8], which consists of RRAM cells integrated in multilayered VNAND-like structure; these are simple and cost-effective 3D processes to achieve high memory density (Fig. 13.3). Figure 13.4a shows a simple neural circuit where synapses are emulated by 1TmR VRRAM pillars which all share a transistor per level. One transistor per pillar is used to select a single synapse composed of m RRAM in connected in parallel. The nonselected pillars are left floating to cancel the sneak path currents. This configuration is ideal in reading mode since each input spike activates the corresponding transistor pillar and is integrated into every output neuron automatically. The sequential programming of the cells composing a single synapse is not possible without back-end selector; therefore, extrinsic stochasticity cannot be used. Only the internal switching probability can be adopted to program the cells: all the cells in one pillar are programmed together, and SET/RESET conditions are tuned to control the probability to switch the memory. The pillar transistor has to be oversized to have on every cell of the VRRAM pillar the desired SET voltage drop (V_{SET}) to achieve SET switching probability independently of the other cells' resistance status (LRS or HRS). Figure 13.4b reports the voltage drop on a given cell (violet in Fig. 13.4b), as a function on the oversize factor of the pillar access transistor (n) for different states of the other cells in the pillar (all in LRS, all in HRS and 50% in LRS and 50% in HRS). The number of levels is fixed, $m = 3$. It clearly appears that even if an oversize factor of $n = 2$ is the minimal sizing to obtain a positive set voltage ($V_{TE}-V_{BE} > 0$), it leads to a voltage variation V_D on the selected cell of up to 80 mV depending on the state of the others cells. Such difference in the voltage seen by the cell would translate in a different switching probability for the same programming pulse. To reduce this *context effect* and guarantee the desired voltage drop on the cells, whatever the context, the oversize factor must be increased above 30 (in this case, a voltage difference of 5 mV is estimated independently on the state of the other cells) [8].

13.2 Example of Convolutional Neural Network for Image Classification with Supervised Learning

Over the last past decade, software implementations of machine learning algorithms have been deployed in many cloud-centric applications such as speech recognition, face identification, photo search, etc. As this rise of machine learning applications continues, it is foreseeable to fabricate embedded neuromorphic circuits thereby eliminating the latency of cloud access and avoiding the large energy cost per bit transmission through the cloud.

The convolutional neural networks (CONV) are among the most promising architectures for embedded applications due to their relatively small memory footprint. In this chapter, we explore the potentialities of RRAM-based CONV architectures. Two different visual pattern recognition applications are demonstrated: the standard MNIST handwritten digits database benchmark and the German Traffic Sign Recognition Benchmark (GTSRB) database. We evaluated the feasibility and performances of such a network in the recognition phase with synaptic weights computed with classical backpropagation algorithms. To this purpose, we used the event-driven "Xnet" simulator [9].

The organization of convolutional layers in convolutional neural networks is originally inspired by the structure of the visual system in mammals [10, 11, 12]. Software implementations of CONVSs were applied with great success in applications such as traffic sign recognition [13], the analysis of biological images

[14], and the detection of faces, complex text, pedestrians on the streets, and human bodies in natural images [15, 16, 17]. A major recent practical success of software implementations of CONVSs is very high since it is the face recognition software proposed by Facebook [18]. The power consumption to perform convolution operations is computationally expensive in the CONV implementation on CPUs and GPUs. This hinders their integration in portable devices. In recent years, dedicated system on chip (SoC) solutions and FPGA platforms have been used to implement these networks to increase performances while decreasing their power consumption. A hardware implementation of CNNs based on the RRAM devices can further improve the power efficiency. We proposed the use of RRAM synapses presented in Sect. 13.1 to store the kernel features [19, 20].

The proposed CONV architecture is composed of a feature extraction module (made of two cascaded convolutional layers) and a classification module (made of two fully connected layers); see Fig. 13.5. While in the fully connected classification module the neurons of a given layer are connected to every neuron of the previous layer by a large number of synapses, in convolutional layers, a small set of synapses (constituting several kernels) are shared among different neurons to connect layer *N* and *N-1* through a convolutional operation. A convolutional layer is composed of several (output) feature maps, each of them being connected to some or all the feature maps of the previous layer through a convolutional kernel. The kernel corresponds to a feature that has to be localized in the input image. In a layer, each feature map contains the results of the convolution of some or all the input maps (which are the output feature maps of the previous layer), each of them with a different convolution kernel. The output feature map contains information about the locations where the kernel features are present in the input map. The extraction module therefore transforms the input image into a simpler set of feature maps. The classification module connects the obtained set of feature maps to the output neuron layer. Each output neuron is associated to a category (the 10-digit categories for the MNIST database and the 43 German Traffic Signs for the GTSRB database): an output neuron spikes when the image presented to the input of the network belongs to its category. In the first convolutional layer, each output feature map is connected to the input image through a convolutional kernel; 16 and 32 feature maps are used in the first convolutional layer for the MNIST and GTSRB applications, respectively. For the second convolutional layer, we adopted a non-complete connection scheme; each output feature map is connected only to some of the input features maps through a convolutional kernel [21]. For MNIST, 16 input feature maps are connected to 24 output feature maps through 90 kernels, while for the complex GTSRB, 32 input feature maps are connected to 48 output feature maps through 186 kernels. To produce smaller output volume data, it is possible to increase the stride with which the kernel is sided on the input data or to add a max-pooling layer [21]. For the MNIST and GTSRB, the stride is 2 and 1, respectively. In the case of GTSRB network, the down-sampling operation is performed by adding a max-pooling layer. The estimated size of the OxRAM array needed to implement the CNNs is 600 kb for MNIST and 1 Mb for GTSRB, respectively.

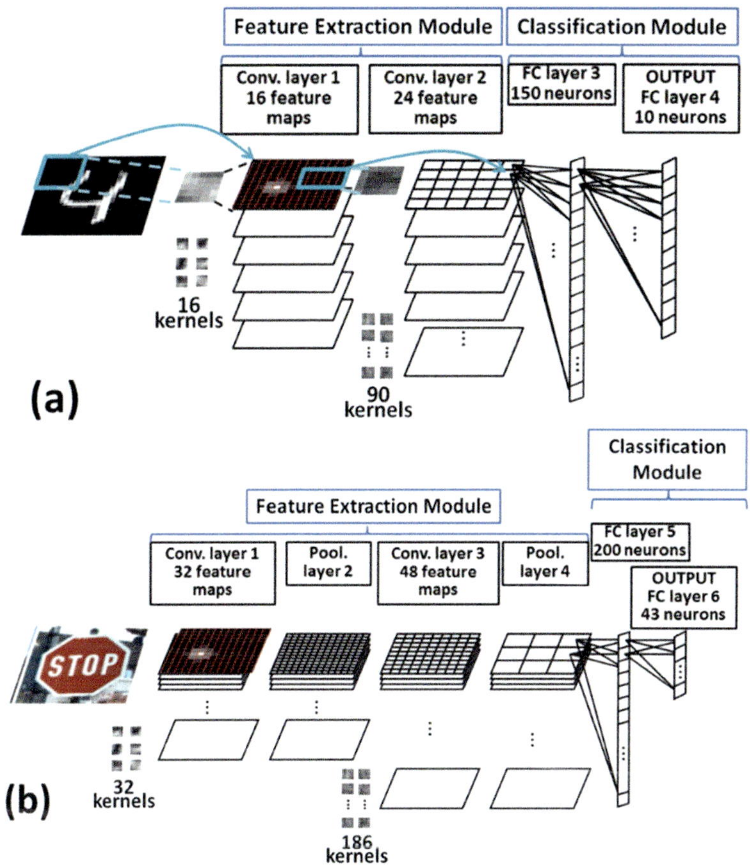

Fig. 13.5 CNN architecture for (**a**) handwritten digit recognition (MNIST database) and (**b**) traffic sign recognition (GTSRB database). For the second convolutional layer, a non-complete connection scheme (each output feature map is connected only to some of the input features maps) is adopted. The max-pooling layers in the GTSRB network reduce the size of the feature maps by a factor 2, thus reducing the complexity of the network. The stride is 2 and 1 for the MNIST and GTSRB networks, respectively (Adapted from Garbin et al. [20])

In order to study the impact of the OxRAM electrical performances and reliability on the network, we fully characterized a 16 kbit OxRAM demonstrator integrated into a 28 nm CMOS digital test-chip (Fig. 13.6) [22]. OxRAM devices feature a metal-insulator-metal (MIM) structure composed of a 5 nm thick HfO_2 layer sandwiched between a Ti top electrode and a TiN bottom electrode. A bitcell is composed of 1 transistor-1 resistor (1T1R) structure. Figure 13.6c reports the cumulative distributions of low-resistance state (LRS) and high-resistance state (HRS) extracted from the 16 kb OxRAM array statistics. No correction code or smart programming algorithms have been used. All the network simulations

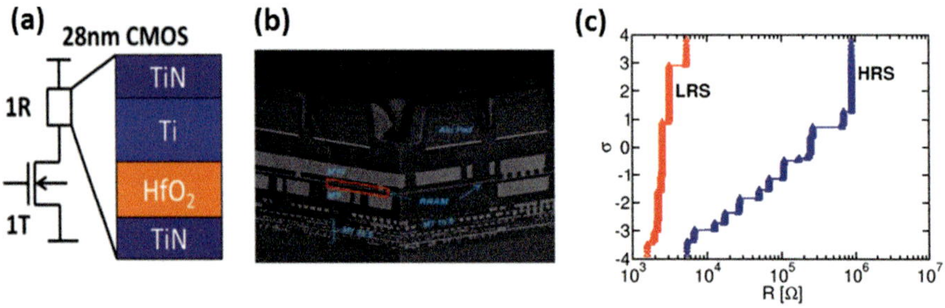

Fig. 13.6 (**a**) 1T1R bitcell schematic, (**b**) SEM cross section of CMOS 28 nm stack including HfO_2-based OxRAM cells, and (**c**) cumulative distributions of LRS and HRS for 16 kbit demonstrator. Distributions are cut at 1 M due to lower limit in current sensing (Adapted from Benoist et al. [22])

Fig. 13.7 Recognition success as a function of the number *n* of parallel OxRAM devices used to implement an equivalent synapse (Fig. 13.1) (Adapted from Garbin et al. [20])

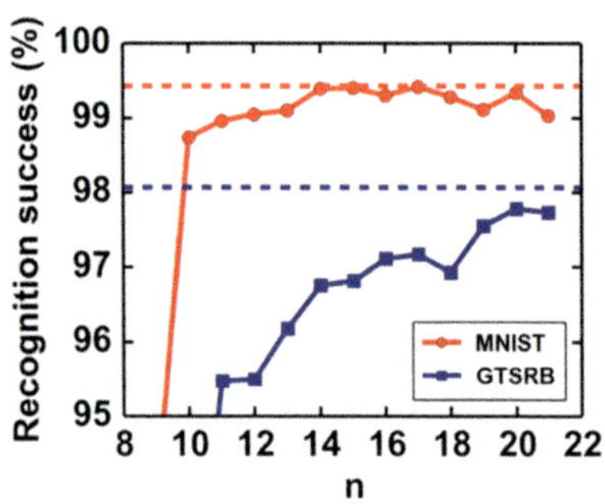

presented in the following take into account the real LRS and HRS distributions presented in Fig. 13.6c.

To implement a synapse in hardware, *n* OxRAM devices connected in parallel are used as shown in Fig. 13.1. In order to define the optimal value of the synaptic weight resolution, i.e., the number (*n*) of OxRAM devices needed per synapse, simulations have been performed on both MNIST and GTSRB databases using the CNN architectures presented in Fig. 13.5. Figure 13.7 shows the simulation results in terms of recognitions success as a function of the value of OxRAM cells per synapse, *n*. The recognition success improves as *n* increases, for *n* higher than 12 the maximum network performance greater than 99% is reached for MNIST. The blue curve in Fig. 13.7 reports the recognition success for the GTSRB database. As this application poses a more complex task, a higher number of cells per synapse, here n~20, is needed. More complex application tasks are more demanding in terms of number of cells per synapse.

13.3 Example of Fully Connected Neural Network with Unsupervised Learning for Real-Time Decoding (Spike Sorting) of Biological Data

RRAM devices coupled with spike coding scheme are a key element to enable the implementation of unsupervised learning with minimal hardware footprint and low power consumption. Embedding neuromorphic learning into low-power devices could enable the design of future sensory processing of *autonomous systems*. An autonomous system can be a simple one, such as a sensory processing system based on environmental sensors or biosensors, or more complex ones, such as a brain-machine interface making decisions based on real-time online processing of in vivo recorded biological signals, or even humanoid robot making decisions and producing behaviors based on the outcome of sophisticated auditory or visual processing. In the following, we will present an alternative approach to perform spike sorting of complex brain signals using RRAM-based neural networks. This approach coupled with brain-machine interfaces (BMIs) may enable the design of autonomous implantable devices for rehabilitation purposes [23, 24].

Neural recording systems are a central component of brain-machine interfaces. The effectiveness and accuracy of BMI scale with the ability to precisely collect the brain signals, triggered, for example, when a person intends to move. This implies the recording of the simultaneous activity of a large population of neurons [25, 26] (Fig. 13.8). Advanced microelectrode arrays [27] allow recording of several hundreds of neurons at sampling rates of typically 20–40 kHz. Indeed, this generates massive data that need to be transmitted to remote systems for further processing or data analysis. To build neural prosthetic devices that can extract information from very large numbers of neurons and decode them in situ without transmitting this information offline, it is necessary to develop additional ultra low-power processing stages to extract relevant information before transmit it to remote

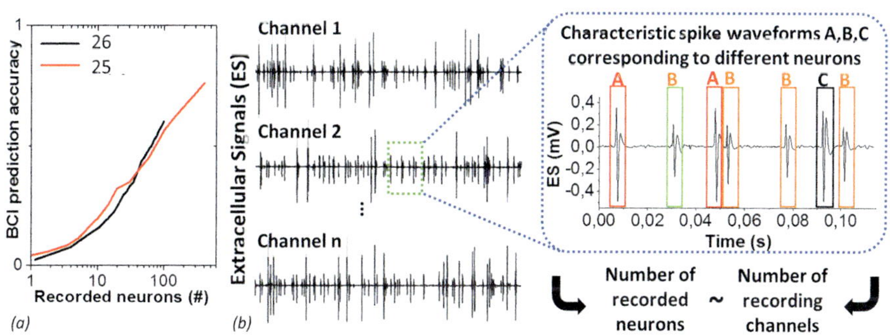

Fig. 13.8 (**a**) Accuracy of BCI for neural prosthesis applications as a function of the number of recorded neurons (Adapted from [25, 26]). (**b**) Example of extracellular electrical signals recorded from *n* channels (microelectrodes) and zoom in showing three distinct spike shapes (*A*, *B*, and *C*) corresponding to three different neurons (Adapted from Werner et al. [23])

computing or actuating modules. To this purpose, here we propose a RRAM-based spiking neural network which converts the recorded bio-signals into asynchronous digital events and classifies them online (spike sorting, Fig. 13.8b). Current spike-sorting methodologies are mainly based on matched filtering and statistical clustering techniques [28], where the electrical waveform is analyzed against a temporally sliding template and a spike is identified whenever the two are found to match within some tolerance. Thus, the standard spike-sorting approach consists of spike detection, feature extraction (typically principal components analysis, PCA), and classification. Algorithms of this type have been implemented in commercial software [29]; however, they present several limitations, as they need user supervision (manual tuning of the parameters), and they are power hungry and therefore not adapted for an embedded application. Therefore, most of the neural signal processing is performed via offline software on desktop computers or GPU. The offline processing is not optimal because it does not allow for real-time processing in closed-loop applications (e.g., in BCI [30, 31]) or real-time data compression prior to wireless transmission with reasonable power consumption in case of high channel counts. New spike-sorting approaches are required to address the described drawbacks of state-of-the-art techniques. In the following, we will describe a RRAM-based neuromorphic system targeting to perform real-time spike sorting with nanowatt-level power consumption and reasonable spike-sorting performances.

In order to illustrate the validity of the proposed spike-sorting methodology, we measured the extracellular activity from in vitro crayfish nerves recorded simultaneously with intracellular data of one motor or sensory neuron of the T5 ganglion (see Fig. 13.9) [32, 33]. In these data, the extracellular signal (ES) contains two different spike shapes (labeled as Spike A and Spike B) corresponding to two different neurons. The spikes simultaneously observed in the intracellular signal (IS) correlate with the activity of Spike A in the extracellular signal. Therefore, the intracellular signal activity can be used as the ground truth to assess the spike-sorting capability of our system for the detection of Spike A in the extracellular signal data. The entire dataset duration comprises 681 s and is called CF1 subsequently.

We assume that different spike shapes observed in the extracellular signal exhibit distinct representations in the time-frequency domain as shown, for example, in Fig. 13.10 for "Spike A" and "Spike B" which can serve as finger prints for the identification of these spike shapes. This approach allows to trace the activity of single neurons. Figure 13.10 shows the topological view of the two-layer fully connected neural network (FCNN) architecture (implemented in the event-driven simulator "Xnet" [9]) designed to extract, learn, and recognize different spike shapes from biological extracellular signal. The topmost layer above the FCNN represents the frequency band-pass filtering used to convert the extracellular recording into a useful input for the FCNN. Thus, the normalized continuous extracellular signal is encoded by 32 frequency band-pass filters. The 32 filtered signals are then full-wave rectified and presented to the FCNN input layer of 32 neurons where the analog continuous signals are converted into spikes which

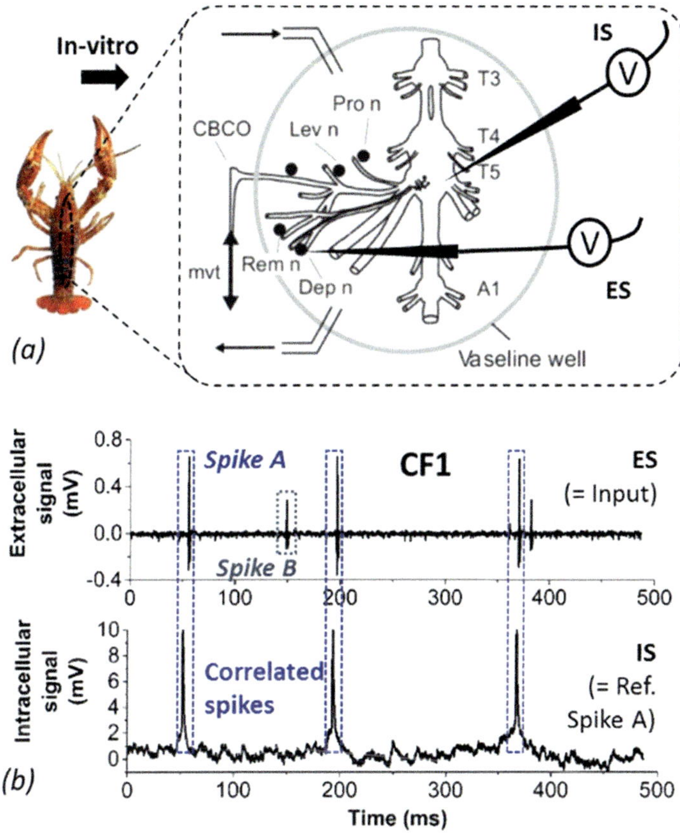

Fig. 13.9 Experiment to obtain real biological data. The extracellular signal reflects Spike A and B and in the following is called CF1. The intracellular activity reflects only Spike A, and it can be used as the ground truth to assess the spike-sorting capability for the detection of Spike A (Adapted from Werner et al. [23])

are then propagated along the synapses to the FCNN output layer of five neurons. The neurons of both layers are described by the leaky integrate fire (LIF) model [34], and they are fully connected by 32 × 5 excitatory synapses. The firing event of an output neuron indicates that the spike inspected in the input signal (Spike A or B in the example of Fig. 13.9) belongs to the specific class corresponding to this output neuron. To minimize delay time between input (spike event in the extracellular signal) and output (corresponding output neuron of SNN spikes), we used second order butterworth band-pass filters. Moreover, the filters have been defined to resolve and distinguish different spike signals with duration of about 1–2 ms and with an interspike time of tens of ms. We adopted 32 filters distributed on a linear range between 100 and 2000 Hz (the frequency spike spectrum does not exceed 3000 Hz [35]). This filter bank can be used on different spiking neural data as demonstrated in the following (Fig. 13.12). By using the band-pass filter approach

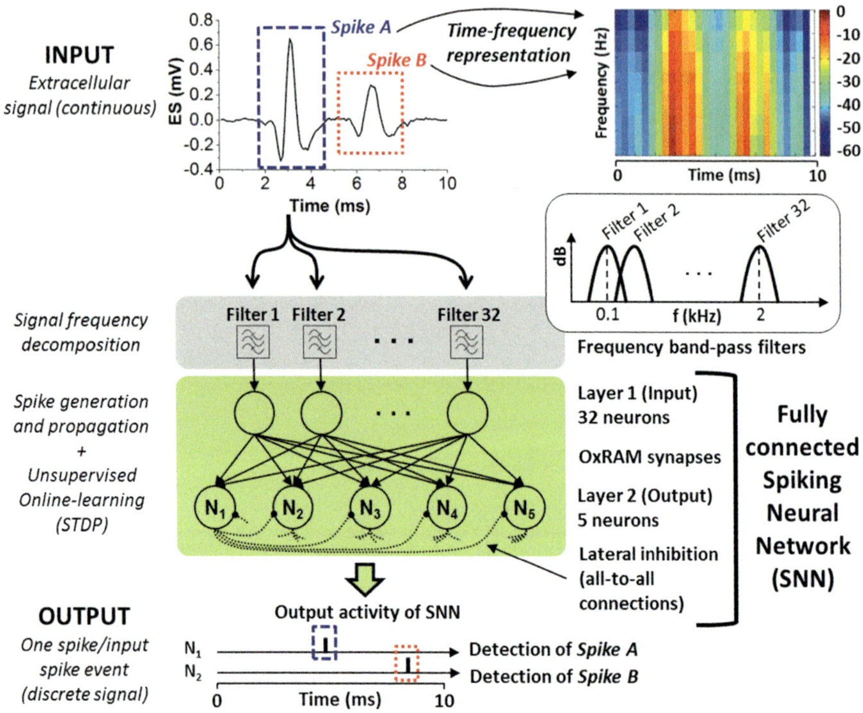

Fig. 13.10 Functional schematic of FCNN. The extracellular signal is fed through 32 frequency band-pass filters which are connected to the FCNN. Synapses are based on HfO_2-based OxRAM devices. Output neurons become selective to different input spikes shapes (Adapted from Werner et al. [23])

to encode spiking data, the FCNN does receive strong input signals if a spike is observed in the input data, whereas rather low-frequency signals are not able to excite the network sufficiently. Thus, no dedicated method to remove low-frequency noise is required, and spike detection is inherently implemented. As shown in Fig. 13.10, the number of input neurons corresponds to the number of filters. The number of output neurons determines the maximum number of spike classes that the FCNN is able to classify. Therefore, we can classify a maximum of five spike classes in this case. A sufficiently high number of output neurons has to be chosen so that every spike shape contained in the extracellular data can be assigned to one output neuron, i.e., the number of output neurons has to be at least as high as the (a priori unknown) number of spike shapes in the extracellular signal. The goal is that every spike shape will be learned and recognized by one of the output neurons whereas nonselective neurons remain silent, i.e. the number of spiking output neurons indicates the number of spike classes. To this purpose, lateral inhibition is implemented with recurrent inhibitory synapses across the

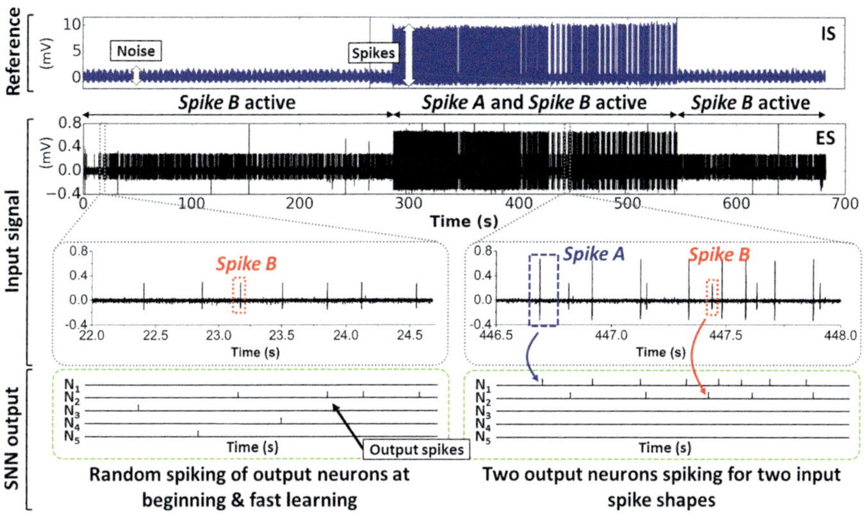

Fig. 13.11 Schematic illustration of the learning phase for the Fully Connected Neural Network (Fig. 13.10) applied on the CF1 (Fig. 13.9) data (Adapted from Werner et al. [23])

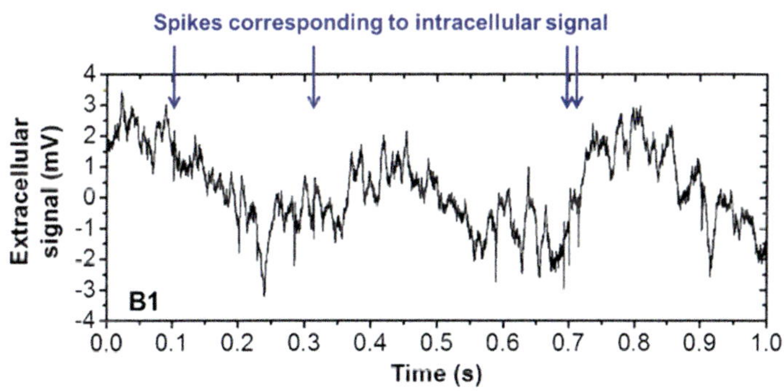

Fig. 13.12 Sequence of real biological spiking data used for verification of the spike-sorting system, in vivo from implanted rat hippocampus [36]. Intracellular recording was simultaneously obtained and provide the ground truth for valid quantification of the spike recognition rate for the labeled spikes (*blue arrows*) (Adapted from Werner et al. [24])

output layer to prevent the neurons from simultaneous spiking (i.e., to avoid the neurons to learn identical inference classes).

One of the key challenges for spike-sorting algorithms is the real-time functionality for a priori unknown data. This objective requires an online learning algorithm and specifically a synaptic latency lower than the duration of biological spikes (approx. 1–2 ms). Simplified probabilistic STDP and 20 binary OxRAM per synapses (Fig. 13.1) are adopted to fulfill these requirements. Extrinsic probability tuning a pseudorandom number generator is adopted ($p_{\mathrm{SET}} = 0.071$, $p_{\mathrm{RESET}} = 0.047$, $\Delta T = 7.7$ ms, see Fig. 13.1b). The low synaptic latency is achieved thanks to the OxRAM fast switching synapses (few tens of ns).

The complete spike-sorting system consisting of band-pass filters and FCNN was simulated with the "Xnet" (event-driven) simulator. Figure 13.10 illustrates schematically the unsupervised learning response of our FCNN for the treatment of the crayfish data (CF1, Fig. 13.9). Initially (0–285 s), only Spike B is present in the input data. The FCNN output, i.e., the firing patterns of the five output neurons N_1-N_5, is completely random. Thanks to the introduced lateral inhibition, one output neuron, here N_2, becomes gradually selective to Spike B. Then (285–545 s), also Spike A is observed in the input signal. In this period, N_1 starts to spike predominantly when the Spike A appears, while N_2 continues to fire for Spike B. The remaining output neurons N_3, N_4, and N_5 are rather silent. At the end of the test case (545–681 s), only Spike B is present. As expected, only N_2 shows activity whereas N_1, N_3, N_4 and N_5 are inactive.

In order to quantify the recognition rate of Spike A, we correlated the activity of N_1 with the intracellular signal (IS in Fig. 13.9). A Spike A event is considered to be recognized by N_1 if N_1 spikes within 20 ~ ms after the Spike A event. The recognition rate was calculated as the ratio of recognized spikes to the total number of Spike A events (truth from IS data) in a given time interval (fixed to 10 s). The system reached its mean spike recognition rate of 85.5% after 15 s (corresponding to 50 Spike A events), calculated starting from the first occurrence of Spike A in the input signal.

We tested our spike-sorting FCNN with respect to its applicability on other more complex neural spiking data (Fig. 13.12). We used another dataset recorded from anesthetized (in vivo) rat hippocampus (publicly available online provided by the Buszaki lab [36, 37]). It features simultaneous recording of extra- and intracellular signals and are in the following referred to as B1. As before in the case of CF1, we use the intracellular recording as a ground truth for the quantification of the recognition rate. B1 comprises a strongly increased background noise level with respect to CF1. Without changing the parameters of our FCNN, the recognition rate 70% for B1. This result confirms that thanks to the STDP learning rule, the proposed network can be used on different biological data without tuning parameters. State-of-the-art spike-sorting algorithms based on spike detection, feature extraction, and clustering (i.e., standard methodology) achieve recognition rates around 90% on the dataset B1 [38] and therefore outperform our proposed approach in terms of accuracy. However, the reported method does not incorporate a spike detection step but uses previously extracted and aligned spike waveforms for the classification. Moreover, the mathematical algorithm is rather complex. For this reason, the standard approach is impractical for real-time applications with low power consumption.

13.4 Conclusions

In this chapter, we proposed resistive RAM (RRAM) cells connected in parallel, thereby providing synaptic analog behavior as artificial synapsis in neuromorphic architectures. The vertical RRAM technology is presented as a possible solution to gain area by realizing one pillar per synapse. We presented two applications that can best exploit the properties of neuromorphic RRAM architectures. The first is visual pattern recognition with supervised learning using convolutional neural networks for embedded systems and portable devices. The proposed solution, with respect to typical software implementation on CPUs and GPUs, allows to decrease the power consumption without degrading performances. Recognition success rates higher than 99% and 97% have been demonstrated for the MNIST and GTSRB networks, respectively. These results are similar to the state-of-the-art recognition success rates obtained with formal CNN models, implemented with floating-point precision synapses. These success rates are reached using 12 and 20 RRAM devices per synapse for the MNIST and GTSRB applications, respectively. The second application is real-time spike sorting of complex brain signals using a fully connected neural network with unsupervised learning. We showed that the proposed network allows for fast adaptation to new input data (it is not needed to know a priori the number of spike shapes to classify and when they appear in the input data). The network has been tested on different sets of real biological spiking data, and functionality was proven for all datasets without parameter tuning. In contrast to standard spike-sorting techniques, the proposed approach offers several advantages, e.g., no power-consuming CPU or GPU are needed, and no parameters (e.g., threshold level for spike detection) have to be optimized manually as a function of the input data. This approach coupled with brain-machine interfaces (BMIs) may enable the design of prosthetic devices that require to extract and decode in situ information from very large numbers of neurons without transmitting this information offline.

References

1. E. Vianello, O. Thomas, G. Molas, O. Turkyilmaz, N. Jovanovic, D. Garbin, G. Palma, M. Alayan, C. Nguyen, J. Coignus, B. Giraud, T. Benoist, M. Reyboz, A. Toffoli, C. Charpin, F. Clermidy, L. Perniola. Resistive memories for ultra-low-power embedded computing design, in *IEEE International Electron Devices Meeting (IEDM)* (2014)
2. J. Bill, R. Legenstein, A compound memristive synapse model for statistical learning through stdp in spiking neural networks. Front. Neurosci. **8**, 412 (2014)
3. D. Garbin, E. Vianello, O. Bichler, Q. Rafhay, C. Gamrat, G. Ghibaudo, B. DeSalvo, L. Perniola, HfO_2–based OxRAM devices as synapses for convolutional neural networks. IEEE Trans. Electron Devices **62**, 2494–2501 (2015)
4. D. Roclin, O. Bichler, C. Gamrat, J-O. Klein, Sneak paths effects in CBRAM memristive devices arrays for spiking neural networks, in *IEEE/ACM International Symposium on Nano-scale Architectures (NANOARCH)* (2014), p. 15–18

5. P. Y. Simard, D. Steinkraus, J. C. Platt, Best practices for convolutional neural networks applied to visual document analysis, in *IEEE International Conference on Document Analysis and Recognition* (2003), pp. 958–963
6. G. Piccolboni, G. Molas, J. M. Portal, R. Coquand, M. Bocquet, D. Garbin, E. Vianello, C. Carabasse, V. Delaye, C. Pellissier, T. Magis, C. Cagli, M. Gely, O. Cueto, D. Deleruyelle, G. Ghibaudo, B. De Salvo, L. Perniola, Investigation of the potentialities of Vertical Resistive RAM (VRRAM) for neuromorphic applications, in *IEEE International Electron Devices Meeting (IEDM)* (2015)
7. M. Suri, O. Bichler, D. Querlioz, G. Palma, E. Vianello, D. Vuillaume, C. Gamrat, B. -De-Salvo, CBRAM devices as binary synapses for lowpower stochastic neuromorphic systems: auditory (Cochlea) and visual (Retina) cognitive processing applications, in *IEEE International Electron Devices Meeting (IEDM)* (2012)
8. G. Piccolboni, M. Parise, G. Molas, A. Levisse, J. M. Portal, R. Coquand, C. Carabasse, M. Bernard, A. Roule, J. P. Noel, B. Giraud, M. Harrand, C. Cagli, T. Magis, E. Vianello, B. De Salvo, G. Ghibaudo, L. Perniola, Vertical CBRAM (V-CBRAM): from experimental data to design perspectives, in *IEEE International Memory Workshop (IMW)* (2016)
9. O. Bichler, D. Querlioz, S. J. Thorpe, J.-P. Bourgoin, C. Gamrat, Unsupervised features extraction from asynchronous silicon retina through Spike-Timing-Dependent Plasticity, in *International Joint Conference on Neural Networks (IJCNN)* (2011), pp. 859–866
10. D.H. Hubel, T.N. Wiesel, Receptive fields, binocular interaction and functional architecture in the cats visual cortex. J. Physiol. **160**, 106–154 (1962)
11. D.J. Felleman, D.C. Van Essen, Distributed hierarchical processing in the primate cerebral cortex. Cereb. Cortex **1**, 1–47 (1991)
12. K. Fukushima, Artificial vision by multi–layered neural networks: neocognitron and its advances. Neural Netw. **37**, 103–119 (2013)
13. D. Ciresan, U. Meier, J. Masci, J. Schmidhuber, Multi–column deep neural network for traffic sign classification. Neural Netw. **32**, 333–338 (2012)
14. F. Ning, D. Delhomme, Y. LeCun, F. Piano, L. Bottou, P.E. Barbano, Toward automatic phenotyping of developing embryos from videos. IEEE Trans. Image Process. **14**, 1360–1371 (2005)
15. P. Sermanet, K. Kavukcuoglu, S. Chintala, Y. LeCun, Pedestrian detection with unsupervised multi–stage feature learning, in *IEEE Conference on Computer Vision and Pattern Recognition (CVPR)* (2013)
16. R. Vaillant, C. Monrocq, Y. Le Cun, A convolutional neural network hand tracker. IEEE Proc. Vis. Image Signal Proces. **141**, 245–250 (1994)
17. C. Garcia, M. Delakis, Convolutional face finder: a neural architecture for fast and robust face detection. IEEE Trans. Pattern Anal. Mach. Intell. **26**, 1408–1423 (2004)
18. Y. Taigman, M. Yang, M. Ranzato, L. Wolf, "Deepface: closing the gap to human-level performance in face verification, in *IEEE Conference on Computer Vision and Pattern Recognition (CVPR)* (2014), pp. 1701–1708
19. D. Garbin, O. Bichler, E. Vianello, Q. Rafhay, C. Gamrat, L.Perniola, G. Ghibaudo, B. -De-Salvo, Variability–tolerant convolutional neural network for pattern recognition applications based on OxRAM synapses, in *IEEE International Electron Devices Meeting (IEDM)* (2014)
20. D. Garbin, E. Vianello, O. Bichler, M. Azzaz, Q. Rafhay, P. Candelier, C. Gamrat, G. Ghibaudo, B. DeSalvo, L. Perniola, On the impact of OxRAM–based synapses variability on convolutional neural networks performance, in *IEEE/ACM International Symposium on Nanoscale Architectures (NANOARCH)* (2015), pp. 193–198
21. Y. LeCun, L. Bottou, Y. Bengio, P. Haffner, Gradient-based learning applied to document recognition. Proc. IEEE **86**, 2278–2324 (1998)
22. A. Benoist, S. Blonkowski, S. Jeannot, S. Denorme, J. Damiens, J. Berger, P. Candelier, E. Vianello, H. Grampeix, J.F. Nodin, E. Jalaguier, L. Perniola, B. Allard, 28 nm advanced

CMOS resistive RAM solution as embedded non-volatile memory, in *IEEE International Reliability Physics Symposium (IRPS)* (2014)
23. T. Werner, D. Garbin, E. Vianello, O. Bichler, D. Cattaert, B. Yvert, B. De Salvo, L. Perniola, Real-time decoding of brain activity by embedded Spiking Neural Networks using OxRAM synapses, in *IEEE International Symposium on Circuits and Systems (ISCAS)* (2016), pp. 2318–2321
24. T. Werner, E. Vianello, O. Bichler, D. Garbin, D. Cattaert, B. Yvert, B. De Salvo, L. Perniola, Spiking neural networks based on OxRAM synapses for real-time unsupervised spike sorting. Front. Neurosci. **10**, 474 (2016)
25. P. J. Ifft, S. Shokur, Z. Li, M. Lebedev, M. Nicolelis, A brain-machine interface enables bimanual arm movements in monkeys. Sci. Transl. Med., vol. 5: 210ra154; 2013.
26. J. Wessberg, C.R. Stambaugh, D.J. Kralik, P.D. Beck, M. Laubach, J.K. Chapin, et al., Real-time prediction of hand trajectory by ensembles of cortical neurons in primates. Nature **408**, 361–365 (2000)
27. M.E. Spira, A. Hai, Multi-electrode array technologies for neuroscience and cardiology. Nat. Nanotechnol. **8**, 83–94 (2013)
28. M. Lewicki, A review of methods for spike sorting: the detection and classification of neural action potentials. Network: Comput. Neural Syst. **9**, R53–R78 (1998)
29. R. Bestel, A.W. Daus, C. Thielemann, A novel automated spike sorting algorithm with adaptable feature extraction. J. Neurosci. Methods **211**, 1686–1178
30. L.R. Hochberg, M.D. Serruya, G.M. Friehs, J.A. Mukand, M. Saleh, A.H. Caplan, A. Branner, D. Chen, R.D. Penn, J.P. Donoghue, Neuronal ensemble control of prosthetic devices by a human with tetraplegia. Nature **442**, 164–171 (2006)
31. L.R. Hochberg, D. Bacher, B. Jarosiewicz, N.Y. Masse, J.D. Simeral, J. Vogel, S. Haddadin, J. Liu, S.S. Cash, P. van der Smagt, J.P. Donoghue, Reach and grasp by people with tetraplegia using a neurally controlled robotic arm. Nature **485**, 372–375 (2012)
32. D. Cattaert, A.E. Manira, Shunting versus inactivation: analysis of presynaptic inhibitory mechanisms in primary afferents of the crayfish. J. Neurosci. **19**, 6079–6089 (1999)
33. D. Cattaert, J.-P. Delbecque, D.H. Edwards, F. Issa, Social interactions determine postural network sensitivity to 5-HT. J. Neurosci. **30**, 5603–5616 (2010)
34. W. Gerstner, W.M. Kistler, *Spiking Neuron Models* (Cambridge University Press, Cambridge, 2002)
35. H. G. Rey, C. Pedreira, R. Q. Quiroga, Past, present and future of spike sorting techniques. Brain Res. Bull. **119**, 106–117 (2015)
36. D. Henze, Z. Borhegyi, J. Csicsvari, A. Mamiya, K.D. Harris, G. Buzsaki, Intracellular features predicted by extracellular recordings in the hippocampus in vivo. J. Neurophysiol. **84**, 390–400 (2000)
37. K.D. Harris, D. Henze, J. Csicsvari, H. Hirase, G. Buzsaki, Accuracy of tetrode spike separation as determined by simultaneous intracellular and extracellular measurements. J. Neurophysiol. **84**, 401–414 (2000)
38. J. Gasthaus, Y. W. Teh, F. Wood, G. Dilan, Dependent Dirichlet process spike sorting, in *Advances in Neural Information Processing Systems (NIPS)* (2008)

ERRATUM TO

Chapter 13
Binary OxRAM/CBRAM Memories for Efficient Implementations of Embedded Neuromorphic Circuits

Elisa Vianello, Thilo Werner, Giuseppe Piccolboni, Daniele Garbin, Olivier Bichler, Gabriel Molas, Jean Michel Portal, Blaise Yvert, Barbara De Salvo, and Luca Perniola

S. Yu (ed.), *Neuro-inspired Computing Using Resistive Synaptic Devices*,
DOI 10.1007/978-3-319-54313-0_13

DOI 10.1007/978-3-319-54313-0_14

The original version of this book was inadvertently published with an incorrect figure in chapter 13. The authors and the editor would like to remove Figure 13.12 (a) as, unfortunately, the recognition rate for the complex biological (CF2) data was portrayed lower than it should be.

The figure 13. 12(a) has been now erased and the text is slightly modified to accommodate this deletion, in the correct version.

The updated online version of the original chapter can be found under
http://dx.doi.org/10.1007/978-3-319-54313-0_13

S. Yu (ed.), *Neuro-inspired Computing Using Resistive Synaptic Devices*,
DOI 10.1007/978-3-319-54313-0_14

Printed in the United States
By Bookmasters